UNDERSTANDING EMOTIONS IN MATHEMATICAL THINKING AND LEARNING

UNDERSTANDING EMOTIONS IN MATHEMATICAL THINKING AND LEARNING

Ulises Xolocotzin Eligio

ELSEVIER

ACADEMIC PRESS
An imprint of Elsevier

Academic Press is an imprint of Elsevier
125 London Wall, London EC2Y 5AS, United Kingdom
525 B Street, Suite 1800, San Diego, CA 92101-4495, United States
50 Hampshire Street, 5th Floor, Cambridge, MA 02139, United States
The Boulevard, Langford Lane, Kidlington, Oxford OX5 1GB, United Kingdom

Notices
Knowledge and best practice in this field are constantly changing. As new research and experience broaden our understanding, changes in research methods, professional practices, or medical treatment may become necessary.

Practitioners and researchers must always rely on their own experience and knowledge in evaluating and using any information, methods, compounds, or experiments described herein. In using such information or methods they should be mindful of their own safety and the safety of others, including parties for whom they have a professional responsibility.

To the fullest extent of the law, neither the Publisher nor the authors, contributors, or editors, assume any liability for any injury and/or damage to persons or property as a matter of products liability, negligence or otherwise, or from any use or operation of any methods, products, instructions, or ideas contained in the material herein.

Library of Congress Cataloging-in-Publication Data
A catalog record for this book is available from the Library of Congress

British Library Cataloguing-in-Publication Data
A catalogue record for this book is available from the British Library

ISBN: 978-0-12-802218-4

Publisher: Nikki Levy
Acquisition Editor: Emily Ekle
Editorial Project Manager: Barbara Makinster
Production Project Manager: Punithavathy Govindaradjane
Cover Designer: Matthew Limbert

Typeset by SPi Global, India

Contents

II

COGNITION AND EMOTION IN MATHEMATICAL ACTIVITY

3. Being in Control

ALEXANDRE BOROVIK

4. Epistemic States of Convincement. A Conceptualization from the Practice of Mathematicians and Neurobiology

MIRELA RIGO-LEMINI, BENJAMÍN MARTÍNEZ-NAVARRO

5. The Impact of Anxiety and Working Memory on Algebraic Reasoning

KELLY TREZISE, ROBERT A. REEVE

III

EMOTIONS IN THE LEARNING AND TEACHING OF MATHEMATICS

IIIA

LEARNERS IN DIFFERENT EDUCATIONAL LEVELS

IIIB

LEARNERS WITH MATHEMATICAL DIFFICULTIES

IIIC

LEARNERS OUT OF THE SCHOOL

IV

THEORETICAL ADVANCES

Contributors

Mike Askew University of the Witwatersrand, Johannesburg, South Africa; Monash University, Melbourne, VIC, Australia

Sophie Batchelor Loughborough University, Loughborough, United Kingdom

Alexandre Borovik University of Manchester, Manchester, United Kingdom

Laurinda Brown University of Bristol, Bristol, United Kingdom

Sara Cordes Boston College, Chestnut Hill, MA, United States

Camilla Gilmore Loughborough University, Loughborough, United Kingdom

Inés M. Gómez-Chacón Universidad Complutense de Madrid, Madrid, Spain

Janet Goodall University of Bath, Bath, United Kingdom

Jennifer Hall Monash University, Melbourne, VIC, Australia

Tracy Helliwell University of Bristol, Bristol, United Kingdom

Paul Hernandez-Martinez Loughborough University, Loughborough, United Kingdom

Michelle Hurst Boston College, Chestnut Hill, MA, United States

Matthew Inglis Loughborough University, Loughborough, United Kingdom

Sarah E. Ives California State University, Sacramento, CA, United States

Sue Johnston-Wilder University of Warwick, Coventry, United kingdom

Clare Lee The Open University, Milton Keynes, United Kingdom

Lyndon C. Martin York University, Toronto, ON, Canada

Benjamín Martínez-Navarro The Center for Research and Advanced Studies of Mexico's National Poly-technical Institute, Mexico City, Mexico

John Mason University of Oxford, Oxford; Open University, Milton Keynes, United Kingdom

Martina Metz University of Calgary, Calgary, AB, Canada

Maria Pampaka The University of Manchester, Manchester, United Kingdom

Robert A. Reeve University of Melbourne, Melbourne, VIC, Australia

David Reid University of Bremen, Bremen, Germany

Mirela Rigo-Lemini The Center for Research and Advanced Studies of Mexico's National Poly-technical Institute, Mexico City, Mexico

Melissa Rodd UCL Institute of Education, London, United Kingdom

Wolff-Michael Roth University of Victoria, Victoria, BC, Canada

Rosemary Russell AR & RR Education Ltd, Poole, United Kingdom

Kelli M. Slaten Georgia Gwinnett College, Lawrenceville, GA, United States

Miwa A. Takeuchi University of Calgary, Calgary, AB, Canada

Jo Towers University of Calgary, Calgary, AB, Canada

Kelly Trezise University of Melbourne, Melbourne, VIC, Australia

Hamsa Venkat University of the Witwatersrand, Johannesburg, South Africa

Margaret Walshaw Massey University, Palmerston North, New Zealand

Ulises Xolocotzin Eligio Centre for Research and Advanced Studies, Mexico City, Mexico

Preface

Ask yourself and the people around you: how do you feel about mathematics? You will find that the question makes sense. It is only natural to invoke a diverse range of emotions in describing our relationship with mathematics. Doing mathematics is clearly an activity that is rich in emotional experiences. And yet, it is rather difficult to explain what the nature and implications of such experiences are. Scholars studying mathematical thinking and learning have traditionally concentrated on cognitive, social, cultural, developmental, technological, and neural factors. It is only fairly recently that they have turned to the study of emotions.

During the last 20 years, research that investigates the ways in which emotions relate to mathematics has expanded rapidly in number, breadth, and depth. Researchers are delivering insights about the ways in which individuals' emotions influence, and are influenced by, the individual and environmental factors involved in using, learning, teaching, and investigating mathematics. These findings are presented in academic events, discussed in book chapters, and reported in academic journals. However, to date no edited book has been published that focuses specifically on the emotional aspects of mathematics.

This volume collects contributions that advance our current understanding of the links between emotions and mathematics. This topic is relevant across disciplines, but opportunities for researchers to become aware of the work done in fields other than their own are lacking and much needed. This book includes contributions from an international group that includes young researchers and leading figures from disciplines such as mathematics education, psychology, and mathematics. The reader will be able to appreciate the theoretical and methodological diversity that is applied across disciplines.

Understanding Emotions in Mathematical Thinking and Learning will be of interest for researchers, graduate students, and teachers. The assembled chapters present information on the current state of the field, novel research trends, innovative takes on established research lines (e.g., mathematics anxiety), and emergent theoretical views.

The chapters are organized into four sections. The first presents an overview of the field. The reader will find a review that describes the origins and development of the research trends that drive the current literature (Chapter 1), and the proposal of a holistic approach that integrates a number of widely used methodologies for investigating the interrelationships between cognition and affect in mathematics (Chapter 2).

The second section covers the interaction between cognition and emotion during mathematical activity, including an account of the emotions felt by mathematicians in their quest to achieve a sense of control over mathematical objects (Chapter 3), a characterization of the epistemic states associated with convincement in mathematics as a kinds of emotion (Chapter 4), and a review of the interactive relationships between mathematics anxiety and cognitive abilities in the context of algebraic problem solving (Chapter 5).

The third section covers some of the ways in which emotions are involved in the learning and teaching of mathematics. This theme has motivated the development of several research lines. This is reflected in the size of this section, which is divided in four parts. The first part deals with the emotional experiences of learners in different educational levels, including a qualitative study on the emotions of elementary school students while doing mathematics (Chapter 6), a mixed-methods study of the role that emotions play in the decline of students' dispositions towards mathematics throughout secondary school (Chapter 7), and a review of the ways in which affective experiences interact with beliefs during the transition to postsecondary mathematics education (Chapter 8).

The second part addresses learners with mathematical difficulties and introduces a psychoanalytical approach for conceptualizing and attending to special needs in the mathematics classroom from a relational perspective (Chapter 9), and outlines the construct of "Mathematical resilience," which entails positive attributes that help learners and teachers deal with negativity towards mathematics (Chapter 10).

The third part deals with learners outside school, and it presents an intervention that has positively changed the emotions experienced by a mother and daughter dyad while doing mathematics at home (Chapter 11), and an investigation on the influence of parents in the emergence of mathematics anxiety during early childhood (Chapter 12). The fourth part presents large-scale interventions for mathematics teachers. One addresses the fear, dislike, and anxiety towards mathematics of primary school teachers with a workshop based on emotional and embodied games and activities (Chapter 13). The other one explores the benefits of writing assignments among preservice elementary teachers who have negative emotions towards mathematics (Chapter 14).

The fourth section presents the underlying concepts and potential applications of recent theoretical advances. The reader will be introduced to a phenomenological approach that combines Dual Systems Theory with the notion of a bicameral brain to incorporate emotions in the analysis of mathematical explorations (Chapter 15), a societal-historical theory to study emotions in general, and in particular mathematics anxiety, in the context of the mathematics classroom (Chapter 16), and an approach to study the decision-making of mathematics teachers that integrates concepts such as emotional orientation and somatic markers (Chapter 17).

The experiences, insights, and findings shared by the contributors to this volume let us see that studying the links between emotions and mathematics is a challenging and fascinating endeavor. Hopefully, this book will help the reader to appreciate the potential of a research domain that is likely to help us achieve a deeper understanding of the foundations of mathematical thinking and learning.

ACKNOWLEDGMENTS

I would like to thank all of the authors for their kind attention and contributions to this project. I am also grateful to the colleagues who helped me to shape this volume with insightful comments and advice.

SECTION I

INTRODUCTION: AN OVERVIEW OF THE FIELD

CHAPTER

1

An Overview of the Growth and Trends of Current Research on Emotions and Mathematics

Ulises Xolocotzin Eligio

Centre for Research and Advanced Studies, Mexico City, Mexico

INTRODUCTION

This chapter describes contemporary research on the relationship between emotions and mathematics. Scholars interested in mathematical thinking and learning avoided studying emotional issues for many years. Consequently, most of what we know about the topic revolves around cognitive, social, and technological issues. However, studies addressing emotional issues are no longer rare. In specialized journals, books, and academic events, it is increasingly common to find research in which the emotions of individuals learning, teaching, performing, and researching mathematics are described, explained, or influenced. Thus, the time seems suitable to reflect on the ways in which the relationship between emotions and mathematics has been investigated.

This chapter offers a birds'-eye view of current research across disciplines. Rather than summarizing key findings and advances, the aim is to identify patterns of growth and current research trends. Emotions are a pervasive component of human experience, and they can relate to mathematics in many different ways. This review identifies the specific links between emotions and mathematics that have attracted the interest of researchers. In this way, we can reflect on the sort of emotional issues and views of emotions that are shaping the current literature. This could be an antecedent to the analyses of results, theories, or research methods. These very important aspects, however, are beyond the scope of the current chapter.

Understanding Emotions in Mathematical Thinking and Learning

http://dx.doi.org/10.1016/B978-0-12-802218-4.00001-7

A couple of clarifications shall be made. First, my use of the term "mathematical activity" refers to activities that produce mathematical knowledge, such as learning, teaching, using, and researching mathematics. Second, contemporary research refers to the work produced during the last two decades.

Specifically, this review aims to answer three research questions:

1. What is the growth pattern of the literature linking emotions to mathematics?
2. In which contexts of mathematical activity have emotions been studied?
3. What are the current trends in the literature linking emotions and mathematics?

The remainder of the chapter is organized as follows. First, there is a brief summary of the historical context of contemporary research. This is followed by a description of the methods employed for collecting and analyzing the literature. Then, the results of the analysis are discussed by answering the research questions presented above. The chapter concludes with a discussion of the current state of the literature.

First Half of the 20th Century: Early Explorations

Nowadays, researchers study a range of emotional phenomena in relation to mathematics. It is worth recalling that emotions were widely considered worth studying until after the second half of the 20th century. Before that, emotions were largely neglected by the dominant approaches to mathematical thinking and learning.

The views of mathematical activity that dominated the first half of the 20th century did not include the belief that emotions could be related to mathematics. For example, Gillette (1901) argued that the abstract nature of mathematical truths makes them distant from mundane experiences. Therefore, mathematics does not provoke any emotional reactions in the layman. Following this rationale, it is logical to conclude that emotions and mathematics are fundamentally separated dimensions of human experience. Views of this kind, however, were about to change soon.

Polya (1945) proposed the integration of emotions in the analysis of mathematical activity, specifically in relation to problem solving. He argued that "teaching to solve problems is education of the will" (p. 94), suggesting that students need to become familiar with the emotional struggle that is required to find a solution. At about the same time, the study of mathematics anxiety appeared and became the first research line that connected mathematics with emotional phenomena.

The construct *Mathematics Anxiety* is rooted in accounts of what was described as "mathemaphobia" (Gough, 1954), or "number anxiety" (Dreger

& Aiken, 1957). Richardson and Suinn (1972) defined mathematics anxiety as the feelings of tension and anxiety that disrupt the manipulation of numbers and the solving of mathematical problems. A wealth of research has been produced in an attempt to understand the origins of math anxiety and alleviate its consequences.

Thorough reviews by Ashcraft (2002), Hembree (1990), and Ma (1999) outline the conclusions of 30 years of research on mathematics anxiety. First, it is clear that mathematics anxiety dampens mathematical performance. However, there is not enough evidence to conclude that poor performance causes anxiety. The effects of mathematics anxiety seem to be more pronounced in males than in females. The negative relationship between anxiety and performance, however, is stable across grade levels and ethnic groups. Individuals who suffer from mathematics anxiety tend to avoid mathematics and learn less when they are exposed to the topic. Nevertheless, it has been proved that anxiety does not relate to overall intelligence. Research on mathematics anxiety is constantly expanding. Extensive reviews of the state of the art can be consulted in Suárez-Pellicioni, Núñez-Peña, and Colomé (2016) and Chang and Beilock (2016).

This chapter concentrates on literature outside the mathematics anxiety domain. This research line played a crucial role in emotions being recognized as an inherent component of mathematical activity. Moreover, scholars working on the topic produce large amounts of literature—so much that it can be considered to be a standalone research domain that requires a complete volume for its review. However, mathematics anxiety is only one way in which emotions and mathematics can be related.

Anxiety mainly refers to the experience of fear. There are many more emotions that can be experienced in relation to mathematics. Also, performance is the main concern in mathematics anxiety research. Although this is important, emotions are likely to influence a wider range of issues involved in the process and outcome of mathematical activity.

A variety of emotional issues other than mathematics anxiety have attracted the attention of researchers. Below we will see that the amount of literature that stems from these interests is growing steadily. Before we continue, it is worth remembering where the current trends in emotion research originate. The following section describes the historical context in which contemporary research lines emerged.

Second Half of the 20th Century: Contemporary Approaches Emerge

The Generalized Turn to Emotions

The modern interest for understanding the links between emotions and mathematics appeared in the context of a generalized turn to the study of emotions. Towards the end of the 20th century and after decades of neglect, emotions started to regain the attention of disciplines such as psychology, philosophy, and neuroscience.

The renovated interest for emotions resulted in the development of conceptual advances that established the foundations of contemporary emotion research. These include, for instance, the outline of information-processing mechanisms that trigger emotional responses (Frijda, 1986; Ortony, Clore, & Collins, 1990), the characterization of evolutionarily defined universal basic emotions (Ekman, 1992), the turn to the social and cultural dimensions of emotions (Harré, 1989; Parkinson, 1996), and the groundbreaking hypothesis that people's emotions and their corresponding bodily sensations lie at the core of human rationality (Damasio, 1994).

Works started to appear showing that emotions and rational thinking are not separated, but closely intertwined. Examples of this include accounts of the role that emotions play during scientific and mathematical discovery (Burton, 2004; Thagard, 2002). The study of emotions is nowadays a well-established and interdisciplinary research domain that transcends its original niche to reach technological fields. Concepts such as "affective computing" (Picard, 1997) and "emotional design" (Norman, 2004) created awareness of the need to consider users' emotions. Nowadays, the capacity to detect and respond to people's emotions is as a requirement for designing everyday technology.

The turn to the study of emotions resonated among mathematical activity scholars. Below we review the developments that set the ground from which the modern approaches to study the relationship between emotions and mathematics emerged.

The Beginning of Contemporary Research

The widespread interest for studying emotions was embraced by researchers interested in mathematical activity. Around the 1990s, theoretical and empirical investigations started to appear that addressed a range of emotional phenomena, especially in mathematics education.

The book *Affect and Mathematical Problem Solving* by McLeod and Adams (1989) summarized the then state of the art in the study of affect in mathematics education (Hannula, 2014), and it is arguably one of the most influential works on the topic (Zan, Brown, Evans, & Hannula, 2006). The ideas proposed in this book challenged the dominant perspective on mathematical problem solving, namely cognitive science. The authors claimed that affective factors such as beliefs, attitudes, and emotions play critical roles in mathematical problem solving, and they introduced concepts that became central in the years to come. These included, for example, student confidence; interest and curiosity; the need to integrate emotion with mathematical cognition; metacognition; the personal, social, and cultural determinants of the emotional experience of mathematics; emotions in the classroom; and teachers' emotions.

McLeod and Adams summarized a period of pioneering research that introduced affective issues. Shortly after this, a theoretical article by Goldin (2000) pioneered the study of specific momentary emotional states emerging during mathematical problem solving. This work proposed a dynamic model of emotional states that can either support the realization of a solution—for example, curiosity, puzzlement, and pleasure—or inhibit useful cognitive activity—for example, frustration, anxiety, and fear.

Mathematics education has played an important role in advancing the study of emotions in relation to mathematics. A number of the ideas that drive the current literature were introduced in the first special issue covering affect in mathematics education, published in *Educational Studies in Mathematics* in 2006.

The editors explained that emotions were the least frequently investigated of the three affective constructs introduced by McLeod and Adams (1989) (Zan et al., 2006). The articles in this special issue introduced innovative perspectives and constructs, such as socioconstructivism (Eynde, Corte, & Verschaffel, 2006); somatic markers (Brown & Reid, 2006); social semiotics and psychoanalysis (Evans, Morgan, & Tsatsaroni, 2006); meta-affect (DeBellis & Goldin, 2006); and the links between motivation, goals, and emotions (Hannula, 2006). The articles collected in this special issue have become key references, as is indicated by their record of 136 citations.

METHOD

Databases Searched and Search Terms

The data were collected from searches in databases such as PsychInfo, EBSCO Host, Math Educ, IEEE, and ERIC, as well as in citation databases such as Web of Science and Scopus. The term entered in the search engines was "(Emotions OR Feelings OR Moods) AND Mathematics." When available in the interface, specifications were made for searching the term in the title, abstract, and keywords.

The term "affect" was not used in order to avoid the return of a large amount of irrelevant entries. The words affect and "emotion" are often used indistinctively (e.g., Forgas, 2008). However, affect is commonly employed as an umbrella term that includes constructs other than emotions, such as beliefs, attitudes, motivation, and self-efficacy, among others (e.g., Pepin & Roesken-Winter, 2015). Since publications that use the term affect for referring to emotions are likely to include both terms in the keywords or abstract (e.g., Hannula, 2014), one might argue that the probability of missing relevant entries after excluding the term is likely to be low.

Selection Criteria for Including Articles

In addition to the search term, additional criteria were used for narrowing the search to identify empirical research, published in peer-reviewed journals until 2015 and written in English. The focus on empirical research, both qualitative and quantitative, favored the intention of providing information on the trends in the study of the relationship between emotions and mathematics. Theoretical works are certainly valuable but, given their intrinsically abstract nature, they are less informative of the populations, contexts, and specific themes that attract the interest of researchers.

Regarding the focus on journal articles, it is fair to recognize that a considerable amount of research is reported in conference proceedings, book chapters, and project reports. However, these outlets tend to present preliminary and/or exploratory studies, from which further developments are reported in journal articles. Publications written in English are selected because they are the most numerous and accessible for wider audiences—it is common for researchers from non-English speaking countries to publish in English-written journals.

Data Screening

The search returned 459 results in total. The collected entries were screened in order to collate those matching the selection criteria. Research linking emotions and mathematics is not widespread, yet these words are widely used. Therefore, it was not surprising that the search returned a large number of irrelevant entries, 361. These entries were rejected for at least one the following reasons:

Not written in English. The abstract or keywords were written in English, but the actual content was written in other language; for example, see Botella (2012).

Nonempirical content. These included items such as introductions to special issues, editorials, or theoretical articles. Examples include works by Davis (2007), Dowker, Ashcraft, and Krinzinger (2012), and Rodd (2006).

Not addressing emotions or mathematics. Some articles used words related to emotions—for example, feelings, moods, etc.—or referred to mathematical activity in the abstract or keywords, but their content did not address either in any specific manner. For example, some articles used the word emotions, but their actual content did not approach a specific emotional issue (e.g., Rüppel, Liersch, & Walter, 2015; Weiland & Yoshikawa, 2013; Weis, Heikamp, & Trommsdorff, 2013). Others included measures with emotional content, for example, socioemotional skills, but said very little about emotions (e.g., Duncan et al., 2007; Hagborg, 1990).

Another set of articles mentioned the word "mathematics" but did not address this subject specifically. For example, studies of "academic emotions" that involved research on a range of subjects, for example, STEM subjects, did not discuss mathematics (e.g., Bieg, Goetz, & Hubbard, 2013; Flunger, Pretsch, Schmitt, & Ludwig, 2013).

Not studying relationships between emotions and mathematics. Studies of this kind investigated the effects of the stress or anxiety caused by doing mathematics on a physiological or cognitive variable. They reported experiments in which mathematical activity, e.g., an arithmetic test, was used as an experimental stimulus for inducing anxiety or stress. Responses to the mathematical stimulus were then compared to responses to a control stimulus, e.g., a reading test (Papousek, Paechter, & Lackner, 2011; Radstaak, Geurts, Brosschot, Cillessen, & Kompier, 2011; Vlemincx, Taelman, De Peuter, Van Diest, & Van Den Bergh, 2011).

Mathematical analysis of emotional data. The search showed articles that reported computational techniques for the modeling of emotions or the analysis of emotion-related data (e.g., Adamatzky, 2003; Costantini & Perugini, 2014). These were excluded since they do not address the relationship between emotions and mathematics.

Data Analysis

After the initial screening, 98 entries were selected for analysis. These articles were diverse in terms of the themes, methods, populations, and overarching views on emotions. Therefore, the analytical strategy consisted of organizing the articles into categories that were useful for addressing the research questions. For example, data used for determining the growth pattern of the literature linking emotions to mathematics (RQ1) required a publication breakdown by year. To determine the contexts of mathematical activity in which have emotions been studied (RQ2), information was needed about the populations and scenarios addressed in each article. Additionally, discovering trends in the literature linking emotions and mathematics (RQ3) required a review of the contents in each article—examining their research questions, theories, objectives, and methods, in order to construct categories of research trends.

RESULTS

Fig. 1 illustrates how the articles were organized in categories according to three dimensions, namely year of publication, research context, and research trends. The following section presents a description of the different categories in order to answer the research questions. Some examples are given to illustrate the content of the different categories. The complete list of articles and their corresponding cross-classification can be consulted in Appendix A.

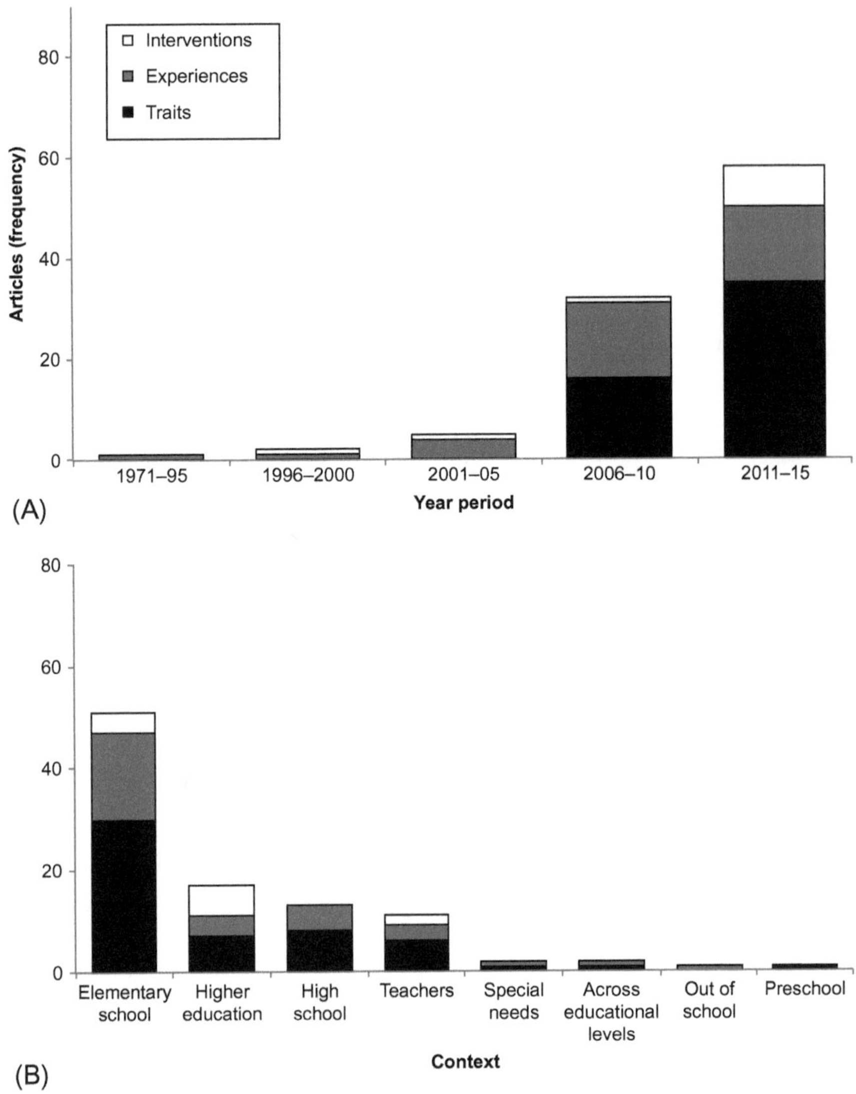

FIG. 1 Number of articles reported in each research theme—intervention, experience, and trait—by year period (A) and context (B).

What Is the Growth Pattern of the Literature Linking Emotions to Mathematics?

After an initial inspection of the articles' publication years, it was observed that the literature growth could be best described in five periods. The first one, ranging from 1971 to 1995, captured the oldest entry from 1975 and illustrated the state of affairs before the advent of current research. The remaining 5-year periods described the development of contemporary literature during the last 20 years.

Fig. 1A shows the distribution of articles in each period, with a breakdown by research theme that is discussed in a subsequent section. It is striking that before 1995, only one published article reported an empirical investigation linking emotions to mathematics. This article presents the case of a patient with autism who experienced positive emotions in relation to certain prime numbers (Park & Youderian, 1975). I believe that this illustrates very well that, except for the case of mathematics anxiety, the role of emotions in mathematical activity was largely neglected at that time.

Seven articles were published in the 1996–2005 period, which can be considered the beginning of contemporary literature. The incipient interest about emotions could have been motivated by two key events, namely the turn to emotions observed across the humanities and social sciences and the publication of "Affect and Mathematical Problem Solving" by McLeod and Adams (1989).

The articles published during this period indicate that the topic of emotions started to gain a place within the study of mathematical thinking and learning. Researchers began addressing phenomena that is especially relevant to doing mathematics, such as the *aha!* experience (Liljedahl, 2005), and introduced pioneering theoretical models for studying the emotions experienced while learning mathematics (Gómez-Chacón, 2000; Kleine, Goetz, Pekrun, & Hall, 2005). Although small in number, there were also some articles that addressed emotional issues related to mathematics from a cognitive psychology perspective (e.g., Albert & Hayward, 2002).

The data shows that the literature started flourishing about a decade ago, in 2006. There were 32 articles published in the period ranging from 2006 to 2010, which is a remarkable growth in comparison to what was published in the previous period. Moreover, this period not only brought an increase in the volume of literature, but also a remarkable diversification of the approaches and topics addressed.

Research from mathematics education had an important development during this period, which include the publication of the first journal special number on affect edited by Zan et al. (2006). This was certainly a milestone considering that the articles in this issue introduced novel research lines in mathematics education. At the same time, studies that addressed the mechanisms underlying the emotional experience of mathematics from a cognitive psychology perspective also increased in number (e.g., Goetz, Frenzel, Hall, & Pekrun, 2008; Johns, Inzlicht, & Schmader, 2008).

Two other important developments occurred in this period. First, articles reporting methodological developments were published, including, for instance, the improvement of questionnaires for assessing emotional orientations (Frenzel, Thrash, Pekrun, & Goetz, 2007c), and the introduction of less traditional methods relying on drawings (Rule & Harrell, 2006; Zambo & Zambo, 2006). Second, the links between emotions and mathematics started to be explored in out-of-school scenarios (Else-Quest, Hyde, & Hejmadi, 2008; Gyllenhaal, 2006).

The 2011–15 period has been the most productive one, since 58 articles were published, representing 59.18% of the literature. Considering the previous periods, it is possible to appreciate a remarkable growth in the literature. Some research lines became firmly established. For example, there were theoretical developments that articulated emotions with other affective constructs, for example, attitudes (Di Martino & Zan, 2011). At the same time, other theoretical proposals appeared (Lewis, 2013). Also, there was an increase in the interest for studying specific emotions (Daschmann, Goetz, & Stupnisky, 2011; Tulis & Ainley, 2011) and categories of emotions that are particularly relevant for the production of mathematical knowledge, in other words, epistemic emotions (Muis, Psaradellis, Lajoie, Di Leo, & Chevrier, 2015). The accelerated expansion of the literature is indicative of a thriving research domain. Below we discuss the contexts of mathematical activity in which emotions have been studied.

In Which Contexts of Mathematical Activity Have Emotions Been Studied?

The population addressed in the articles indicated the contexts of mathematical activity in which researchers have decided to study emotions. In this way, research contexts were relatively straightforward to identify. A first inspection showed that virtually all of the articles were conducted in formal education contexts. With this in mind, nine categories were defined from the data, including: preschool, elementary school (K-12, elementary and secondary school), high school, higher education (including undergraduate and graduate), learners out of school (including adult education), learners with special needs, teacher education, and research across educational levels.

Fig. 1B shows the distribution of articles conducted in each context. Of the articles, 53.12% (51) report investigations in elementary school. The following contexts in decreasing frequency are the following: higher education, 17.34% (17); high school, 12.24% (12); and teacher education, 11.22% (11). The least frequent categories were special education and research across educational levels, with each having 2.04% (2), and out-of-school and preschool, with 1.02% (1) each.

One can only wonder why most research is conducted in elementary and secondary school scenarios. This probably reflects the fact that mathematics education research has traditionally concentrated on primary education, although postsecondary education, teacher education, and preschool education have gained attention (Niss, 2004). Another reason might be practical; more than half of the articles (32) conducted in elementary school report quantitative studies with large samples that sometimes reached the thousands (e.g., Costa & Faria, 2015; Frenzel, Goetz, Luedtke, Pekrun, & Sutton, 2009; Frenzel, Pekrun, & Goetz, 2007a, 2007b; Goetz, Preckel, Pekrun, & Hall, 2007).

The proportion of articles with large samples is smaller in studies involving higher education (3), high school (3), and teacher education (4). Articles with large samples identify patterns of relationships between affective constructs, by analyzing questionnaire data with statistical techniques that are very sensitive to the sample size. Arguably, conditions that favor this kind of methodology, such as the accessibility to a population that is somehow homogeneous in their mathematics instruction, are easier to obtain in primary school scenarios.

Categorizing the contexts addressed in the literature permitted the identification of the kinds of populations that require more attention. It is certainly positive that higher education, high school, and teacher education are starting to attract the interest of researchers. However, mathematics-related emotional issues among learners in preschool, individuals with special needs, and learners out of school have barely been discussed. The study of emotional issues with these populations could be seen as quite challenging. For example, one might think of the potential challenges involved in studying children's emotional experiences. However, there is evidence to suggest that methods that are widely used with adults, for example, self-reports, are also reliable for young children (Durbin, 2010).

There is also need for more longitudinal or cross-sectional studies that analyze emotional issues across educational levels. These designs can reveal whether the relationship between emotions and mathematics is reciprocal and/or changes over time. Finally, it is worth mentioning the populations that are yet to be addressed, such as professional mathematicians and adult learners. Emotional issues are irrelevant in these contexts; Burton (2004) explained that positive emotions are an essential component of professional mathematical practice, and the pioneering work by Evans (2000) showed that adults' mathematical activities are always emotional.

What Are the Current Trends in the Literature Linking Emotions and Mathematics?

In order to answer this question, the content of each article was reviewed in order to identify similarities and differences in the type of emotional issues addressed, theoretical frameworks, and research methods. These characteristics revealed distinctive ways of relating emotions to mathematics, and they are categorized in three research trends: emotional traits, emotional states, and emotional interventions. A second order of categories was defined within each research trend, which described different groups of articles determined by their specifics themes. Table 1 shows the frequencies of articles within each research trend and theme. What these trends entail and the sort of themes addressed within each trend are described below.

TABLE 1 Frequencies of Articles Organized by Specific Themes Within Research Trends

Research theme	Description	*f*
Emotional traits		
Emotions as outcomes	Aim to identify the sources of individual differences in emotional orientations, usually using differences in affective constructs (e.g., goals, values, self-efficacy, identity), and contextual variables such as academic domain or classroom environment	31
Emotions as predictors	Assess whether emotional orientations can predict a range of outcome variables, mainly performance but also other affective constructs such as attitudes and self-regulation	17
Trait research methods	Report the development of questionnaires for assessing emotional orientations	4
Emotions as predictors and outcomes	Study the dynamics of the relationship between emotional orientations and other affective constructs and performance. Involves cross-sectional and longitudinal studies	2
Emotional experiences		
Emotion-cognition	Explore the links between emotional states and cognitive mechanisms during mathematical tasks	14
Emotion-environment	Explore the ways in which emotional states relate to cultural and social factors during mathematical tasks	10
Descriptions of emotional states	Study the qualities of the emotional states experienced during mathematical tasks	7
State research Methods	Report methodological tools for exploring individuals' emotional states during mathematical tasks	1
Emotional interventions		
Nonemotional instruction	Report interventions involving training in nonemotional skills expected to impact individuals' emotions	2
Emotional instruction	Report the effects of instruction on skills that are thought to directly impact individuals' emotions, e.g., metacognition and emotion control	6
Emotional resources	Introduce educational resources designed to improve individuals' emotional experiences while doing mathematics	4

(Continued)

Emotional Traits

Research following this trend approaches the relationship between emotions and mathematics from an individual differences perspective. This trend is most discussed, representing 55.10% (54) of the literature. This approach capitalizes on the fact that people differ in their dispositions to feel certain emotions. Trait differences in emotional reactivity increase the probability of experiencing trait-congruent emotions. For instance, individuals with higher trait anxiety are more likely to feel anxious. Individual differences in emotion orientation can also explain why individuals react differently to similar situations (Revelle & Scherer, 2009).

This kind of research treats emotions as traits. From this point of view, individuals have an emotional orientation towards mathematics, which is stable and determined by individual differences in affective factors such as attitudes, goals, beliefs, motivation, etc. Studies in this research trend follow the hypothesis that emotional orientations and their interactions with affective and cognitive constructs are reliable predictors of mathematics performance.

It is difficult to explain why this trend is predominant. It might be due to the fact that most of the research on emotions and mathematics has been conducted in school contexts. In mathematics education, the study of emotional phenomena intends to predict students' behavior (Di Martino & Zan, 2011). Diversity is the rule rather than the exception in the mathematics classroom, and emotional orientations can explain why some students engage with mathematics in more enjoyable and successful ways than others.

The study of emotional traits is growing steadily. Fig. 1A shows that the first articles to discuss this appeared in the 2006–10 period (16), and the amount rapidly expanded during the 2011–15 period (35). Also, this research trend has presence in most of the mathematical activity contexts, except for the out-of-school category. Four themes were identified within this research trend: emotional outcomes, emotional predictors, trait research methods, and emotions as outcomes and predictors.

Emotional outcomes (31) represent most of the emotional traits literature, and the aim is to identify the sources of individual differences in emotional orientations towards mathematics. Pekrun's control-value theory (CVT), (Pekrun, 1992) is invoked by many of these works (12). CVT provides a framework for studying *academic emotions* and *achievement emotions*, including pride, hope, anger, frustration, and boredom. These emotions are caused by appraisals of control and values (Pekrun, 2006).

Articles that use CVT support the predictions of this theory; this means, for example, that achievement and academic emotions in relation to mathematics can be predicted by individual differences in goals (Jang & Liu, 2012), values (Frenzel, Pekrun, & Goetz, 2007b), self-concepts (Goetz et al., 2008), and perceived control and value in the learning environment (Frenzel, Pekrun, & Goetz, 2007a).

Other articles framed by CVT investigate the domain specificity of academic emotions, reporting, for example, that these are more closely related within the domain of mathematics than in other domains (Goetz, Frenzel, Pekrun, Hall, & Luedtke, 2007).

Articles that do not rely on CVT (19) offer diverse theoretical perspectives. These include, for example, studies linking emotional traits with attitudes (Di Martino & Zan, 2011), self-efficacy (Putwain & Remedios, 2014), and implicit theories of emotions (Romero, Master, Paunesku, Dweck, & Gross, 2014).

Emotional predictors (17) assess the extent to which emotional orientations predict individual differences from a range of variables. Most studies are designed to predict mathematics achievement (e.g., Johns et al., 2008; Kyttälä & Björn, 2010). However, other outcome variables are also studied, including, for example, the image of mathematics (Lane, Stynes, & O'Donoghue, 2014) and attitudes towards learning technologies (Baya'a & Daher, 2013; Lee & Yuan, 2010).

Although the usage of CVT is not as frequent as in the *emotional outcomes* theme, some works refer to constructs derived from this theory—for instance, academic and achievement emotions (Bailey, Taasoobshirazi, & Carr, 2014; Villavicencio & Bernardo, 2013). Also, emotional orientations are related to a variety of constructs, including self-regulation (Chatzistamatiou, Dermitzaki, & Bagiatis, 2014; Cho & Heron, 2015) and emotional intelligence (Costa & Faria, 2015), for example.

Research methods (4). All of these articles report the development of questionnaires for assessing emotional constructs derived from CVT, such as achievement emotions (Frenzel, Thrash, et al., 2007c; Peixoto, Mata, Monteiro, Sanches, & Pekrun, 2015), as well as specific emotions such as boredom (Daschmann et al., 2011) and interest (Wininger, Adkins, Inman, & Roberts, 2014).

Emotions as predictors and outcomes (2). These articles study the dynamics of mutual influence between emotional orientations and variables such as mathematics performance (Ahmed, van der Werf, Kuyper, & Minnaert, 2013), beliefs, and flow experience (Stephanou, 2011).

Emotional States

Articles following this trend represent the second largest proportion of the literature, 32.65%, (32). These works investigate how emotions are experienced during mathematical activity. Unlike the articles in the *emotional traits* themes, these articles do not try to explain stable emotional orientations towards mathematics. Instead, their aim is to investigate the determinants of emotional states and the implications that such states have for the process and outcome of mathematical activity.

Fig. 1A shows that this research trend has always been present in the literature. There was a significant increase in the 2006–10 period to reach 15 articles. The number of articles in the 2011–15 period was also 15. This indicates that there is interest for this research line, but it is not having the accelerated growth of the emotional traits one. Nevertheless, its presence across contexts is worth noting. Four specific themes were identified within this research trend, including studies addressing interactions between emotion-cognition and emotion-environment, descriptions of emotional states, and research methods.

Emotion-cognition interactions (14). These articles focus on internal processes, addressing a range of ways in which emotions might interact with cognitive mechanisms during mathematical activity. The majority (8) study problem-solving situations, and report a number of interesting findings. For example, negative emotions derive from both beliefs and difficulties for deploying key cognitive abilities required for solving a problem, such as processing symbols, structuring a problem, and thinking flexibly (Gómez-Chacón, 2000). Mathematically sophisticated learners tend to experience more diverse emotional states during problem-solving (Allen & Carifio, 2007), and epistemic emotions influence self-regulation and performance positively (Muis et al., 2015).

Emotion-cognition interactions have also been investigated in mathematical activities other than problem solving. Some studies describe the positive relationship between abstract reasoning and positive emotions in test situations (Goetz, Preckel, et al., 2007). Studies of emotion-cognition interactions in the classroom investigate teachers' efforts to regulate their emotions (Sutton, Mudrey-Camino, & Knight, 2009), and the benefits of aha! experiences (Liljedahl, 2005). There is one laboratory study showing that emotions influence the perception of quantities (Droit-Volet, 2013).

Emotion-environment interactions (10). These articles investigate the links between individuals' emotional states and some of the social and cultural processes involved in mathematical activity. Half of these articles address everyday classroom practices. They identify some determinants of students' emotional states, including, for instance, teachers' understandability, illustration, pace, and difficulty (Goetz, Luedtke, Nett, Keller, & Lipnevich, 2013), and their responses to incorrect answers (Tainio & Laine, 2015; Tulis, 2013). Similarly, some studies identify environmental determinants of teachers' emotional states, including students' motivation and discipline (Becker, Keller, Goetz, Frenzel, & Taxer, 2015), and access to professional and personal support for coping with stress (Mujtaba & Reiss, 2013).

Some articles in this theme approach emotions from a perspective that coordinates cognitive, affective, and social factors. These articles study issues such as the emotions associated with discursive positioning during problem solving (Daher, 2015; Evans et al., 2006) and the emotional expressions employed for constructing identity in a mathematical task (Heyd-Metzuyanim & Sfard, 2012). A couple of studies investigate girls' emotional reactions to perceived gender unfairness and its negative consequences for performance during mathematical testing (Keller & Dauenheimer, 2003; Mangels, Good, Whiteman, Maniscalco, & Dweck, 2012).

Descriptions of emotional states (7). These articles characterize the emotional states experienced during mathematical activity, without discussing cognitive or environmental factors. Their specific aims are diverse. For example, testing whether Pekrun's model for permanent emotional orientations towards mathematics can also be applied to describe emotional states during specific situations (Dettmers et al., 2011; Kleine et al., 2005).

Other studies describe the emotions felt by individuals in out-of-school mathematical activities, such as doing homework (Else-Quest et al., 2008) and visiting mathematics exhibits (Gyllenhaal, 2006). Finally, other articles focus on rarely addressed emotions, such as aesthetic experience (Brinkmann, 2009) or ideas that have great potential for systematically explaining the emotional experience of mathematics, such as flow theory (Schweinle, Turner, & Meyer, 2008) and the cognitive structure of emotions theory (Martínez-Sierra & García González, 2014).

State research methods (1). Only one article was dedicated to introducing a methodological tool for studying the experience of emotions during mathematical tasks. This study reported the potential of using drawings to determine children's feelings about mathematics (Zambo & Zambo, 2006).

Emotional Interventions

Articles following this trend are the least frequent, 12.5% (12). These works investigate ways of intentionally influencing individuals' emotional states to improve the experience and outcome of mathematical activity. Fig. 1A shows that this trend has had a small presence of representation since the 1996–2000 period, with modest growth in the 2011–15 period. The trend is growing, albeit slowly. Fig. 1B shows that these studies have been conducted in elementary schools, higher education, and with teachers. There is much diversity in the theoretical and methodological approaches of the reported interventions; however, three research themes are identifiable within this trend.

Nonemotional instruction. These works report interventions that involve instruction in nonemotional skills that are expected to have an impact on individuals' emotions—for instance, training on parental involvement to generate positive feelings towards mathematics homework (Van Voorhis, 2011), or the provision of an inquiry learning environment to facilitate positive emotions towards mathematics (Dogan, 2012).

Emotional instruction. These articles implement training in a range of skills that are thought to be useful for gaining control over one's own emotions. These include, for instance, metacognition for student teachers and for university students who struggle with mathematics (Zan, 2000; Caballero, Blanco, & Guerrero, 2011), control of intrusive thoughts in a mathematics student (Albert & Hayward, 2002), and mindfulness for engineering students in a calculus course (Bellinger, DeCaro, & Ralston, 2015).

Emotional resources. Some articles report the effects of educational materials designed to positively impact students' emotions. Some examples include the usage of humor in paper-based materials to generating interest (Matarazzo, Durik, & Delaney, 2010), the implementation of feedback to create positive emotions during computerized assessment (Liu et al., 2015), and the development of adaptive tutoring systems that detect and react to students' emotions while learning mathematics (Arroyo, Burleson, Tai, Muldner, & Woolf, 2013; Arroyo et al., 2014).

CONCLUSIONS

The data shows that current research regarding the relationships between emotions and mathematics is thriving. Over the last two decades there has been a significant growth in the amount of empirical studies, addressing a range of emotional phenomena but mostly focusing on educational settings. The literature seems to follow at least three research trends. The most prominent one, which is rapidly growing, focuses on the causes and consequences of individual differences in emotional orientations towards mathematics. A second trend that has gained presence but is not expanding studies the ways in which emotional states interact with cognitive, social, and environmental factors during mathematical activity. The third trend, which is just emerging, involves the exploration of instruction and resources designed for improving mathematical activity by intentionally influencing emotions.

Perhaps researchers feel attracted to the study of emotional orientations because these orientations are part of the system of constructs that compose the affective domain. These constructs are firmly established, considering both the longstanding history of their conceptual foundations and the large amount of data that supports their validity and usefulness. Moreover, one might argue that the stability of emotional orientations allow for more stable predictions of learners' behavior. There is a firm ground in which to stand for researchers interested in this way of linking emotions to mathematics. In the future, important progress is expected for this research line.

More research is necessary for understanding the ways in which emotional states interact with cognitive, social, and environmental factors during mathematical activity. Very little is known about the way in which the temporary emotional states that seem to be inherent to the experience of doing mathematics participate in the production of mathematical knowledge. The amount of literature can be seen as little, considering the recognized relevance of transitory emotional states during mathematical tasks (Goldin, 2014; Hannula, 2015).

One reason for the slow growth of this research trend might be the relative lack of adequate research methods for accessing individuals' emotions in "real time" while doing mathematical tasks. This is not an exclusive problem of the researchers studying mathematical thinking and learning. The problem of accessing individuals' emotions is inherited from the larger context of disciplines that study emotional phenomena. Additionally, there is a shortage of theories for conceptualizing the relationship between emotional states and the cognitive and social processes that are activated during mathematical activity. Thus, future research should address both conceptual and methodological challenges for developing this research line.

It is not surprising that articles reporting interventions are still so rare. After all, only 20 years have passed since the study of emotions started flourishing. On the one hand, one could say that the development of interventions for influencing emotions will flourish until more basic research lines, in the study of emotional traits and states for instance, achieve more mature stages. On the other hand, at least in educational research, it is quite common for theoretical developments to be sustained in applied research. Perhaps one way to advance our understanding of the links between mathematics and emotions is by implementing more intervention studies.

The research on emotions and mathematics is definitively growing. However, it is worth noting that the current scope does not cover all of the possible relations between emotions and mathematics. First, more research is needed for attending forms of mathematical activity other than learning and teaching in school scenarios. For example, we know very little about the emotional issues involved in the mathematical activity of young learners, adults out of school, and professional mathematicians.

The current literature can be expanded in depth and breadth by integrating a wider range of schools of thought. Diversity in conceptual frameworks will precipitate the development of more research lines to follow, and the achievement of new discoveries. I would like to mention a couple of examples; one refers to embodied cognition, which has been embraced with enthusiasm among scholars in fields such as mathematics education, yet very little research has been conducted for addressing emotions from this perspective (Brown & Reid, 2006).

A second example refers to cognitive psychology, which has a long-standing tradition in the study of mathematical thinking and learning. To date, this approach has concentrated on low-level cognitive mechanisms such as number processing (Droit-Volet, 2013). It is well documented that emotions interact with higher-level mechanisms such as reasoning (Blanchette & Richards, 2010). This interaction might have important implications for mathematical activity, but these are yet to be explored.

Before concluding, it is worth mentioning some limitations of the current review. First, every effort was made to capture a representative sample of the literature on emotions and mathematics. However, given the focus on journal articles outside the math anxiety domain, a number of relevant works on the topic were not included (e.g., Radford, 2015; Roth & Walshaw, 2015). Second, this research offers an overview of the literature. The intention is to reflect on the current directions of research. Future complementary reviews will be required to analyze the theories, methods, and main findings of the literature.

To conclude, current research in emotions and mathematics is growing rapidly. Future research should examine a wider range of perspectives in order to extend and complement the current views. This is a formidable endeavor, since the study of emotions implies huge conceptual and practical challenges. However, this should not discourage us. We are just starting to appreciate the potential benefits of understanding the way in which emotions work during mathematical activity.

APPENDIX A ARTICLES REVIEWED AND THEIR CLASSIFICATIONS BY YEAR PERIOD, RESEARCH CONTEXT, AND RESEARCH TREND

Article	Research trend	Context
Period 1971–95		
Park, D., & Youderian, P. (1975). Light and number: Ordering principles in the world of an autistic child. *Journal of Autism and Childhood Schizophrenia*, *4*(4), 313–323.	Emotional states	Special needs

Continued

Article	Research trend	Context
Period 1996–2000		
Gómez-Chacón, I. M. (2000). Affective influences in the knowledge of mathematics. *Educational Studies in Mathematics*, *43*(2), 149–168.	Emotional states	Higher education
Zan, R. (2000). A metacognitive intervention in mathematics at university level. *International Journal of Mathematical Education in Science & Technology*, *31*(1), 143–150.	Emotional intervention	Higher education
Period 2001–06		
Efklides, A., & Petkaki, C. (2005). Effects of mood on students' metacognitive experiences. *Learning and Instruction*, *15*(5), 415–431.	Emotional states	Elementary education
Kleine, M., Goetz, T., Pekrun, R., & Hall, N. (2005). The structure of students' emotions experienced during a mathematical achievement test. *ZDM—International Journal on Mathematics Education*, *37*(3), 221–225.	Emotional states	Elementary education
Keller, J., & Dauenheimer, D. (2003). Stereotype threat in the classroom: Dejection mediates the disrupting threat effect on women's math performance. *Personality and Social Psychology Bulletin*, *29*(3), 371–381.	Emotional states	High school
Liljedahl, P. G. (2005). Mathematical discovery and affect: The effect of AHA! experiences on undergraduate mathematics students. *International Journal of Mathematical Education in Science and Technology*, *36*(2–3), 219–234.	Emotional states	Higher education
Albert, I., & Hayward, P. (2002). Treatment of intrusive ruminations about mathematics. *Behavioural and Cognitive Psychotherapy*, *30*(2), 223–226.	Emotional intervention	Higher education

Article	Research trend	Context
Period 2006–10		
Frenzel, A. C., Goetz, T., Luedtke, O., Pekrun, R., & Sutton, R. E. (2009). Emotional transmission in the classroom: Exploring the relationship between teacher and student enjoyment. *Journal of Educational Psychology*, *101*(3), 705–716.	Emotional traits	Elementary education
Frenzel, A. C., Pekrun, R., & Goetz, T. (2007). Girls and mathematics—A "hopeless" issue? A control-value approach to gender differences in emotions towards mathematics. *European Journal of Psychology of Education*, *22*(4), 497–514.	Emotional traits	Elementary education
Frenzel, A. C., Thrash, T. M., Pekrun, R., & Goetz, T. (2007). Achievement emotions in Germany and China—A cross-cultural validation of the academic emotions questionnaire-mathematics. *Journal of Cross-Cultural Psychology*, *38*(3), 302–309.	Emotional traits	Elementary education
Frenzel, A. C., Pekrun, R., & Goetz, T. (2007). Perceived learning environment and students' emotional experiences: A multilevel analysis of mathematics classrooms. *Learning and Instruction*, *17*(5), 478–493.	Emotional traits	Elementary education
Goetz, T., Cronjaeger, H., Frenzel, A. C., Ludtke, O., & Hall, N. C. (2010). Academic self-concept and emotion relations: Domain specificity and age effects. *Contemporary Educational Psychology*, *35*(1), 44–58.	Emotional traits	Elementary education
Goetz, T., Frenzel, A. C., Hall, N. C., & Pekrun, R. (2008). Antecedents of academic emotions: Testing the internal/external frame of reference model for academic enjoyment. *Contemporary Educational Psychology*, *33*(1), 9–33.	Emotional traits	Elementary education

Continued

Article	Research trend	Context
Goetz, T., Frenzel, A. C., Pekrun, R., & Hall, N. C. (2006). The domain specificity of academic emotional experiences. *Journal of Experimental Education*, *75*(1), 5–29.	Emotional traits	Elementary education
Goetz, T., Frenzel, A. C., Pekrun, R., Hall, N. C., & Luedtke, O. (2007). Between- and within-domain relations of students' academic emotions. *Journal of Educational Psychology*, *99*(4), 715–733.	Emotional traits	Elementary education
Stevens, T., Olivrez, A., & Hamman, D. (2006). The role of cognition, motivation, and emotion in explaining the mathematics achievement gap between Hispanic and White students. *Hispanic Journal of Behavioral Sciences*, *28*(2), 161–186.	Emotional traits	Elementary education
Kyttala, M., & Bjorn, P. M. (2010). Prior mathematics achievement, cognitive appraisals and anxiety as predictors of Finnish students' later mathematics performance and career orientation. *Educational Psychology*, *30*(4), 431–448.	Emotional traits	High school
Lee, C.-Y., & Yuan, Y. (2010). Gender differences in the relationship between Taiwanese adolescents' mathematics attitudes and their perceptions toward virtual manipulatives. *International Journal of Science and Mathematics Education*, *8*(5), 937–950.	Emotional traits	High school
Johns, M., Inzlicht, M., & Schmader, T. (2008). Stereotype threat and executive resource depletion: Examining the influence of emotion regulation. *Journal of Experimental Psychology: General*, *137*(4), 691–705.	Emotional traits	Higher education
Smith, C. A., & Kirby, L. D. (2009). Relational antecedents of appraised problem-focused coping potential and its associated emotions. *Cognition and Emotion*, *23*(3), 481–503.	Emotional traits	Higher education

Article	Research trend	Context
Preckel, F., Goetz, T., & Frenzel, A. (2010). Ability grouping of gifted students: Effects on academic self-concept and boredom. *British Journal of Educational Psychology, 80*(3), 451–472.	Emotional traits	Special needs
Hodgen, J., & Askew, M. (2007). Emotion, identity and teacher learning: Becoming a primary mathematics teacher. *Oxford Review of Education, 33*(4), 469–487.	Emotional traits	Teachers
Kunter, M., Tsai, Y.-M., Klusmann, U., Brunner, M., Krauss, S., & Baumert, J. (2008). Students' and mathematics teachers' perceptions of teacher enthusiasm and instruction. *Learning and Instruction, 18*(5), 468–482.	Emotional traits	Teachers
Näring, G., Briët, M., & Brouwers, A. (2006). Beyond demand-control: Emotional labour and symptoms of burnout in teachers. *Work and Stress, 20*(4), 303–315.	Emotional traits	Teachers
Allen, B. D., & Carifio, J. (2007). Mathematical sophistication and differentiated emotions during mathematical problem solving. *Journal of Mathematics and Statistics, 3*(4), 163–167.	Emotional states	Elementary education
DeBellis, V. A., & Goldin, G. A. (2006). Affect and meta-affect in mathematical problem solving: A representational perspective. *Educational Studies in Mathematics, 63*(2), 131–147.	Emotional states	Elementary education
Else-Quest, N. M., Hyde, J. S., & Hejmadi, A. (2008). Mother and child emotions during mathematics homework. *Mathematical Thinking & Learning, 10*(1), 5–35.	Emotional states	Elementary education
Evans, J., Morgan, C., & Tsatsaroni, A. (2006). Discursive positioning and emotion in school mathematics practices. *Educational Studies in Mathematics, 63*(2), 209–226.	Emotional states	Elementary education

Continued

Article	Research trend	Context
Goetz, T., Preckel, F., Pekrun, R., & Hall, N. C. (2007). Emotional states during test taking: Does cognitive ability make a difference? *Learning and Individual DIfferences*, *17*(1), 3–16.	Emotional states	Elementary education
Schweinle, A., Turner, J., & Meyer, D. (2008). Understanding young adolescents' optimal experiences in academic settings. *Journal of Experimental Education*, *77*(2), 125–143.	Emotional states	Elementary education
Zambo, D., & Zambo, R. (2006). Using thought bubble pictures to assess students' feelings about mathematics. *Mathematics Teaching in the Middle School*, *12*(1), 14–21.	Emotional states	Elementary education
Hannula, M. S. (2006). Motivation in mathematics: Goals reflected in emotions. *Educational Studies in Mathematics*, *63*(2), 165–178.	Emotional states	High school
Op't Eynde, P., De Corte, E., & Verschaffel, L. (2006). "Accepting emotional complexity": A socio-constructivist perspective on the role of emotions in the mathematics classroom. *Educational Studies in Mathematics*, *63*(2), 193–207.	Emotional states	High school
Brown, L., & Reid, D. A. (2006). Embodied cognition: Somatic markers, purposes and emotional orientations. *Educational Studies in Mathematics*, *63*(2), 179–192.	Emotional states	Higher education
Gyllenhaal, E. D. (2006). Memories of math: Visitors' experiences in an exhibition about calculus. *Curator: The Museum Journal*, *49*(3), 345–364.	Emotional states	Out of school
Sutton, R. E., Mudrey-Camino, R., & Knight, C. C. (2009). Teachers' emotion regulation and classroom management. *Theory Into Practice*, *48*(2), 130–137.	Emotional states	Teachers

Article	Research trend	Context
Brinkmann, A. (2009). Mathematical beauty and its characteristics—A study on the students' points of view. *Montana Mathematics Enthusiast*, *6*(3), 365–380.	Emotional states	Across educational levels
Matarazzo, K. L., Durik, A. M., & Delaney, M. L. (2010). The effect of humorous instructional materials on interest in a math task. *Motivation and Emotion*, *34*(3), 293–305.	Emotional intervention	Higher education
Boylan, M. (2009). Engaging with issues of emotionality in mathematics teacher education for social justice. *Journal of Mathematics Teacher Education*, *12*(6), 427–443.	Emotional intervention	Teachers
Period 2011–15		
Stephanou, G. (2014). Feelings towards child-teacher relationships, and emotions about the teacher in kindergarten: Effects on learning motivation, competence beliefs and performance in mathematics and literacy. *European Early Childhood Education Research Journal*, *22*(4), 457–477.	Emotional traits	Preschool
Ahmed, W., van der Werf, G., Kuyper, H., & Minnaert, A. (2013). Emotions, self-regulated learning, and achievement in mathematics: A growth curve analysis. *Journal of Educational Psychology*, *105*(1), 150–161.	Emotional traits	Elementary education
Bieg, M., Goetz, T., & Lipnevich, A. A. (2014). What students think they feel differs from what they really feel—Academic self-concept moderates the discrepancy between students' trait and state emotional self-reports. *PLOS ONE*, *9*(3).	Emotional traits	Elementary education
Buff, A. (2014). Enjoyment of learning and its personal antecedents: Testing the change-change assumption of the control-value theory of achievement emotions. *Learning and Individual Differences*, *31*, 21–29.	Emotional traits	Elementary education

Continued

Article	Research trend	Context
Buff, A., Reusser, K., Rakoczy, K., & Pauli, C. (2011). Activating positive affective experiences in the classroom: ``Nice to have{"} or something more? *Learning and Instruction*, *21*(3), 452–466.	Emotional traits	Elementary education
Costa, A., & Faria, L. (2015). The impact of emotional intelligence on academic achievement: A longitudinal study in Portuguese secondary school. *Learning and Individual Differences*, *37*, 38–47.	Emotional traits	Elementary education
Daschmann, E. C., Goetz, T., & Stupnisky, R. H. (2011). Testing the predictors of boredom at school: Development and validation of the precursors to boredom scales. *British Journal of Educational Psychology*, *81*(3), 421–440.	Emotional traits	Elementary education
de Corte, E., Depaepe, F., Eynde, P. O., Verschaffel, L., Corte, E., Depaepe, F., … Verschaffel, L. (2011). Students' self-regulation of emotions in mathematics: An analysis of meta-emotional knowledge and skills. *ZDM—International Journal on Mathematics Education*, *43*(4), 483–495.	Emotional traits	Elementary education
Frenzel, A. C., Pekrun, R., Dicke, A.-L., & Goetz, T. (2012). Beyond quantitative decline: Conceptual shifts in adolescents' development of interest in mathematics. *Developmental Psychology*, *48*(4), 1069–1082.	Emotional traits	Elementary education
Galla, B. M., & Wood, J. J. (2012). Emotional self-efficacy moderates anxiety-related impairments in math performance in elementary school-age youth. *Personality and Individual Differences*, *52*(2), 118–122.	Emotional traits	Elementary education
Goetz, T., Frenzel, A. C., Luedtke, O., & Hall, N. C. (2011). Between-domain relations of academic emotions: Does having the same instructor make a difference? *Journal of Experimental Education*, *79*(1), 84–101.	Emotional traits	Elementary education

Article	Research trend	Context
Goetz, T., Haag, L., Lipnevich, A. A., Keller, M. M., Frenzel, A. C., & Collier, A. P. M. (2014). Between-domain relations of students' academic emotions and their judgments of school domain similarity. *Frontiers in Psychology*, *5*.	Emotional traits	Elementary education
Goetz, T., Nett, U. E., Martiny, S. E., Hall, N. C., Pekrun, R., Dettmers, S., & Trautwein, U. (2012). Students' emotions during homework: Structures, self-concept antecedents, and achievement outcomes. *Learning and Individual Differences*, *22*(2, SI), 225–234.	Emotional traits	Elementary education
Jang, L. Y., & Liu, W. C. (2012). 2 x 2 Achievement goals and achievement emotions: A cluster analysis of students' motivation. *European Journal of Psychology of Education*, *27*(1), 59–76.	Emotional traits	Elementary education
Peixoto, F., Mata, L., Monteiro, V., Sanches, C., & Pekrun, R. (2015). The achievement emotions questionnaire: Validation for pre-adolescent students. *European Journal of Developmental Psychology*, *12*(4), 472–481.	Emotional traits	Elementary education
Putwain, D., & Remedios, R. (2014). Fear appeals used prior to a high-stakes examination: What makes them threatening? *Learning and Individual Differences*, *36*, 145–151.	Emotional traits	Elementary education
Roesken, B., Hannula, M. S., & Pehkonen, E. (2011). Dimensions of students' views of themselves as learners of mathematics. *ZDM—International Journal on Mathematics Education*, *43*(4), 497–506.	Emotional traits	Elementary education
Romero, C., Master, A., Paunesku, D., Dweck, C. S., & Gross, J. J. (2014). Academic and emotional functioning in middle school: The role of implicit theories. *Emotion*, *14*(2), 227–234.	Emotional traits	Elementary education

Continued

Article	Research trend	Context
Stephanou, G. (2011). Students' classroom emotions: Socio-cognitive antecedents and school performance. *Electronic Journal of Research in Educational Psychology*, *9*(1), 5–48.	Emotional traits	Elementary education
Swanson, J., Valiente, C., Lemery-Chalfant, K., Bradley, R. H., & Eggum-Wilkens, N. D. (2014). Longitudinal relations among parents' reactions to children's negative emotions, effortful control, and math achievement in early elementary school. *Child Development*, *85*(5), 1932–1947.	Emotional traits	Elementary education
Tulis, M., & Ainley, M. (2011). Interest, enjoyment and pride after failure experiences? Predictors of students' state-emotions after success and failure during learning in mathematics. *Educational Psychology*, *31*(7), 779–807.	Emotional traits	Elementary education
Wininger, S. R., Adkins, O., Inman, T. F., & Roberts, J. (2014). Development of a student interest in mathematics scale for gifted and talented programming identification. *Journal of Advanced Academics*, *25*(4), 403–421.	Emotional traits	Elementary education
Bailey, M., Taasoobshirazi, G., & Carr, M. (2014). A multivariate model of achievement in geometry. *Journal of Educational Research*, *107*(6), 440–461.	Emotional traits	High school
Burić, I., & Sorić, I. (2012). The role of test hope and hopelessness in self-regulated learning: Relations between volitional strategies, cognitive appraisals and academic achievement. *Learning and Individual Differences*, *22*(4), 523–529.	Emotional traits	High school
Kim, C., Park, S. W., & Cozart, J. (2014). Affective and motivational factors of learning in online mathematics courses. *British Journal of Educational Technology*, *45*(1), 171–185.	Emotional traits	High school

Article	Research trend	Context
Lane, C., Stynes, M., & O'Donoghue, J. (2014). The image of mathematics held by Irish post-primary students. *International Journal of Mathematical Education in Science & Technology*, *45*(6), 879–891.	Emotional traits	High school
Ozgün-Koca, S. A., & Şen, A. I. (2011). Evaluation of beliefs and attitudes of high school sudents towards science and mathematics courses. *Journal of Turkish Science Education*, *8*(1), 42–60.	Emotional traits	High school
Winberg, T. M., Hellgren, J. M., & Palm, T. (2014). Stimulating positive emotional experiences in mathematics learning: Influence of situational and personal factors. *European Journal of Psychology of Education*, *29*(4), 673–691.	Emotional traits	High school
Cho, M.-H., & Heron, M. L. (2015). Self-regulated learning: The role of motivation, emotion, and use of learning strategies in students' learning experiences in a self-paced online mathematics course. *Distance Education*, *36*(1), 80–99.	Emotional traits	Higher education
Lewis, G. (2013). Emotion and disaffection with school mathematics. *Research in Mathematics Education*, *15*(1), 70–86.	Emotional traits	Higher education
Niculescu, A. C., Tempelaar, D., Leppink, J., Dailey-Hebert, A., Segers, M., & Gijselaers, W. (2015). Feelings and performance in the first year at university: Learning-related emotions as predictors of achievement outcomes in mathematics and statistics. *Electronic Journal of Research in Educational Psychology*, *13*(3), 431–462.	Emotional traits	Higher education
Stanisavljevic, D., Trajkovic, G., Marinkovic, J., Bukumiric, Z., Cirkovic, A., & Milic, N. (2014). Assessing attitudes towards statistics among medical students: Psychometric properties of the Serbian version of the survey of attitudes towards statistics (SATS). *PLOS ONE*, *9*(11).	Emotional traits	Higher education

Continued

Article	Research trend	Context
Villavicencio, F. T., & Bernardo, A. B. I. (2013). Positive academic emotions moderate the relationship between self-regulation and academic achievement. *British Journal of Educational Psychology*, *83*(2), 329–340.	Emotional traits	Higher education
Baya'a, N., & Daher, W. (2013). Mathematics teachers' readiness to integrate ICT in the classroom. *International Journal of Emerging Technologies in Learning*, *8*(1), 46–52.	Emotional traits	Teachers
Chatzistamatiou, M., Dermitzaki, I., & Bagiatis, V. (2014). Self-regulatory teaching in mathematics: Relations to teachers' motivation, affect and professional commitment. *European Journal of Psychology of Education*, *29*(2), 295–310.	Emotional traits	Teachers
Kunter, M., Frenzel, A., Nagy, G., Baumert, J., & Pekrun, R. (2011). Teacher enthusiasm: Dimensionality and context specificity. *Contemporary Educational Psychology*, *36*(4), 289–301.	Emotional traits	Teachers
Di Martino, P., & Zan, R. (2011). Attitude towards mathematics: A bridge between beliefs and emotions. *ZDM—International Journal on Mathematics Education*, *43*(4), 471–482.	Emotional traits	Across educational levels
Daher, W. (2015). Discursive positionings and emotions in modelling activities. *International Journal of Mathematical Education in Science and Technology*, *46*(8), 1149–1164.	Emotional states	Elementary education
Dettmers, S., Trautwein, U., Lüdtke, O., Goetz, T., Frenzel, A. C., & Pekrun, R. (2011). Students' emotions during homework in mathematics: Testing a theoretical model of antecedents and achievement outcomes. *Contemporary Educational Psychology*, *36*(1), 25–35.	Emotional states	Elementary education

Article	Research trend	Context
Goetz, T., Luedtke, O., Nett, U. E., Keller, M. M., & Lipnevich, A. A. (2013). Characteristics of teaching and students' emotions in the classroom: Investigating differences across domains. *Contemporary Educational Psychology*, *38*(4), 383–394.	Emotional states	Elementary education
Heyd-Metzuyanim, E., & Sfard, A. (2012). Identity struggles in the mathematics classroom: On learning mathematics as an interplay of mathematizing and identifying. *International Journal of Educational Research*, *51–52*, 128–145. The University of Haifa, Israel.	Emotional states	Elementary education
Muis, K. R., Psaradellis, C., Lajoie, S. P., Di Leo, I., & Chevrier, M. (2015). The role of epistemic emotions in mathematics problem solving. *Contemporary Educational Psychology*, *42*, 172–185.	Emotional states	Elementary education
Tainio, L., & Laine, A. (2015). Emotion work and affective stance in the mathematics classroom: The case of IRE sequences in Finnish classroom interaction. *Educational Studies in Mathematics*, *89*(1), 67–87.	Emotional states	Elementary education
Tulis, M. (2013). Error management behavior in classrooms: Teachers' responses to student mistakes. *Teaching and Teacher Education*, *33*, 56–68.	Emotional states	Elementary education
Tulis, M., & Fulmer, S. M. (2013). Students' motivational and emotional experiences and their relationship to persistence during academic challenge in mathematics and reading. *Learning and Individual Differences*, *27*, 35–46.	Emotional states	Elementary education
Martínez-Sierra, G., & García González, M. S. D. S. (2014). High school students' emotional experiences in mathematics classes. *Research in Mathematics Education*, *16*(3), 234–250.	Emotional states	High school

Continued

Article	Research trend	Context
Mangels, J. A., Good, C., Whiteman, R. C., Maniscalco, B., & Dweck, C. S. (2012). Emotion blocks the path to learning under stereotype threat. *Social Cognitive and Affective Neuroscience*, *7*(2), 230–241.	Emotional states	Higher education
Droit-Volet, S. (2013). Emotion and magnitude perception: Number and length bisection. *Frontiers in Neurorobotics*. 7(24)	Emotional states	Higher education
Becker, E. S., Keller, M. M., Goetz, T., Frenzel, A. C., & Taxer, J. L. (2015). Antecedents of teachers' emotions in the classroom: An intraindividual approach. *Frontiers in Psychology*, *6*.	Emotional states	Teachers
Mujtaba, T., & Reiss, M. (2013). Factors that lead to positive or negative stress in secondary school teachers of mathematics and science. *Oxford Review of Education*, *39*(5), 627–648.	Emotional states	Teachers
Arroyo, I., Woolf, B. P., Burelson, W., Muldner, K., Rai, D., & Tai, M. (2014). A multimedia adaptive tutoring system for mathematics that addresses cognition, metacognition and affect. *International Journal of Artificial Intelligence in Education*, *24*(4), 387–426.	Emotional intervention	Elementary education
Arroyo, I., Burleson, W., Tai, M., Muldner, K., & Woolf, B. P. (2013). Gender differences in the use and benefit of advanced learning technologies for mathematics. *Journal of Educational Psychology*, *105*(4), 957–969.	Emotional intervention	Elementary education
Dogan, H. (2012). Emotion, confidence, perception and expectation case of mathematics. *International Journal of Science and Mathematics Education*, *10*(1), 49–69.	Emotional intervention	Elementary education
Van Voorhis, F. L. (2011). Adding families to the homework equation: A longitudinal study of mathematics achievement. *Education and Urban Society*, *43*(3), 313–338.	Emotional intervention	Elementary education

Article	Research trend	Context
Bellinger, D. B., DeCaro, M. S., & Ralston, P. A. S. (2015). Mindfulness, anxiety, and high-stakes mathematics performance in the laboratory and classroom. *Consciousness and Cognition*, *37*, 123–132.	Emotional intervention	Higher education
Kim, C., & Hodges, C. B. (2012). Effects of an emotion control treatment on academic emotions, motivation and achievement in an online mathematics course. *Instructional Science*, *40*(1), 173–192.	Emotional intervention	Higher education
Liu, C.-J., Huang, C.-F., Liu, M.-C., Chien, Y.-C., Lai, C.-H., & Huang, Y.-M. (2015). Does gender influence emotions resulting from positive applause feedback in self-assessment testing? Evidence from Neuroscience. *Educational Technology & Society*, *18*(1), 337–350.	Emotional intervention	Higher education
Caballero, A., Blanco, L. J., & Guerrero, E. (2011). Problem solving and emotional education in initial primary teacher education. *Eurasia Journal of Mathematics, Science and Technology Education*, *7*(4), 281–292.	Emotional intervention	Teachers

References

Adamatzky, A. (2003). Affectons: Automata models of emotional interactions. *Applied Mathematics and Computation, 146*(2–3), 579–594.

Ahmed, W., van der Werf, G., Kuyper, H., & Minnaert, A. (2013). Emotions, self-regulated learning, and achievement in mathematics: A growth curve analysis. *Journal of Educational Psychology, 105*(1), 150–161.

Albert, I., & Hayward, P. (2002). Treatment of intrusive ruminations about mathematics. *Behavioural and Cognitive Psychotherapy, 30*(2), 223–226.

Allen, B. D., & Carifio, J. (2007). Mathematical sophistication and differentiated emotions during mathematical problem solving. *Journal of Mathematics and Statistics, 3*(4), 163–167.

Arroyo, I., Burleson, W., Tai, M., Muldner, K., & Woolf, B. P. (2013). Gender differences in the use and benefit of advanced learning technologies for mathematics. *Journal of Educational Psychology, 105*(4), 957–969.

Arroyo, I., Woolf, B. P., Burelson, W., Muldner, K., Rai, D., & Tai, M. (2014). A multimedia adaptive tutoring system for mathematics that addresses cognition, metacognition and affect. *International Journal of Artificial Intelligence in Education, 24*(4), 387–426.

Ashcraft, M. H. (2002). Math anxiety: Personal, educational, and cognitive consequences. *Current Directions in Psychological Science, 11*(5), 181–185.

Bailey, M., Taasoobshirazi, G., & Carr, M. (2014). A multivariate model of achievement in geometry. *Journal of Educational Research, 107*(6), 440–461.

Baya'a, N., & Daher, W. (2013). Mathematics teachers' readiness to integrate ICT in the classroom. *International Journal of Emerging Technologies in Learning, 8*(1), 46–52.

Becker, E. S., Keller, M. M., Goetz, T., Frenzel, A. C., & Taxer, J. L. (2015). Antecedents of teachers' emotions in the classroom: An intraindividual approach. *Frontiers in Psychology, 6*, 635.

Bellinger, D. B., DeCaro, M. S., & Ralston, P. A. S. (2015). Mindfulness, anxiety, and high-stakes mathematics performance in the laboratory and classroom. *Consciousness and Cognition, 37*, 123–132.

Bieg, M., Goetz, T., & Hubbard, K. (2013). Can I master it and does it matter? An intraindividual analysis on control-value antecedents of trait and state academic emotions. *Learning and Individual Differences, 28*, 102–108.

Blanchette, I., & Richards, A. (2010). The influence of affect on higher level cognition: A review of research on interpretation, judgement, decision making and reasoning. *Cognition & Emotion, 24*(4), 561–595.

Botella, J. M. (2012). Is there a relationship in primary education between affective factors in mathematics and academic performance? *Estudios Sobre Educacion, 23*, 141–155.

Brinkmann, A. (2009). Mathematical beauty and its characteristics—A study on the students' points of view. *Montana Mathematics Enthusiast, 6*(3), 365–380.

Brown, L., & Reid, D. A. (2006). Embodied cognition: Somatic markers, purposes and emotional orientations. *Educational Studies in Mathematics, 63*(2), 179–192.

Burton, L. L. 2004. *Mathematicians as enquirers: Learning about learning mathematics: Vol. 34.*

Caballero, A., Blanco, L. J., & Guerrero, E. (2011). Problem solving and emotional education in initial primary teacher education. *Eurasia Journal of Mathematics, Science and Technology Education, 7*(4), 281–292.

Chang, H., & Beilock, S. L. (2016). The math anxiety-math performance link and its relation to individual and environmental factors: A review of current behavioral and psychophysiological research. *Current Opinion in Behavioral Sciences, 10*, 33–38.

Chatzistamatiou, M., Dermitzaki, I., & Bagiatis, V. (2014). Self-regulatory teaching in mathematics: Relations to teachers' motivation, affect and professional commitment. *European Journal of Psychology of Education, 29*(2), 295–310.

Cho, M.-H., & Heron, M. L. (2015). Self-regulated learning: The role of motivation, emotion, and use of learning strategies in students' learning experiences in a self-paced online mathematics course. *Distance Education, 36*(1), 80–99.

Costa, A., & Faria, L. (2015). The impact of emotional intelligence on academic achievement: A longitudinal study in Portuguese secondary school. *Learning and Individual Differences, 37*, 38–47.

Costantini, G., & Perugini, M. (2014). Generalization of clustering coefficients to signed correlation networks. *PLoS ONE, 9*(2), 1–10.

Daher, W. (2015). Discursive positionings and emotions in modelling activities. *International Journal of Mathematical Education in Science and Technology, 46*(8), 1149–1164.

Damasio, A. 1994. *Descartes' error: Emotion, reason, and the human brain*. New York, NY: G.P. Putnam's Sons.

Daschmann, E. C., Goetz, T., & Stupnisky, R. H. (2011). Testing the predictors of boredom at school: Development and validation of the precursors to boredom scales. *British Journal of Educational Psychology, 81*(3), 421–440.

Davis, R. B. (2007). Emotion and thought. *Mathematics Teaching in the Middle School, 12*(9), 522–529.

DeBellis, V. A., & Goldin, G. A. (2006). Affect and meta-affect in mathematical problem solving: A representational perspective. *Educational Studies in Mathematics, 63*(2), 131–147.

Dettmers, S., Trautwein, U., Lüdtke, O., Goetz, T., Frenzel, A. C., & Pekrun, R. (2011). Students' emotions during homework in mathematics: Testing a theoretical model of antecedents and achievement outcomes. *Contemporary Educational Psychology, 36*(1), 25–35.

Di Martino, P., & Zan, R. (2011). Attitude towards mathematics: A bridge between beliefs and emotions. *ZDM—International Journal on Mathematics Education, 43*(4), 471–482.

Dogan, H. (2012). Emotion, confidence, perception and expectation case of mathematics. *International Journal of Science and Mathematics Education, 10*(1), 49–69.

Dowker, A., Ashcraft, M., & Krinzinger, H. (2012). The development of attitudes and emotions related to mathematics. *Child Development Research, 2012*, 1–3, article ID 238435 http://dx.doi.org/10.1155/2012/238435.

Dreger, R. M., & Aiken, L. R. (1957). The identification of number anxiety in a college population. *Journal of Educational Psychology, 48*(6), 344.

Droit-Volet, S. (2013). Emotion and magnitude perception: Number and length bisection. *Frontiers in Neurorobotics, 7*, 24.

Duncan, G. J., Dowsett, C. J., Claessens, A., Magnuson, K., Huston, A. C., Klebanov, P., et al. (2007). School readiness and later achievement. *Developmental Psychology, 43*(6), 1428–1446.

Durbin, C. E. (2010). Validity of young children's self-reports of their emotion in response to structured laboratory tasks. *Emotion, 10*(4), 519.

Ekman, P. (1992). An argument for basic emotions. *Cognition & Emotion, 6*(3), 169–200.

Else-Quest, N. M., Hyde, J. S., & Hejmadi, A. (2008). Mother and child emotions during mathematics homework. *Mathematical Thinking & Learning, 10*(1), 5–35.

Evans, J. (2000). *Adults' mathematical thinking and emotions. A study of numerate practices.* London: Routledge/Falmer.

Evans, J., Morgan, C., & Tsatsaroni, A. (2006). Discursive positioning and emotion in school mathematics practices. *Educational Studies in Mathematics, 63*(2), 209–226.

Eynde, P. O., Corte, E. D., & Verschaffel, L. (2006). Accepting emotional complexity: A socio-constructivist perspective on the role of emotions in the mathematics classroom. *Educational Studies in Mathematics, 63*(2), 193–207.

Flunger, B., Pretsch, J., Schmitt, M., & Ludwig, P. (2013). The role of explicit need strength for emotions during learning. *Learning and Individual Differences, 23*, 241–248.

Forgas, J. P. (2008). Affect and cognition. *Perspectives on Psychological Science, 3*(2), 94–101.

Frenzel, A. C., Goetz, T., Luedtke, O., Pekrun, R., & Sutton, R. E. (2009). Emotional transmission in the classroom: Exploring the relationship between teacher and student enjoyment. *Journal of Educational Psychology, 101*(3), 705–716.

Frenzel, A. C., Pekrun, R., & Goetz, T. (2007a). Perceived learning environment and students' emotional experiences: A multilevel analysis of mathematics classrooms. *Learning and Instruction, 17*(5), 478–493.

Frenzel, A. C., Pekrun, R., & Goetz, T. (2007b). Girls and mathematics—A "hopeless" issue? A control-value approach to gender differences in emotions towards mathematics. *European Journal of Psychology of Education, 22*(4), 497–514.

Frenzel, A. C., Thrash, T. M., Pekrun, R., & Goetz, T. (2007c). Achievement emotions in Germany and China—A cross-cultural validation of the academic emotions questionnaire-mathematics. *Journal of Cross-Cultural Psychology, 38*(3), 302–309.

Frijda, N. H. (1986). *The emotions*. Cambridge/New York/Paris: Ed. de la Maison des Sciences de l'Homme and Cambridge University Press.

Gillette, J. M. (1901). The relation of emotion to mathematical belief. *Psychological Review, 8*(6), 602–606.

Goetz, T., Frenzel, A. C., Hall, N. C., & Pekrun, R. (2008). Antecedents of academic emotions: Testing the internal/external frame of reference model for academic enjoyment. *Contemporary Educational Psychology, 33*(1), 9–33.

Goetz, T., Frenzel, A. C., Pekrun, R., Hall, N. C., & Luedtke, O. (2007a). Between- and within-domain relations of students' academic emotions. *Journal of Educational Psychology, 99*(4), 715–733.

Goetz, T., Luedtke, O., Nett, U. E., Keller, M. M., & Lipnevich, A. A. (2013). Characteristics of teaching and students' emotions in the classroom: Investigating differences across domains. *Contemporary Educational Psychology, 38*(4), 383–394.

Goetz, T., Preckel, F., Pekrun, R., & Hall, N. C. (2007b). Emotional experiences during test taking: Does cognitive ability make a difference? *Learning and Individual Differences*, *17*(1), 3–16.

Goldin, G. A. (2000). Affective pathways and representation in mathematical problem solving. *Mathematical Thinking and Learning*, *2*(3), 209–219.

Goldin, G. A. (2014). Perspectives on emotion in mathematical engagement, learning, and problem solving. In R. Pekrun & L. Linnenbrink-Garcia (Eds.), *International handbook of emotions in education* (pp. 391–414). New York, NY: Routledge.

Gómez-Chacón, I. M. (2000). Affective influences in the knowledge of mathematics. *Educational Studies in Mathematics*, *43*(2), 149–168.

Gough, M. F. (1954). Why failures in mathematics? Mathemaphobia: Causes and treatments. *The Clearing House: A Journal of Educational Strategies, Issues and Ideas*, *28*(5), 290–294.

Gyllenhaal, E. D. (2006). Memories of math: Visitors' experiences in an exhibiton about calculus. *Curator: The Museum Journal*, *49*(3), 345–364.

Hagborg, W. J. (1990). The Revised Behavior Problem Checklist and severely emotionally disturbed adolescents: Relationship to intelligence, academic achievement, and sociometric ratings. *Journal of Abnormal Child Psychology*, *18*(1), 47–53.

Hannula, M. S. (2006). Motivation in mathematics: Goals reflected in emotions. *Educational Studies in Mathematics*, *63*(2), 165–178.

Hannula, M. S. (2014). Affect in mathematics education. In S. Lerman (Ed.), *Encyclopedia of mathematics education* (pp. 23–27). Dordrecht: Springer Netherlands.

Hannula, M. S. (2015). Emotions in problem solving. In *Selected regular lectures from the 12th international congress on mathematical education* (pp. 269–288).

Harré, R. (1989). An outline of the social constructionist viewpoint. In R. Harré (Ed.), *The social construction of emotions* (pp. 2–13). Oxford/New York: Blackwell Publishing.

Hembree, R. (1990). The nature, effects, and relief of mathematics anxiety. *Journal for Research in Mathematics Education*, *21*(1), 33–46.

Heyd-Metzuyanim, E., & Sfard, A. (2012). Identity struggles in the mathematics classroom: On learning mathematics as an interplay of mathematizing and identifying. *International Journal of Educational Research*, *51–52*, 128–145.

Jang, L. Y., & Liu, W. C. (2012). 2x2 Achievement goals and achievement emotions: A cluster analysis of students' motivation. *European Journal of Psychology of Education*, *27*(1), 59–76.

Johns, M., Inzlicht, M., & Schmader, T. (2008). Stereotype threat and executive resource depletion: Examining the influence of emotion regulation. *Journal of Experimental Psychology: General*, *137*(4), 691–705.

Keller, J., & Dauenheimer, D. (2003). Stereotype threat in the classroom: Dejection mediates the disrupting threat effect on women's math performance. *Personality and Social Psychology Bulletin*, *29*(3), 371–381.

Kleine, M., Goetz, T., Pekrun, R., & Hall, N. (2005). The structure of students' emotions experienced during a mathematical achievement test. *ZDM—International Journal on Mathematics Education*, *37*(3), 221–225.

Kyttälä, M., & Björn, P. M. (2010). Prior mathematics achievement, cognitive appraisals and anxiety as predictors of finnish students' later mathematics performance and career orientation. *Educational Psychology*, *30*(4), 431–448.

Lane, C., Stynes, M., & O'Donoghue, J. (2014). The image of mathematics held by Irish post-primary students. *International Journal of Mathematical Education in Science & Technology*, *45*(6), 879–891.

Lee, C.-Y., & Yuan, Y. (2010). Gender differences in the relationship between Taiwanese adolescents' mathematics attitudes and their perceptions toward virtual manipulatives. *International Journal of Science and Mathematics Education*, *8*(5), 937–950.

Lewis, G. (2013). Emotion and disaffection with school mathematics. *Research in Mathematics Education*, *15*(1), 70–86.

Liljedahl, P. G. (2005). Mathematical discovery and affect: The effect of AHA! experiences on undergraduate mathematics students. *International Journal of Mathematical Education in Science and Technology*, *36*(2–3), 219–234.

Liu, C.-J., Huang, C.-F., Liu, M.-C., Chien, Y.-C., Lai, C.-H., & Huang, Y.-M. (2015). Does gender influence emotions resulting from positive applause feedback in self-assessment testing? Evidence from neuroscience. *Educational Technology & Society*, *18*(1), 337–350.

Ma, X. (1999). A meta-analysis of the relationship between anxiety toward mathematics and achievement in mathematics. *Journal for Research in Mathematics Education*, *30*(5), 520–540.

Mangels, J. A., Good, C., Whiteman, R. C., Maniscalco, B., & Dweck, C. S. (2012). Emotion blocks the path to learning under stereotype threat. *Social Cognitive and Affective Neuroscience*, *7*(2), 230–241.

Martínez-Sierra, G., & García González, M.S.D.S. (2014). High school students' emotional experiences in mathematics classes. *Research in Mathematics Education*, *16*(3), 234–250.

Matarazzo, K. L., Durik, A. M., & Delaney, M. L. (2010). The effect of humorous instructional materials on interest in a math task. *Motivation and Emotion*, *34*(3), 293–305.

McLeod, D. B., & Adams, V. M. 1989. *Affect and mathematical problem solving: A new perspective*. New York: Springer-Verlag.

Muis, K. R., Psaradellis, C., Lajoie, S. P., Di Leo, I., & Chevrier, M. (2015). The role of epistemic emotions in mathematics problem solving. *Contemporary Educational Psychology*, *42*, 172–185.

Mujtaba, T., & Reiss, M. (2013). Factors that lead to positive or negative stress in secondary school teachers of mathematics and science. *Oxford Review of Education*, *39*(5), 627–648.

Niss, M. (2004). Key issues and trends in research on mathematical education. In H. Fujita, Y. Hashimoto, B. R. Hodgson, P. Y. Lee, S. Lerman, & T. Sawada (Eds.), *Proceedings of the ninth international congress on mathematical education: 2000 Makuhari Japan* (pp. 37–57). Dordrecht: Springer Netherlands.

Norman, D. A. 2004. *Emotional design: Why we love (or hate) everyday things*. New York, NY: Basic Books.

Ortony, A., Clore, G. L., & Collins, A. 1990. *The cognitive structure of emotions*. New York: Cambridge University Press.

Papousek, I., Paechter, M., & Lackner, H. K. (2011). Delayed psychophysiological recovery after self-concept-inconsistent negative performance feedback. *International Journal of Psychophysiology*, *82*(3), 275–282.

Park, D., & Youderian, P. (1975). Light and number: Ordering principles in the world of an autistic child. *Journal of Autism and Childhood Schizophrenia*, *4*(4), 313–323.

Parkinson, B. (1996). Emotions are social. *The British Psychological Society*, *87*, 663–683.

Peixoto, F., Mata, L., Monteiro, V., Sanches, C., & Pekrun, R. (2015). The achievement emotions questionnaire: Validation for pre-adolescent students. *European Journal of Developmental Psychology*, *12*(4), 472–481.

Pekrun, R. (1992). The impact of emotions on learning and achievement: Towards a theory of cognitive/motivational mediators. *Applied Psychology*, *41*, 359–376.

Pekrun, R. (2006). The control-value theory of achievement emotions: Assumptions, corollaries, and implications for educational research and practice. *Educational Psychology Review*, *18*(4), 315–341.

Pepin, B., & Roesken-Winter, B. (2015). *From beliefs to dynamic affect systems in mathematics education. Exploring a mosaic of relationships and interactions*. Switzerland: Springer International Publishing.

Picard, R. W. (1997). *Affective computing*. Cambridge, MA: MIT Press.

Polya, G. 1945. *How to solve it: A new aspect of mathematical method*. Princeton, NJ: Princeton University Press.

Putwain, D., & Remedios, R. (2014). Fear appeals used prior to a high-stakes examination: What makes them threatening? *Learning and Individual Differences*, *36*, 145–151.

Radford, L. (2015). Of love, frustration, and mathematics: A cultural-historical approach to emotions in mathematics teaching and learning. In B. Pepin & B. Rösken-Winter (Eds.), *From beliefs to dynamic affect systems in mathematics education. Exploring a mosaic of relationships and interactions* (pp. 25–49). New York/Heidelberg/London: Springer International Publishing.

Radstaak, M., Geurts, S. A. E., Brosschot, J. F., Cillessen, A. H. N., & Kompier, M. A. J. (2011). The role of affect and rumination in cardiovascular recovery from stress. *International Journal of Psychophysiology*, *81*(3), 237–244.

Revelle, W., & Scherer, K. R. (2009). Personality and emotion. In D. Sander & K. R. Scherer (Eds.), *The Oxford companion to emotion and the affective sciences*. New York and Oxford: Oxford University Press.

Richardson, F. C., & Suinn, R. M. (1972). The mathematics anxiety rating scale: Psychometric data. *Journal of Counseling Psychology*, *19*(6), 551.

Rodd, M. (2006). Commentary: Mathematics, emotion and special needs. *Educational Studies in Mathematics*, *63*(2), 227–234.

Romero, C., Master, A., Paunesku, D., Dweck, C. S., & Gross, J. J. (2014). Academic and emotional functioning in middle school: The role of implicit theories. *Emotion*, *14*(2), 227–234.

Roth, W.-M., & Walshaw, M. (2015). Rethinking affect in education from a societal-historical perspective: The case of mathematics anxiety. *Mind, Culture, and Activity*, *22*(3), 217–232.

Rule, A. C., & Harrell, M. H. (2006). Symbolic drawings reveal changes in preservice teacher mathematics attitudes after a mathematics methods course. *School Science and Mathematics*, *106*(6), 241–258.

Rüppel, F., Liersch, S., & Walter, U. (2015). The influence of psychological well-being on academic success. *Journal of Public Health (Germany)*, *23*(1), 15–24.

Schweinle, A., Turner, J., & Meyer, D. (2008). Understanding young adolescents' optimal experiences in academic settings. *Journal of Experimental Education*, *77*(2), 125–143.

Stephanou, G. (2011). Students' classroom emotions: Socio-cognitive antecedents and school performance. *Electronic Journal of Research in Educational Psychology*, *9*(1), 5–48.

Suárez-Pellicioni, M., Núñez-Peña, M. I., & Colomé, À. (2016). Math anxiety: A review of its cognitive consequences, psychophysiological correlates, and brain bases. *Cognitive, Affective, & Behavioral Neuroscience*, *16*(1), 3–22.

Sutton, R. E., Mudrey-Camino, R., & Knight, C. C. (2009). Teachers' emotion regulation and classroom management. *Theory Into Practice*, *48*(2), 130.

Tainio, L., & Laine, A. (2015). Emotion work and affective stance in the mathematics classroom: The case of IRE sequences in Finnish classroom interaction. *Educational Studies in Mathematics*, *89*(1), 67–87.

Thagard, P. R. (2002). The passionate scientist: Emotion in scientific cognition. In P. Carruthers, S. P. Stich, & M. Siegal (Eds.), *The cognitive basis of science*. Cambridge, UK: Cambridge University Press.

Tulis, M. (2013). Error management behavior in classrooms: Teachers' responses to student mistakes. *Teaching and Teacher Education*, *33*, 56–68.

Tulis, M., & Ainley, M. (2011). Interest, enjoyment and pride after failure experiences? Predictors of students' state-emotions after success and failure during learning in mathematics. *Educational Psychology*, *31*(7), 779–807.

Van Voorhis, F. L. (2011). Adding families to the homework equation: A longitudinal study of mathematics achievement. *Education and Urban Society*, *43*(3), 313–338.

Villavicencio, F. T., & Bernardo, A. B. I. (2013). Positive academic emotions moderate the relationship between self-regulation and academic achievement. *British Journal of Educational Psychology*, *83*(2), 329–340.

Vlemincx, E., Taelman, J., De Peuter, S., Van Diest, I., & Van Den Bergh, O. (2011). Sigh rate and respiratory variability during mental load and sustained attention. *Psychophysiology*, *48*(1), 117–120.

Weiland, C., & Yoshikawa, H. (2013). Impacts of a prekindergarten program on children's mathematics, language, literacy, executive function, and emotional skills. *Child Development*, *84*(6), 2112–2130.

Weis, M., Heikamp, T., & Trommsdorff, G. (2013). Gender differences in school achievement: The role of self-regulation. *Frontiers in Psychology, 4*, 442.

Wininger, S. R., Adkins, O., Inman, T. F., & Roberts, J. (2014). Development of a student interest in mathematics scale for gifted and talented programming identification. *Journal of Advanced Academics, 25*(4), 403–421.

Zambo, D., & Zambo, R. (2006). Using thought bubble pictures to assess students' feelings about mathematics. *Mathematics Teaching in the Middle School, 12*(1), 14–21.

Zan, R. (2000). A metacognitive intervention in mathematics at university level. *International Journal of Mathematical Education in Science & Technology, 31*(1), 143–150.

Zan, R., Brown, L., Evans, J., & Hannula, M. S. (2006). Affect in mathematics education: An introduction. *Educational Studies in Mathematics, 63*(2), 113.

CHAPTER

2

Appraising Emotion in Mathematical Knowledge: Reflections on Methodology

Inés M. Gómez-Chacón
Universidad Complutense de Madrid, Madrid, Spain

INTRODUCTION

This chapter poses methodological questions around the evaluation of emotion in learning environments against a backdrop of the interaction between emotion and cognition. For most of the 20th century, research on emotions was conducted primarily by philosophers and psychologists, whose studies were based on a clearly defined suite of "theories of emotion" that continue to drive research in the area (Lewis, Haviland-Jones, & Barrett, 2008). In the 1980s, however, mathematics researchers such as McLeod (1992) pioneered the study of emotional response in mathematics, clearing the way for the implementation of a more systematic program beginning in the 1990s (e.g., DeBellis & Goldin, 2006; Leder, Pehkonen, & Törner, 2002; McLeod & Adams, 1989; Pepin & Rösken-Winter, 2015; Schoenfeld, 1992). Most of these studies focused on relatively stable affective traits such as attitudes, beliefs, and values, whereas emotion has been studied much less intensely (DeBellis & Goldin, 2006; Evans, 2000, 2003; Gómez-Chacón, 2000a, 2000b; Hannula, 2002; Op 't Eynde, De Corte, & Verschaffel, 2001).

Although interest in the matter is growing, it continues to be a minority field. For instance, judging from the keywords, only four of the 20 chapters of a book on recent progress in the affective dimension in mathematics teaching (edited by Pepin & Rösken-Winter, 2015) focus on emotion: Radford (2015), Di Martino and Zan (2015), Gómez-Chacón (2015), and Hannula (2015).

http://dx.doi.org/10.1016/B978-0-12-802218-4.00002-9

The initial premise is that emotional experience is difficult to detect because emotions are constructs (i.e., conceptual entities that cannot be directly measured) with fuzzy boundaries and with substantial interindividual variability in terms of expression and experience.

Aware of the pitfalls inherent in such complexity and consequently proceeding with all due precaution, this chapter poses questions that the author believes may drive the discipline in a forward direction.

More explicitly, it addresses both elements on which a consensus has been reached and others still open to debate and in need of further research, as summarized below.

1. Emotions are generally accepted to be significant elements in the key stages that determine success in solving a problem. In the late 1990s and early 2000s, certain experts (e.g., Goldin, 2000; Gómez-Chacón, 2000a, 2000b) coined the phrase "affective pathways" to describe how the typical patterns of emotional states lead to success or failure in solving problems and in the long run form part of a more stable system of attitudes, beliefs, and values held in connection with mathematics. Other more recently posed questions, which remain open, include the determination of affective pathways relating to the social and contextual dimensions of emotion and cognition, cognitive development along positive pathways, and the effect of metaemotion in task performance (Gómez-Chacón, 2011, 2012, 2015).
2. A number of affect-related phenomena (such as intellectual courage, astonishment, curiosity, interest, wonder, surprise, self-assuredness, doubt, fear of the unknown, joy of verification and of knowing) have been labeled "epistemic emotions or feelings" (Muis, Psaradellis, Lajoie, Di Leo, & Chevrier, 2015; Pekrun & Stephens, 2012). Several studies have explored how some of these epistemic emotions are beneficial for learning when they can be resolved through the use of appropriate learning strategies or via new lines of research, in which experts look for a better understanding of the role that emotions play in self-regulated learning.
3. The influence to which "individual" emotion is subject is highly cultural, with culture playing an instrumental role in shaping "inner" experience. Schoenfeld (2015) observed that in the context of emotion, the term "culture" needs to be "unpacked" (p. 400) and broached not from the assumption that cultures are uniform, but rather from more dynamic conceits such as those put forward by identity theories (Lave & Wenger, 1991). Studies that conceive of mathematical learning as a form of social participation (Abreu, 2002; Abreu, Crafter, Gorgorio, & Prat, 2013; Evans & Wedege, 2004; Gómez-Chacón & Figueral, 2007) or common cognition on learning mathematics as an interplay of mathematizing and identifying

(Heyd-Metzuyanim & Sfard, 2012) have revealed a need to address the plurality of meanings, appraisals, legitimacies, and identities that coexist in a classroom as well as the social management of that plurality. In such studies, the mathematics student is someone whose personal meanings do not always concur with the classroom-legitimized meanings, someone influenced by environmental appraisals and constantly reconstructing his or her identity in line with the activities in which he or she participates.

Many aspects of the aforementioned issues have been identified. This chapter draws from empirical data collected in earlier research (Gómez-Chacón, 1997, 2000a, 2000b) to take a closer look at items 1 and 3 above. The aim is to methodologically and theoretically determine affective pathways from a holistic approach to emotional experience and the study of the person in context. The aforementioned study was chosen because (a) it addresses a range of issues that are commonly studied separately and (b) it was conducted from two theoretical perspectives (the social-cognitive and the theory of identity approaches) and a combination of methods. The discussion is preceded by a section on emotion.

SYSTEMIC APPROACH TO THE STUDY OF EMOTIONS

Emotion is very likely one of the most basic of the objects of debate under the umbrella of affect. Researchers studying the psychology of emotion have used different approaches without agreeing on what an emotion actually is. Nonetheless, opinions have converged in certain respects. First, emotions are agreed to be connected to personal goals and value systems. Second, emotions are associated with physiological reactions and differ from the nonemotional part of cognition. Third, emotions are viewed from a functional standpoint; in other words, they play a leading role in human beings' adaptive dimension (Evans, Hannula, Philippou, & Zan, 2003; Mandler, 1989).

Cognitivist (social-cognitive) and constructivist trends prevail in the literature on the theory underlying the study of emotion. Emotion is explained as the interruption of a plan and as the result of a series of cognitive processes: assessment of a situation, attribution of causality, evaluation of expectations, and objectives and conformity to social rules.

Traditional theories of emotion appear to be characterized by polarization (Scherer, 2000): innate or constructed, subjective or social, static or dynamic. In the social-constructivist theory of learning, the aim is to characterize emotion in ways that entail the interaction between these poles. The study referred to here drew from the systemic conception introduced by Mascolo, Harkins, and Harakal (2000: 127) for the development of affect and motivation in mathematics and is used in mathematics education research by authors such as Op 't Eynde and De Corte (2002), Gómez-Chacón (2006), and Gómez-Chacón and Figueral (2007). Mascolo, Harkins, and Harakal proposed a systemic approach to emotional development built on the following four assertions.

1. Emotional states and experiences comprise multicomponent processes.
2. Emotional experience emerges through the mutual regulation of system components over time and within contexts.
3. Sensitive contexts, which constitute one such system component, not only act autonomously and self-adjust but are subject to changes in the social context.
4. Emotional experience self-organizes in a series of more or less stable models (or attractors), ranked by degree of variation.

This systemic approach consequently views emotion as a multicomponent process comprising systems of appraisal, affect, and action (Mascolo et al., 2000). The coordinated interaction of all these systems is regarded here as instrumental to and constitutive of emotional experience.

The approach adopted in the study was guided by three fundamentals.

1. In the first fundamental, *the emphasis on the emotional process*, emotions are characterized as episodes in which the systems of appraisal, affect, and behavior (action) coact dynamically. Emotions emerge via the interactions among the cognitive (appraisal), affective (autonomic nervous system and monitor system), and conative (motor and motivational) realms. Emotions are not, then, automatic biological responses, but rather a product of social interaction, in which emotional responses are evaluated and managed based on an individual's ideologies and principles. Such management is not limited to expression; it extends to emotional experience. Emotions are the product of people's evaluations and reevaluations based on experience, their support network, and social confrontation in the face of given stimuli.
2. *The systemic approach is not chaotic but governed by an ontology of specific contexts.* Emotions tend to self-organize in a finite number of models, in other words, basic emotions. Nonetheless, interindividual differences clearly exist. Emotions are not automatic responses or the consequence of the triggering of physiological processes; they are the complex result of learning, social influence, and individual interpretation.
3. *From the systemic perspective emotional experience is sited within a social-historic context.* Emotions are a social-cognitive construct, that is, their origin and reality are social and they are built on attitudes and transitory social roles, experienced as passions based on the language and morals of a given culture.

One important aspect in research on affect is understanding the interaction between it and cognition. In approaches such as Goldin's (2000), affect is interpreted as a representation system parallel to the cognitive system involved in problem solving. Other approaches stress the social dimension. From the social-constructivist perspective, the first aspect defined by that interaction is social context (Op 't Eynde et al., 2001). The theories on analysis of practical discourse focus on the positions adopted in a social context (Evans, 2000; Gómez-Chacón, 2011). In the Vygotskian school, emotions constitute a dimension in the zone of proximal development (Nelmes, 2003) merged with thought and the relationships among reasoning, action, and emotion (Radford, 2015; Roth & Radford, 2011). Other approaches are adopted in the fields of neuroscience and mathematical thought (Schlöglmann, 2002) and in studies conducted from the perspective of self-regulation (Malmivuori, 2001) or psychoanalytical theory (Evans, 2000; Evans & Tsatsaroni, 1996).

In the research discussed here, cognition-affect interaction is broached from a social-constructivist perspective and from the dimension of social identity. A number of related notions are discussed below.

STUDY OF EMOTIONAL EXPERIENCE: A HOLISTIC APPROACH

The traditional theories of emotion (expressions, embodiments, cognitive appraisal, and social constructs), which prevailed for several decades, determined specific methodologies. Russell (2003) contended that theories differ because they actually address different factors, not necessarily because they present different views of the same conceit. He posited that if "emotion" is actually a "heterogeneous cluster of loosely related events, patterns and dispositions, then these diverse theories might each concern a somewhat different subset of events or different aspects of those events."

In much of the present author's prior research, the study of emotional experience has been broached holistically for the following reasons.

- The objects of study are not reduced to variables; rather, they are viewed as a whole. The nature of social reality is regarded not as a static, objective, idealized construct with an existence independent of subjects, as in positivist theory, but as a dynamic construct articulated by a group of subjects who interact in a setting.
- The aim is to understand people within their own reference framework. Context and situation are inseparable: one cannot be researched without the other.

The research proposed in this chapter is patterned on the triangle formed by the cognitive conditions, emotional dimension, and social context present in adult learning processes, specifically in the context of social exclusion. The approach adopted is illustrated with data gathered over two academic years from a group of 23 17- to 19-year-old vocational training students. All participants had experienced academic failure and were at risk of social exclusion.

The research design combined ethnographic and case study techniques and reflections on the activity at hand. Data were collected using a number of sources and procedures: questionnaires, individual interviews with students, semistructured group interviews, interviews with the teacher, classroom observations, field notes, audio recordings, and students' answers to problems.

The case studies focused primarily on the local affective dimension and simple problem-solving scenarios (methodologies described in Gómez-Chacón (2000a, 2000b) and Gómez-Chacón and Figueral (2007)). The global affective dimension and complex scenarios served as the backdrop for determining the origin of subjects' emotional reactions. The settings were a mathematics classroom and a joinery (woodworking) apprenticeship. The spontaneous expression of beliefs and emotional reactions toward mathematics in these settings were subsequently confirmed in interviews and subjects' evaluations of their own progress, with remarks on past and recent positive and negative experiences around learning mathematics. The following attempts were made when interpreting the data:

- to establish and describe significant relationships between cognition and affect (local and global affect, two structures defined in a later section);
- to explore the origin of affective reactions and describe changes observed in subjects;
- to determine whether students' social identity was a reference for understanding their behavior and emotional reactions in a mathematics classroom.

METHODOLOGICAL CONSIDERATIONS IN THE INTERRELATIONSHIPS BETWEEN COGNITION AND AFFECT IN MATHEMATICS

The elements studied included (1) the problem-solvers' emotional state and (2) the position of the zoom when observing the cognition-affect relationship in context.

Problem-Solvers' Emotional State

The affective states of problem solvers can be described in many ways. Different authors (McLeod, 1992; McLeod & Adams, 1989) have described different emotional states by emphasizing and describing some factors such as magnitude and direction, duration of emotional reactions, level of awareness of the emotions that influence problem solving, and degree of control over emotional reactions. These factors constitute a reasonable beginning for an analysis of the characteristics of emotional states and their relationship to problem solving. However, we consider that in addition to these characteristics we need to know more about the ways in which these factors interact with the different types of cognitive processes, the different types of context, and the differing beliefs that the individual holds. In this paper not only do we highlight the need to make the combined nature of cognition, motivation, and emotion on mathematical work in dynamic affect systems explicit, but we present an analytical model based on the structures of (local and global) affect. Our major claim is the following: "if we would like to understand cognitive-affect interplay in the individual when acting 'in the moment,' it is necessary to know the individual on a different level (individual, group, society, holistic view of the individual), and there are some aspects which model the dynamic of cognitive-affect which could be captured (explained and modeled) in detail using the constructs: structures of (local and global) affect."

In our case we seek to confirm specific patterns, specific cases, and simple rules that give a typical structure or rhythmic sequences in this interaction between cognition and affect in the individual and explore whether these same patterns are extended to a group of individuals. The structure definition that we use here is dynamic (dynamic system), and it corresponds to the predominant use of the term today, in which a structure is considered as a system of transformations, including variance addition invariance.

In this chapter *local affect cognition structure* is understood to mean changes in feelings or emotional reactions occurring when solving mathematical problems in the classroom. In this structure, studied in simple scenarios (such as problem-solving stages or the discovery of errors), relationships were established between emotional reactions and cognitive processes in different stages of the mathematical activity (*local affect-cognition structure*). The types of interaction between emotions and the cognitive system (interruptions, deviations, or cognitive shortcuts), which may be expressed via different pathways, arise in the local structure.

Global affect is the combination of all an individual's local affect pathways, which continually feed into self-concept, along with personal beliefs about mathematics and how it is learned. Global affect was explored in complex scenarios; in other words, subjects were viewed in their sociocultural context and their interaction with others. In such scenarios, learning mathematics was regarded as one of the building blocks in subjects' social identity and emotional reactions were contextualized in the social realities from which they stemmed. In the evaluation of subjects' emotions, two pathways were taken into consideration in the cognitive and affective dimensions of learning mathematics. One represented the information on emotional reactions that impact conscious processing (*local affect-cognition structure*). The second addressed the sociocultural influences on individuals and how that information is internalized and shapes their belief systems (*global affect-cognition structure*).

Position of the Zoom When Observing the Cognition-Affect Relationship in Context

As a methodological consideration it is important where we set the zoom when observing the cognition-affect relationship. We consider that the structure and the function of cognition and emotions are inseparable in all behavior. However, in order to characterize the affective and cognitive components we establish categories for viewing specificities in each of them and the use thereof in macroanalysis data and microanalysis ("The Local Dimension of Emotion" and "The Global Dimension of Emotion" sections).

In this study we use the term "cognitive" broadly. On the one hand it refers to the extensive use of processes of evaluation (cognitive appraisal), and on the other, it refers to the characterization of the subjects' personal meanings of the cognitive dimension of the heuristic that acts in the solving of problems. Certain distinctions between "knowledge" and "evaluation/appraisal" are essential in our study.

The need for this differentiation arises not only from our studies but from studies that have expressed the need to make a distinction between knowledge and appraisal (Lazarus, 1991). Also we can speak of the difference between mathematical knowledge in general and contextual mathematical knowledge. The former includes establishing attitudes, beliefs, and intuitions about oneself; the latter is active in a particular situation or with a specific content whose impact on emotion can be very different. Cognitive appraisal refers to the generation of personal meanings and how the valuation of that knowledge makes it potentially emotional—in other words, how the situation globally affects relevance for the person, in relation to the goals, and to resource management.

In the identification of both local and global affect-cognitive structure we focus on the influences of mathematics process, metacognitive and directive processes, students' beliefs, and causal attribution of success or failure. By focusing on these processes we seek to understand the movement between the mathematization processes and subjectivation processes by the subject during the mathematical activities in context. In context some emotions involve others, inducing interpersonal interactions and identity of individual. Local affect-cognitive structure allowed us to make a very valuable microscopic study of the actions and behaviors of subjects (in terms that we describe in "The Local Dimension of Emotion" section in the study of the local affect). However, using this local approach of emotion, there was something that was still not explained and that surpassed the established categories. This made us think about the intrinsic value and dynamism of the person in his or her historical and social reality. For this, "The Global Dimension of Emotion" section we reread the data, this time on the basis of symbolic interaction and on the perspective of social identity. This section discusses a case study of a student's emotions, some of which can only be understood if viewed in a scenario of the *subject's adjustment to school in which he negotiates his contextual identity*. Adrián, the student in question, adopted self-assertive behavior and defense mechanisms: his strategies tended toward differentiation, singularity, nonconformity, systematic opposition to all that school meant to him, and a tendency toward negativism.

Some emotions and their interaction with the cognitive realm cannot be comprehended at the individual level only and must be studied as a social process in conjunction with perspectives that view emotion as cognitive evaluation or physiological expression.

THE LOCAL DIMENSION OF EMOTION

As noted in "Study of Emotional Experience: A Holistic Approach" section, this study focused on different levels of globality and contextualization in the exploration of emotions during problem solving. Each of these levels called for applying tools and methods in different ways and interrelating conceptual theories to interpret the data.

For the local affect, cognitive establishing patterns of interaction cognition and affect requires analysis at the microscopic level of individuals. In "Study of Emotional Experience: A Holistic Approach" section we present the results from a group of 23 17- to 19-year-olds who are taking part in vocational training. This research was carried out in a natural context, tracking people for 2 years during mathematics classroom sessions. The research's aim is to describe the significant pathways that the interaction of affect and cognition seem to follow in each subject. For each individual of the case study we try to give an answer to these questions: Faced with the task, what is the initial affective attitude? What are the causes of the interruptions (gaps or jumps) in the affect-cognition interaction? How are they articulated with the problem-solving process? What are the most frequent emotional reactions? How can the cognitive and affective tendencies be cataloged?

Specific Theoretical and Methodological Issues for Local Affect

The starting point for the study of local affect-cognitive structure was a cognitive understanding of emotion in which the key element, the replies to a self-evaluation questionnaire, were subsequently compared and expanded with classroom recordings and an interview. From that cognitive vantage point for this local affect structure, emotional reactions were viewed as rapidly changing feelings experienced consciously or pre- or unconsciously when solving mathematical problems.

Mathematics classroom observations focused on the items explained in "Problem-solvers' Emotional State" and "Position of the Zoom When Observing the Cognition-Affect Relationship in Context" sections above: emotional snapshots during problem solving, the cognitive demands involved in performing tasks and learning, metacognitive and metaaffective processes, and classroom interactions were instrumental to the signs or indications from which subjects' emotions could be described and corroborated. Audio recordings and field notes or both were used in data collection. Case studies focused on local affect and simple scenarios. The origin of subjects' emotional reactions was determined by classroom observation, subsequently confirmed in the interview and subjects' evaluations of their own progress. One of the tools used to diagnose emotional reactions and subjects' self-evaluation, the *emotional graph*,[1] is described below.

Questionnaire: The Emotional Graph

The questionnaire consisted of six questions, three on feelings and emotional reactions and three on transferal to learning in the workshop and everyday life (Table 1). Students were asked to fill it in after every problem. In this questionnaire they are asked graph their feelings and reactions as they did each the problem. The aim was to gather information on their affective reactions (magnitude, direction, awareness, and control of emotions) and the origin of such feelings (interaction between affective and cognitive factors). Explicit information on magnitude, direction, and awareness was gleaned from students' plots of their emotions along with their notes on the cognitive demands involved in doing each problem.

Moreover, the self-evaluation data furnished on students' questionnaires (Adrián's is given in Fig. 1 by way of example) were supplemented with, and compared to, classroom observations and recordings and the information gathered during interviews. The interviews were designed primarily to confirm the classroom observations, especially individuals' most explicit and recurrent emotional reactions. They aimed to enhance individuals' awareness of the origin, regulation, and control of their emotions.

[1] This tool was chosen because it has been used in a number of studies (McLeod, Metzger, & Craviotto, 1989), so it afforded a certain assurance of validity and reliability. A similar tool with fewer limitations is the Problem Mood Map (Gómez-Chacón, 2000a, 2000b).

TABLE 1 "Emotional graph" Questionnaire

Name				Date
1. How did you feel after finishing the problem?				
Very satisfied	Satisfied	Unsatisfied	Very unsatisfied	
2. Briefly explain your reply.				
3. Graph your feelings and reactions as you did the problem.				
4. Did the problem remind you of situations outside the classroom (such as at home or in public places)? Briefly explain your reply.				
5. Will what you learned with this problem help you in your everyday life?				
6. What suggestions would you make to expand the problem?				

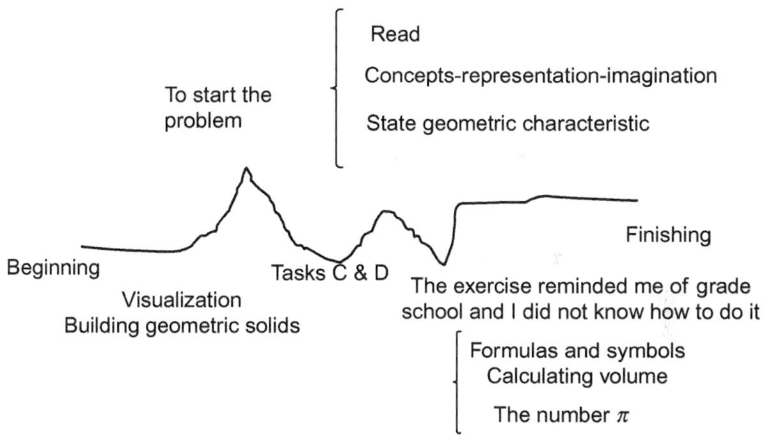

FIG. 1 Case Study: Adrián. Emotional Graph for the Activity: Packaging Geometry

The findings from the classroom observations and recordings and the problem-solving protocol were subsequently pooled and synthesized. Table 2 reproduces part of the protocol for analyzing classroom session observations (see extract of the session in the Appendix). Each row describes one emotional episode, the explanatory unit of analysis (emotional reaction unit, ERU). These units, determined by the stage in the process that the student experienced a given emotion, are characterized by magnitude, direction, level of awareness and control of emotions, and interaction with cognitive processes.

A study of students' emotional graphs and analysis of their classroom sessions yielded answers to the following questions. What caused

TABLE 2 Template for Analyzing Adrián's Data in Session S13

Student's Indication of Positive Emotion on the Graph	Signs of Positive Emotion Recorded by the Researcher	Problem-Solving Stages	Interactions	Metacognition	Metaemotion	Cognitive Processes	Signs of Negative Emotion Recorded by the Researcher	Student's Indication of Negative Emotion on the Graph
ERU29		Processing/ solving problem	Progress with the teacher's cognitive support Positive learning environment			Definition of mathematical objects. Deduction of formula for area of circle Explanation of the process	Bewilderment	
ERU30	Receptive	Processing/ skepticism	Teacher's cognitive support Positive learning environment	Control and regulation		Review of process		
ERU31		Processing/ solving problem	Teacher's cognitive support			Deduction of formula for area of circle Volume of cylinder		
ERU32		Processing					Arguments, Adrián, and MH	

ERU33		Processing	Cognitive support from a classmate			Deduction of formula for area of circle	Upset In a bad mood	
ERU34		Processing/ solving problem	Negative interaction with others, affective support for others		Regulates, recognizes, identifies emotion	*School experience*	Racks brain Laughs, protests	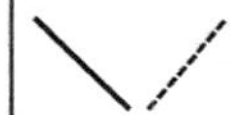 When I finished it reminded me of grade school and I didn't know how to do it
ERU35		Processing	Cognitive support from a classmate Positive learning environment Teacher's cognitive support			Deduction of formula for area	Distraction Bewilderment	
ERU36		Processing/ solving problem	Progress with teacher's cognitive and affective support Classmate's cognitive support		Appraisal and expression of emotion	Deduction of formula for area Exemplification	Resists Protests, does not listen, bewildered, distracted	

interruptions in affect-cognition interactions? What form did such interruptions adopt in problem solving? The direction and magnitude of the emotions and students' awareness of their feelings were revealed as they solved the problem. All this information—in other words, the synthesis of students' notes and researchers' observations—was used to draw a profile of each subject, called an affect-cognition map or pathway.

Selected Findings: Adrián's Affective Pathway, a Case Study

Adrián was the student chosen for this discussion. He is a 17-year-old vocational student in a cabinetmaker workshop. He follows a practical training in the workshop and complementary lessons in mathematics, language, and social skill in order to complete secondary education. He experienced academic failure in elementary school, and he started at secondary school but dropped out due to receiving low marks.

With the data gathered during classroom activities over a 5-month period for a total of 18 mathematics classroom sessions in which a number of educational units were covered, Adrián's emotional reactions were found to have several origins: past learning experiences with mathematics and teachers, knowledge organization and mathematical aptitudes (acquisition and processing of mathematical information, mathematical memory), problems in an area of mathematics he disliked, his mood early in the session, his beliefs about mathematics as an area of knowledge, and his belief that certain personal skills are needed to solve mathematical problems.

On the grounds of session monitoring and the explicit information contained in Adrián's emotional graphs, the following interruptions and changes of direction were identified in his affect-cognition interactions.

Changes in direction from positive to negative arose during the *initial contact* with the problem: when he read and understood the problem; when he read the first page of the problem or the materials he would have to use; with his first overview of the problem. During problem solving, these changes were associated with a lack of understanding of specific mathematics resources (conventions, criteria, methodology); the absence of theoretical and structural knowledge; scant ability to reason with mathematical symbols and spatial relations; pursuit of relationships and connections between the mathematical elements of the problem and the knowledge acquired; ongoing quest for a strategy; explanation, verification, and generalization of the problem; effort required to assimilate changes in the level of difficulty and to consolidate and verbalize what he had learned; his views on mathematics; experiences that reminded him of past school experience.

The negative emotions detected during classroom sessions included rejection, resistance, protests, aggressiveness, anger, bad mood, annoyance, fear, distraction, mental blockage, intellectual stress, bewilderment, boredom, indecision and insecurity, apathy, and indifference.

Changes from negative to positive were likewise attributable to several causes, classified under the following categories. When he used techniques routinely applied in the joinery shop, such as drawing or measuring, he found it easier to grasp the structure of the problem. Other situations included the following: when he remembered mathematical information and was able to draw from and apply it; when he could do all or part of the problem intuitively; when he received cognitive and affective support from the teacher or a classmate; when he became aware of and was able to regulate his emotions; when he could identify and accept errors; when he could move on to the next stage of the problem without assistance or help from his classmates; when he could see his own achievement and skill. Ultimately, this change from negative to positive was determined by his appraisal of the problem.

The affective and cognitive characteristics of the emotional reactions plotted on the emotional graph were consistent with cognitive science findings on the interaction between affective and cognitive representation in problem solving. The case studies (Gómez-Chacón, 1997, 2000a, 2000b) showed, however, that observing and identifying students' change in feelings or emotional reactions during problem solving (local affect) or detecting cognitive processes associated with positive or negative emotions do not suffice to understand students' affective relationships with mathematics. Difficulties in understanding the problem, for instance, or in memory retrieval induce frustration and anxiety, whereas awareness of personal learning progress induces joy and satisfaction. Curiosity, in turn, may improve performance in heuristic processes important for research planning. In addition to detecting significant relationships between cognition and affect and their possible use in mathematics teaching, the data confirmed that students' affective reactions to mathematics must be studied in more complex scenarios (global affect) to be able to contextualize such reactions as part of the social reality in which they arise. The focus must be placed not only on cognitive reasoning but also on evaluation. Contextual values, ideas and practice (culture) must also be taken into consideration, for they establish an order that helps individuals keep their bearings and provides them with a communication code.

THE GLOBAL DIMENSION OF EMOTION

The conclusions drawn from the analysis of the *local affect structure* identified the most crucial contextual factors. The questions posed in this second phase of the study were as follows. What kinds of situations induce students to resist or refuse to learn mathematics? How intense is students' sense of belonging to their social group? What interactions can be identified between belief systems about mathematics, learning the discipline, and strategies for negotiating social identity? Can students' social identity be used as a reference for understanding their behavior and emotional reactions in a mathematics classroom?

These questions were intended to underscore the need to understand the effects of the educational role of the social system, the sociocultural interface, and personal affect on learning. The discussion that follows focuses on messages constantly received by students about the social importance of learning and knowing mathematics. Their self-concept as learners is related to their attitudes, their view of the mathematics universe, their belief system about mathematical learning and their social identity. It also has a heavy impact on how they conceive of and react to mathematics. When the social identity perspective is adopted, account must be taken of symbolic social relationships.[2]

Specific Theoretical and Methodological Issues for Global Affect

This study adopted the social interaction approach developed around conversation analysis that defines identity as "the set of verbal practices through which persons assemble and display who they are while in the presence of, and in interaction, with others" (Hadden & Lester, 1978: 331).

Global analysis and case study techniques were used to answer the research questions posed. Two pillars were established for global analysis: (a) subjects' status as group members and their own opinion of their identification with the group (its affective status, values, beliefs and attitudes) and (b) subjects' negotiation of their social identity. That was followed by a review of the case studies from the perspective of social identity and beliefs.

The guidelines followed to pursue the two pillars of the analysis (Gómez-Chacón, 2011) are listed below.

(a) Subjects' status as group members and their identification with the group: experience, information, position.
(b) Negotiation of social identity:

[2] This study was also based on interactionist studies stressing identity construction and viewing identities as "identity strategies" (*stratégies identitaires*). Camilleri, Kastersztein, et al. (1990) defined identity strategies as "processes or procedures set into action (consciously or unconsciously) by an agent (individual or group) to reach one or more goals (explicitly stated or situated at an unconscious level); procedures elaborated in function of the interaction situation, that is in function of the different determinations (sociohistorical, cultural, psychological) of this situation" (Camilleri et al., 1990, p. 24).

- behaviors and situations in which identity is presumably conspicuous, compared to the behaviors allegedly expected from them and to the scenarios in which they change their behavior
- negotiation of identity, in connection with when to "negotiate"; the status with which they identify; purpose, identity, and management of inequality in groups tagged with a negative social identity
- the resources available to negotiate identity: identification strategies.

Selected Findings: Adrián, a Case Study

Insight into the complexity of emotional experience is provided below in the in-depth review of Adrián's case, focusing on both the classroom observations and the student's own evaluation on the emotional graph for ERU34. The following is an extract of the interview conducted after the classroom session, expressive of his emotional reactions and beliefs about symbols.

INTERVIEWER	What happened to you while you did the problem? On the graph you show a negative emotion when you work with symbols.
ADRIÁN	Yes, symbols spook me.
INTERVIEWER.	Why?
ADRIÁN	Who knows..., you don't understand them and...
INTERVIEWER	And what do you feel?
ADRIÁN	I think, "oh my," this is ... it'll take me years to figure this out.
INTERVIEWER	Really?
ADRIÁN	Yes.
[Silence]	
INTERVIEWER	What happened today when you were doing the problem? Why did you act like that and say it was the same as when you were in Elementary School?
ADRIÁN	π, I didn't understand it. When I was in grade school it wasn't explained to me, I didn't understand... and I decided to ignore the teacher.
INTERVIEWER	Now, when we were doing the problem, did you identify me with your grade school teacher?
ADRIÁN	You explained it to me. She explained it to everyone, for everyone... and she didn't explain it to me again. She said I just didn't pay attention but the problem was that I didn't understand... Lots of times I'd act out in class, but for the one time I was actually paying attention... She didn't want to... and I wasn't going to run after her... She can shove it. I have my pride.

(Continued)

INTERVIEWER	And why did you react like that now?
ADRIÁN	I thought you wouldn't explain it, like all the other teachers…, that you would explain it for everyone, but not for me specially.
INTERVIEWER	What happened when I explained it to you? How did you feel?
ADRIÁN	I thought, "well good, now I know something."
INTERVIEWER	And when you were explaining it to Mohamed?
ADRIÁN	I knew how to do it, I knew the value of π.
INTERVIEWER	What are your feelings about π?
ADRIÁN	I didn't know what it was, what that figure meant [refers to the meaning of the symbol].
INTERVIEWER	Would you like to find out a little more about π and do other problems so you get to know it better?
ADRIÁN	Yes but some other day.

[Adrián showed considerable interest in overcoming his feelings about π and subsequently did a number of related problems.]

Based on that classroom incident, the teacher could have simply thought that Adrián found it difficult to compute areas and volumes or to grasp the meaning of π. Or she might have tried to explore the unobservable at first blush but which can be traced or monitored. For instance, when the student became aggressive, saying "it's very hard and I just don't get it" or "I didn't understand π here or in (elementary) school." The question that might be posed is whether these cognitive and affective findings are indicative of or reveal the origin of this student's learning impediment and the difficulties experienced with areas and volumes.

In this situation, Adrián's reaction, reliving an earlier school experience evoked by the problem, suggests certain elements that block his learning:

- being confronted again with mathematical content with which he had experienced difficulty
- being reminded of prior negative situations around teachers

The teacher, realizing that some students did not know how to calculate the volume of a cylinder or the area of a circle, tried to describe the process "step-by-step" on the blackboard. She consistently explained the underlying conceits, believing that students, seeing where and how she used the strategies and formulas, would deduce that they could be applied in other cases. This was not the type of approach that Adrián needed. He reacted by reliving a negative school experience: "You explained it to me, she explained it for everyone… I didn't understand… I thought you wouldn't explain it." The teacher did the problem with Adrián by drawing from his own ideas. Here the teacher's role was to steer those ideas toward the solution to the problem and help him use the resources at hand, favoring his involvement in the problem, enabling him to regain his lost self-confidence, and enhancing his awareness of what he was learning.

Some of Adrián's beliefs conditioned his personal situation: "I've never understood math"; "math is very hard, I just don't get it." These were played out through emotional reactions such as aggressive behavior, sorrow, or hatred. His comment on the emotional graph (Table 2) was that he felt dissatisfied after finishing the problem "because I didn't understand anything, especially π and finding the volume of circles (sic!)." On the drawing alongside the downslope he wrote "the problem reminded me of grade school and I didn't know how to do it."

When evoking school experience, Adrián exhibited aggressive behavior in an attempt to avoid the fear of not being acknowledged. The monitoring conducted, during two academic years, revealed key moments in his earlier school experience that impacted him negatively, as in the case discussed above. The effects on learning situations were the following:

- challenges to the teacher's authority
- negative teacher-student or peer group interactions
- mistrustful attitudes toward the teacher
- lack of self-confidence
- resistance to learn mathematical concepts
- resistance to perform tasks that required thinking, such as problem solving

The data showed that Adrián's "flashback" reactions to classroom mathematics, primarily in adjustment to school scenarios, appeared in the following situations: when the problem evoked past experience, when he was confronted with mathematical concepts he had trouble mastering, when the teacher evoked past negative situations, or when the mathematical conceit was not in keeping with his preferred mechanical approach.

Simple vs Complex Scenarios in Affective Structures

The emphasis in the study of the 23 students' cognitive-affective responses in mathematics classroom interaction was on refusals to learn and the lack of basic conditions for learning. Similarities and differences were identified among individuals (Gómez-Chacón, 2011):

1. The *similarities* included a lack of self-confidence in their ability to solve problems, a fear of reliving previous academic failure, and a lack of attention to the task at hand (processes and similar). They exhibited resistance, fear, and insecurity. They preferred problems involving the mechanical application of knowledge. They lacked an understanding of problem-solving strategies and stages. In the information

processing stage, they felt more comfortable with known approaches and were afraid to change. Their hatred of the discipline and their own insecurity conditioned their ability to reason. They required the teacher's cognitive assistance to generalize the application of mathematical objects, relationships, and operations.

2. *Differences* were found in the degree of insecurity in their relationships and interaction with their peers, which depended on their position and experience in the social group, the auxiliary representations used for mathematical reasoning, and their thought processes. Some students used their own informal procedures or strategies acquired by practice or in everyday life. Their experience in the t cabinetmaker workshop helps them understand the structure of the problem.

The characteristics of the situations inducing resistance or refusal differed. Consequently, the use of the sociological term "scenario" to refer to these situations is felt to be justified when ascertaining socio-cultural influences on individuals' assimilation of information and the shaping of their belief systems. Hence the expression "more complex scenario" in the preceding section. A "scenario" is an organized setting in which something happens. It is consequently used to mean what takes place in a given context at a specific time and with known resources. In the presence of similar circumstances, the people involved act in more or less the same ways, given their learned social and individual predisposition.

Behaviors believed to be characteristic of socially disadvantaged teenagers with a record of academic failure were studied, along with their causes, in classroom interactions. The alternatives to those behaviors were also explored. Here, the factors identified could be classified under four headings: *adjustment to school, self-justification, demand for interdependence, and response to messages (or differentiation)* (Table 3).

TABLE 3 Scenarios, Student Behaviors, and Their Effects on Learning

Scenario	Behaviors	Effects on Learning Situations
Adjustment to school	• Evocation of school experience • Aggressiveness and attempts to avoid the fear of nonacknowledgement • Negative memories of key school experience • Mechanical approach to problems and reluctance to learn strategies with unknown quantities	– Opposition to teacher's authority; student-teacher interactions or interactions with peers – Mistrust of teacher – Lack of self-confidence – Resistance to learn mathematical concepts – Resistance to perform tasks requiring thinking, such as problem solving

TABLE 3 Scenarios, Student Behaviors, and Their Effects on Learning—cont'd

Scenario	Behaviors	Effects on Learning Situations
Self-justification	• Justification of opinions, values, skills, preferences, and emotional reactions on the grounds of group membership	– Use of informal structures to solve problems – Preference for oral mathematics – Preference for shared learning – Resistance to develop and advance in mathematical knowledge and procedures – Understanding of teacher's task – Defense mechanisms: mockery and boredom
Demand for interdependence	• Reliance on the strength of the group and fear of doing things in new and unknown ways: "We've always done it that way"	– Student-teacher communication – Student-student and student-group communication – Classroom norms – Classroom management – Resistance to learning mathematical concepts – Subject's perception of curriculum
Response to messages (differentiation)	• Differential external appearance (dress, hairstyle) • Differences with classmates • Defiance in appraisals and beliefs about learning mathematics in practical contexts: "This has nothing to do with mathematics"	– Student-teacher communication – Gesturing in classroom interaction – Interference with the learning environment – Isolation, resistance to teamwork – Preferences and liking for a certain types of activity

CONCLUSIVE ISSUES

In "The Local Dimension of Emotion" and "The Global Dimension of Emotion" sections we have described two studies of the same group of individuals, one referring to their local affect-cognitive structure through a cognitive framework and another focusing on the global affect-cognitive structure from a symbolic and identity interactionist approach.

The conclusions drawn in this study are meant to contribute to proposals for advancing our understanding in the following aspects of the interaction between cognition-affect in mathematics: (a) the conceptualization of structures of this interaction and (b) the necessary synergy between theoretical and methodological frameworks in the research on affect.

In relation to the structures of affect-cognition interaction, we recall that global affect-cognitive structure can be captured when we observe the subject at mezzo level and at macroscopic level. It is a result of different aspects:

1. Summary of the pathways followed by the individual in the local affective dimension.
2. Interactions and sociocultural influences on individuals and how that information is internalized and shapes their belief systems. Two aspects to take into account are the social representations of mathematical knowledge and sociocultural identity of subjects.

In the last section we have seen how in the local observation of affect and cognition it allows the pathways to be determined. One repeated observation of individuals in different activities is that it allows us to establish the various pathways (including motion) that can be represented and set as global affect-cognition structure (described "The Local Dimension of Emotion" section).

Identifying the difficulties that originate from the interaction between cognition and affect involves becoming sensitive to these movements, reflecting on them, and understanding where they come from and where they lead us. Capturing this structure of affect more precisely at the level of research in mathematics education would favor more accurate items in the assessment instruments about cognition and affect of the learning of mathematics.

However, as it was shown "The Global Dimension of Emotion" section there are certain elements that can be only defined in the interaction and the sociocultural influences on the individual.

The development of a person is not linear and cannot be predicted by only identifying local episodes. There are numerous ways that could lead to a person developing a skill or understanding of something. Accepting these principles, it clearly has consequences in the study of human beings: lifetime is not only characterized by regularities and the progressive establishment of regularities, but also by the times when these continuities are interrupted, reoriented, and change.

Although in our understanding of the interplay between cognition and affect I think it is possible to establish regularities. However, capturing these moments of rupture (irregularities) is interesting for several reasons. First, at a theoretical level, the bifurcation points in which the person develops new behaviors are highly significant. Second, at an empirical and methodological level, these episodes or moments naturally constitute "laboratories of change." In an experimental context tasks or questionnaires are often proposed without breaking points. People simply have to respond, and the researcher examines in detail these responses or the processes that guide them. However, this research carried out in a natural context, tracking people—in this case for two years—allowed us to see these interruptions, disruptions, or turning points that required a response, or an adjustment by the individual.

The study of identity processes requires a consideration of the person's subjective experience. It becomes more visible when individuals have to undergo transitions and ruptures between practice contexts. For this reason it becomes more visible in studies from a socially disadvantaged context or studies of ethnic minorities.

EPILOG: OPEN QUESTIONS AND UNRESOLVED ISSUES

This section lists important issues around affect in mathematics education that have not yet been suitably solved by research (this author's included). While many have been broadly discussed in the area of affective science, no clear answers have been forthcoming.

Experience vs expression. The existence of a one-to-one relationship between emotional experience and emotional expression is perhaps the most widespread and potentially problematic assumption.

While it may be tempting to think of emotional experience and expression as two sides of the same coin, recent research suggests that such a conclusion is tenuous at best. Physiological and bodily channels obviously convey information on a person's emotional state. Nonetheless, the exact value of such channels as emotional indicators has yet to be determined.

Intercomponent coherence in emotion. Another assumption made by many researchers is that emotional experience is expressed as a synchronized response involving peripheral physiology, facial expression, and instrumental behavior. The corollary to that assumption is that the signals detected are more intense when several channels are observed. With the exception of prototypically intense expressions of emotional experience, however, research findings show low correlations among the various components of a given emotional episode (Barrett, 2006). The identification of cases where emotional components are synchronized as opposed to those where they are only weakly related is a matter outstanding more effective detection.

Affective structures. Labels and categories of emotions based on two- or three-dimensional models are of crucial importance to researchers. The identification of a suitable level of representation can favor progress. Here two significant structures in a person's affective dimension, *local* and *global affect-cognitive structure,* are proposed given their effectiveness in the research conducted. In natural learning contexts, motivation, cognition, and emotion can be conceptualized and worked with empirically at three or more levels of specificity to explore how the self transitions at those levels.

The interrelationships among cultural, social, and personal systems in students' emotions around mathematical thinking. The data showed the existence of interrelationships among cultural, social, and personal systems in students' emotions around mathematical thinking. The study covered these students' social identity and what mathematics and the learning process meant to them. The findings suggest that new approaches to the affective dimension of mathematics in lifelong learning are possible, at least for groups similar to the one studied (branded with a negative identity). The traits comprising these students' identity in their context can be equated to a network of meanings relevant to that network which will emerge in mathematics learning. Such meanings provide insight into the pursuit of a fuller understanding of persons' global affect, their affective view of and reaction to mathematics and mathematics learning, the way they build their belief systems, and their awareness of that process.

Awareness of the importance of social processes in the explanation of emotion is growing steadily. The aim in this chapter (and in this author's research) is to stress the role of culture and social processes on the one hand and the variety of positions adopted by individuals on the other. Social and cultural factors cannot be construed as a "uniform static package." A dialog emerges around the self. The case studies showed that students' behavior reveals different strategies for identifying and negotiating their social identity and consequently different emotional responses.

The emotional dimension can be studied from the standpoint of that dialog by combining perspectives and methods that seek to describe, with a full wealth of symbols, the meanings of affect in the social reality in which it arises. The frameworks for interpreting emotion have traditionally run the risk of centering more on an individual's cognitive events than on issues around arguments and conflicts in "mathematical practice."

APPENDIX CLASSROOM SESSION

Two of the subject's emotional reactions during the "package geometry" problem, namely fear of symbols and previous school experience as the origin of emotion, are discussed below. Students were asked to do the following problem.

Package geometry

1. After making a poster, a group of teenagers decides to throw a party with food and drink for all. The figure shows the packages that the drinks and snacks came in (glass, plastic, metal, cardboard).

(a) List three geometric characteristics for each type of package.
(b) Compare the packages and jot down the geometric differences you observed.
(c) How many parts do you think would be needed to make each package?
(d) Choose a figure and draw a model for building the package.
(e) Calculate the volume of the packages.
(f) Fill in the "emotional graph" questionnaire.

One of the packages for which the students had to draw a model and calculate the volume was a cylinder. Adrián was faced with the need to know the formula for the area of a circle. The following is a transcript of the classroom session recorded.

[..........]

TEACHER	... Now do the cylinder. Is that right, Adrián? Let's see how you did the cylinder.
ADRIÁN	You open up the cylinder.
TEACHER	You open up the cylinder, and then...
ADRIÁN	You measure... 20 times 15. 20 times 15, like...
MOHAMED	And the lid the same way, of course
TEACHER	One thing...OK? What you're getting is this outside area. If you do that, if you measure this, you measure 20 times this, when I wrap it back up, do you see what I'm measuring? It's this. The outside. I'm not measuring what can fit inside. So to measure the inside, what should we do?

[Background noise. While the teacher explains this to Adrián the other students begin to talk about other things. The teacher calls for order.]

TEACHER	Enough! Please. OK? Otherwise there will be no 5-minute break.. OK? Then, Adrián, if the height is 5...
CARLOS	What a dinky tin... unless it's a pepper shaker or something..
TEACHER	All right, 10. Come on, how would we do it?
CARLOS	Look, the volume is equal to multiplying the width...
JAVIER	What width? There is no width.
TEACHER	Let's see, one thing that we need to know, and that's why Mohamed says he doesn't like it. Let's see. What's the first question we need to ask? To know how to find the area of this? [pointing to the base of the cylinder]. How would we find the area of this circle? How would we find the area? That's where you're stuck. How do you find the area of a circle?
ADRIÁN	Smart-aleck. Since you already know [protesting aggressively].

(Continued)

CARLOS	You mean the circumference.
TEACHER	The circumference is the outer line and the area covered is called a circle.
CARLOS	How do you find it?
TEACHER	How?
CARLOS	Uhhh, π times…
JAVIER	3.1416. I never understood that.
TEACHER	π times r
CARLOS	Divided by…. no, squared
TEACHER	Yes, π time r squared.
JAVIER	Geeeez, what you make us do. She says it's easy, pal.
CARLOS	3.14 times the radius, which is 3…
FERNANDO	I don't know how to do π.
TEACHER	Let's see, Adrián. Do you remember?
ADRIÁN	This all sounds to me like grade school and then I couldn't have cared less… [angrily]
TEACHER	All right, explain what you mean. You couldn't have cared less about school?
ADRIÁN	That when we got to π I cut class.
TEACHER	You cut class when you got to π? Why?
ADRIÁN and FERNANDO	Because it's too hard and I don't get it.
TEACHER	You don't get it? ….

[……][The rest of the class keeps working] Adrián is at an impasse. The teacher shows the class how to calculate the area of a circle on the blackboard. Two students, Mohamed and Juan, interrupt because they don't understand the number π. And Mohamed gets distracted talking about an outing. The teacher tries to get the class's attention.]

TEACHER	Let's see, Mohamed. What's this all about?
MOHAMED	I'm not doing it for one reason, because I don't know how.
TEACHER	But I'm showing you how, teaching you. Let's see, I know you don't know how. That doesn't matter, but I'm teaching you how. Carlos, what's the matter?
CARLOS	Ah! the part that's not the lid, the body, has to have the same diameter as … the lid.
TEACHER	Yes, although the drawing doesn't show that.
CARLOS	Well, that's it.

[They calculate the volume, with Carlos and Javier especially collaborating with the teacher. Juan and Adrián are bewildered. The teacher asks what's wrong and Adrián replies.]

ADRIÁN	A lot… the thing about the circles, circles there, circles there, π: I didn't learn it in grade school or here either [sadly].

TEACHER	Well you are going to learn it here, you'll see, come on, you'll get it. [The teacher explains it to Mohamed and Adrián specifically.] Let's see, Mohamed: I'm going to write it down on this sheet of paper. I want to make this tin. I want to calculate the volume of this tin, OK? To calculate the volume of this tin- Adrián, can you see? Adrián … the volume, can you see how to find the volume? I have to find the area of the bottom
ADRIÁN	But how do you find the area of a …?
TEACHER	The area of the base and I multiply it times the height.
ADRIÁN	I don't get it, the area times the area and…
TEACHER	OK, but just a minute. I've got an idea. When I know the area here at the bottom and I multiply it up here, I get the volume, right? How am I going to calculate this area? This area…
ADRIÁN	Don't get started with that π stuff [resisting].
TEACHER	…It happens to be a circle. But don't worry about having to do π. Let's see.
ADRIÁN	I'll never get this [protesting and discouraged].
TEACHER	I don't want you to close up on me, Adrián. You keep saying you don't get it. OK, I understand, but…
ADRIÁN	Who's making us do this problem [angrily]
TEACHER	Adrián, please. The area of a circle, how do you find it? There's a formula that we studied, right? And what's the formula? If you know the radius of this circle, right? if you know the radius, you know… you have a circle,
ADRIÁN	Thing is, I don't know
TEACHER	But we've said it here, we've used an example, we're working on an example and learning. If you know the radius of the circle, which is the distance from this dot in the centre to here, and if it measures 2, the formula, what's the formula? It's π times the radius squared ($A=\pi r^2$).
ADRIÁN	And what's r?
TEACHER	r is the radius.
ADRIÁN	And what's π?
TEACHER	π is an irrational number and it's equal to 3.14, OK? It's actually equal to a little more, but we only use the first two decimals for convenience, all right?

[The teacher estimates π with a string.]

ADRIÁN	I hate this!
TEACHER	Why, why do you hate it?
ADRIÁN	Because I don't understand it.
TEACHER	But what don't you understand? Let's see…
ADRIÁN	Well that π thing, how you calculate it…

(Continued)

TEACHER	No, you don't have to... π equals 3.14. Instead of writing π you can write 3.14 and that's that. Then how would we find the area? What do we have to multiply 3.14 times? The radius. And how much is the radius?
ADRIÁN	2, times 4 [he calms down and continues the problem less agitated.]
[After he's through I ask Adrián to explain the process to Mohamed with a different example.]	
MOHAMED	I'm not going to know how to, teach.
TEACHER	Let's see, what didn't you understand?
MOHAMED	Well all that.
TEACHER	What's "all that"? Let's see, Adrián, why don't you explain it to him and that way you review it? Yes, because know you know how.
ADRIÁN	I know how?
TEACHER	I can see that you know how.
ADRIÁN	That's what I told you.
TEACHER	C'mon. Let's see. I'm going to be right here and you explain it to him and if... C'mon. Explain it.
MOHAMED	C'mon geeeenius [Mohamed encourages Adrián to explain it to him.]

Acknowledgments

This work has been partially supported by grant from the project *NILS Science and Sustainability* (ES07), UCM-EEA Abel Extraordinary Chair, Individual Mobility of Researchers (ES007-ABEL-IM-2013).

References

Abreu, G. (2002). Mathematics learning in out-of-school contexts: A cultural psychology perspective. In L. D. English (Ed.), *Handbook of international research on mathematics education* (pp. 323–354). Mahwah, NJ: Lawrence Erlbaum Associates, Inc.

Abreu, G., Crafter, S., Gorgorio, N., & Prat, M. (2013). Understanding immigrant students' transitions as mathematical learners from a dialogical self perspective. In *Eighth congress of european research in mathematics education (CERME 8), Antalya, Turkey.*

Barrett, J. L. (2006). Are emotions natural kinds? *Perspectives on Psychological Science, 1*, 28–58.

Camilleri, C., Kastersztein, J., et al. (1990). *Stratégies identitaires*. Paris: Presses Universitaires de France.

DeBellis, V. A., & Goldin, G. A. (2006). Affect and meta-affect in mathematical problem solving: A representational perspective. *Educational Studies in Mathematics, 63*(2), 131–147.

Di Martino, P., & Zan, R. (2015). The construct of attitude in mathematics education. In B. Pepin & B. Rösken-Winter (Eds.), *From beliefs to dynamic systems in mathematics education. Exploring a mosaic of relationships and interactions* (pp. 51–72). Switzerland: Springer.

Evans, J. (2000). *Adults' mathematical thinking and emotions: A study of numerate practices.* London: Routledge Falmer.

Evans, J. (2003). Methods and findings in research on affect and emotion in mathematics education. In *Third conference of the european society for research in mathematics education, 28 February–3 March 2003 in Bellaria, Italy.*

Evans, J., Hannula, M., Philippou, G., & Zan, R. (2003). Thematic working group 2: Affect and mathematical thinking. In M. A. Mariotti (Ed.), *Proceedings of CERME 3: Third conference of the european society for research in mathematics education 28 February–3 March 2003 in Bellaria, Italy.* Published on the web http://www.dm.unipi.it/~didattica/CERME3/proceedings/.

Evans, J., & Tsatsaroni, A. (1996). Linking the cognitive and the affective in educational research: Cognitivist, psychoanalytic and poststructuralist models. *British Educational Research Journal: Special Issue on Poststructuralism and Postmodernism, 21*(3), 347–358.

Evans, J., & Wedege, T. (2004). Motivation and resistance to learning mathematics in a lifelong prospective. In *Study group 6: Adults and lifelong math, 10th ICME, Copenhagen.*

Goldin, G. A. (2000). Affective pathways and representations in mathematical problem solving. *Mathematical Thinking and Learning, 17*(2), 209–219.

Gómez-Chacón, I. M.ª (1997). Procesos de aprendizaje en matemáticas con poblaciones de fracaso escolar en contextos de exclusión social. Las influencias afectivas en el conocimiento de las matemáticas, Tesis Doctoral Universidad Complutense de Madrid. Published by Universidad Complutense de Madrid.

Gómez-Chacón, I.Mª. (2000a). *Matemática emocional. Los afectos en el aprendizaje matemático.* Madrid: Narcea.

Gómez-Chacón, I.Mª. (2000b). Affective influences in the knowledge of mathematics. *Educational Studies in Mathematics, 43*(2), 149–168.

Gómez-Chacón, I.Mª. (2006). Affect, mathematical thinking and intercultural learning: A study on educational practice. In M. M. Bosh (Ed.), *Proceedings of CERME 4: Third conference of the european society for research in mathematics education, Spain (pp. 194–204). European Society for Research in Mathematics Education.*

Gómez-Chacón, I.Mª. (2011). Beliefs and strategies of identity in mathematical learning. In B. Roesken & M. Casper (Eds.), *Current state of research on mathematical beliefs XVII proceedings of the MAVI-17 conference Sept. 17–20, (pp. 74–84). Professional School of Education, Ruhr_universität Bochum, Bochum, German.*

Gómez-Chacón, I.Mª. (2012). Affective pathways and interactive visualization in the context of technological and professional mathematical knowledge. *Nordic Studies in Mathematics Education, 17*(3-4), 57–74.

Gómez-Chacón, I.Mª. (2015). Meta-emotion and mathematical modeling processes in computerized environments. In B. Pepin & B. Rösken-Winter (Eds.), *From beliefs and affect to dynamic systems in mathematics education. Exploring a mosaic of relationships and interactions* (pp. 201–226). Switzerland: Springer.

Gómez-Chacón, I.Mª., & Figueral, L. (2007). Identité et facteur affectifs dans l'apprentissage des mathématiques. *Annales de Didactique et de Sciences Cognitives, IREM Strasbourg, 12*, 117–146.

Hadden, S., & Lester, M. (1978). Talking Identity: The production of 'Self' In interaction. *Human Studies, 1*, 331–356.

Hannula, M. S. (2002). Attitude towards mathematics: Emotions, expectations and values. *Educational Studies in Mathematics, 49*(1), 25–46.

Hannula, M. S. (2015). The relevance of affective systems and social factors: a commentary. In B. Pepin & B. Rösken-Winter (Eds.), *From beliefs to dynamic systems in mathematics education. Exploring a mosaic of relationships and interactions* (pp. 269–277). Switzerland: Springer.

Heyd-Metzuyanim, E., & Sfard, A. (2012). Identity struggles in the mathematics classroom: On learning mathematics as an interplay of mathematizing and identifying. *International Journal of Educational Research, 51–52*, 128–145.

Lazarus, R. (1991). *Emotion and adaptation.* New York, NY: Oxford University Press.

Lave, J., & Wenger, E. (1991). *Situated learning: Legitimate peripheral participation*. Cambridge: Cambridge University Press.

Leder, G., Pehkonen, E., & Törner, G. (Eds.), (2002). *Beliefs: A hidden variable in mathematics education?* Dordrecht: Kluwer Academic Publishers.

Lewis, M., Haviland-Jones, J., & Barrett, L. 2008. Handbook of emotions (3rd ed.). New York: Guilford Press.

Malmivuori, M. L. (2001). *The dynamics of affect, cognition, and social environment in the regulation of personal learning processes: The case of mathematics*. Research report 172 Helsinki: Helsinki University Press.

Mandler, G. (1989). Affect and learning: Causes and consequences of emotional interactions. In D. B. McLeody & V. M. Adams (Eds.), *Affect and mathematical problem solving: A new perspectitve* (pp. 3–19). New York, NY: Springer-Verlag.

Mascolo, M. F., Harkins, D., & Harakal, T. (2000). The dynamic construction of emotions: Varieties of anger. In M. D. Lewis & I. Granic (Eds.), *Emotion, development, and self-organization: Dynamic systems approaches to emotional development* (pp. 125–152). Cambridge, UK: Cambridge University Press.

McLeod, D. B. (1992). Research on affect in mathematics education: A reconceptualization. In D. A. Grows (Ed.), *Handbook of research on mathematics teaching and learning* (pp. 575–596). New York, NY: Macmillan.

McLeod, D. B., & Adams, V. M. (1989). *Affect and mathematical problem solving: A new perspective*. Rotterdam, NL: Springer-Verlag Publishing.

McLeod, D. B., Metzger, W., & Craviotto, C. (1989). Comparing experts' and novices' affective reactions to mathematical problem solving: An exploratory study. In G. Vergnaud (Ed.), *Proceedings of the thirteenth international conference for the psychology of mathematics education, Vol. 2* (pp. 296–303). Paris: Laboratoire de Psychologie du Developpement et de l'Education de l'Enfant.

Muis, K. R., Psaradellis, C., Lajoie, S. P., Di Leo, I., & Chevrier, M. (2015). The role of epistemic emotions in mathematics problem solving. *Contemporary Educational Psychology, 42*, 172–185.

Nelmes, P. (2003). Developing a conceptual framework for the role of emotions in the language of teaching and learning. In *Third conference of the european society for research in mathematics education, 28 February–3 March, in Bellaria, Italy.*

Op 't Eynde, P., & De Corte, E. (2002). Accepting emotional complexity: A component systems approach of emotions in the mathematics classroom. In *Symposium motivation and emotion research in education: Theoretical frameworks and methodological issues at the 2002 annual meeting of the american educational research association, in New Orleans, Louisiana.*

Op 't Eynde, P., De Corte, E., & Verschaffel, L. (2001). What to learn from what we feel? The role of students' emotions in the mathematics classroom. In S. Volet & S. Järvelä (Eds.), *A volume in the EARLI/Pergamon "Advances in Learning and Instruction" series. Motivation in learning contexts: Theoretical and methodological implications* (pp. 149–167).

Pekrun, R., & Stephens, E. J. (2012). Academic emotions. In K. Harris, S. Graham, T. Urdan, S. Graham, J. Royer, & M. Zeidner (Eds.), *Individual differences and cultural and contextual factors. APA educational psychology handbook, Vol. 2* (pp. 3–31). Washington, DC: American Psychological Association.

Pepin, B. & Rösken-Winter, B. (Eds.), (2015). *From beliefs and affect to dynamic systems in mathematics education. Exploring a mosaic of relationships and interactions*. Switzerland: Springer.

Radford, L. (2015). Of Love, frustration and mathematics: A cultural-historical approach to emotion in mathematics teaching and learning. In B. Pepin & B. Rösken-Winter (Eds.), *From beliefs to dynamic systems in mathematics education. Exploring a mosaic of relationships and interactions* (pp. 25–50). Switzerland: Springer.

Roth, W., & Radford, L. (2011). *A cultural historical perspective on teaching and learning*. Rotterdam: Sense Publishers.

Russell, A. (2003). Core affect and the psychological construction of emotion. *Psychological Review, 110,* 145–172.

Scherer, K. R. (2000). Emotions as episodes of subsystem synchronization driven by nonlinear appraisal proceses. In M. D. Lewis & I. Granic (Eds.), *Emotion, development, and self-organization: Dynamic systems approaches to emotional development* (pp. 70–99). New York/Cambridge: Cambrigde University Press.

Schlöglmann, W. (2002). Affect and mathematics learning. In A. D. Cockburn & E. Nardi (Eds.), 3. *Proceedings of the 26th PME, Norwich, UK* (pp. 185–192).

Schoenfeld, A. H. (1992). Learning to think mathematically: Problem solving, metacognition, and sense-making in mathematics. In D. A. Grouws (Ed.), *Handbook of research on mathematics teaching and learning* (pp. 334–370). New York NY: Mac Millan P.C.

Schoenfeld, A. H. (2015). What counts, when? Reflections on beliefs, affect, attitude, orientations, habits of mind, grain size, time scale, context, theory, and method. In B. Pepin & B. Rösken-Winter (Eds.), *From beliefs to dynamic systems in mathematics education. Exploring a mosaic of relationships and interactions* (pp. 395–404). Switzerland: Springer.

SECTION II

COGNITION AND EMOTION IN MATHEMATICAL ACTIVITY

CHAPTER

3

Being in Control

Alexandre Borovik

University of Manchester, Manchester, United Kingdom

NAMING INFINITY

I start my narrative referencing an important episode in history of mathematics—it was analyzed by Graham and Kantor (2009) in their book *Naming Infinity*.[1] The book tells a fascinating story of Russian mathematicians of the early 20th century who discovered a spiritual teaching of *Name Worshiping*, an esoteric stream of Russian Orthodox Christianity, and found in it the strength to do things that were inaccessible to their Western colleagues. They developed the emerging set theory—a challenging and paradoxical branch of mathematics—further and further towards more and more intricate degrees of infinity because they were not afraid to name infinity; after all, their religion not just allowed—demanded from them to name the God.

Naming Infinity is set not so much in the mathematical context as in the realm of the wider spiritual quests of Egorov, Luzin, and their mathematician disciples. Mathematically, Egorov and Luzin followed the great French school of analysis of the time; philosophically, they were guided by Father Pavel Florensky, famous Russian Orthodox theologian, philosopher, and polymath. Among other things, Florensky was an electrical engineer and a prominent theoretician of the Russian Symbolism movement in arts. Moscow mathematics was surrounded by a tangled knot of religious, philosophical, ideological, political, and esthetic currents and undercurrents of prerevolutionary and revolutionary Russia.

There is one aspect of the name worshiping/set theory conjunction that especially attracts my attention: a touching childishness of Florensky's

[1] A very important review of the book is Sinkevich (2012).

http://dx.doi.org/10.1016/B978-0-12-802218-4.00003-0

and Luzin's outlook—their somewhat naive but open and sincere view of the world. This was very much part of *Zeitgeist*:

> Только детские книги читать,
> Только детские думы лелеять ...
> **(О. Мандельштам)**[2]

This could be disputed, but one might as well say that Luzin's sincere childishness contributed to his political infantilism that caused so much trouble to him in his later life.

Luzin and his mathematician friends were brave as only children could be brave. Children name the world around them—for them, it is a way of controlling the world. They are happy to use names suggested by adults. But if they have not heard an appropriate name, they are not afraid to invent the names of their own.

As soon as a child encounters mathematics, an ideal world starts to grow inside of his or her mind—the world of mathematics—and a child strives to control this world. Here, children also need names, and again, they are not afraid to make new ones.

QUEST FOR CONTROL

I systematically collect my mathematician colleagues' testimonies about challenges in their early learning of mathematics (and I now have hundreds of the stories). A common thread in the stories is the same as in *Naming Infinity*: children need to *control* the concepts, objects, and structures they face in mathematics, and they achieve the control by *naming* mathematical entities.

As we shall soon see, controlling ideal objects is a highly emotional affair. My correspondents frequently told me that they suffered more from their inability to communicate their difficulties to adults than from the mathematical difficulties as such—they had no shared language with adults around them.

When I started to collect the childhood stories of mathematics (analyzed in detail in my forthcoming book, *The Shadows of the Truth* (Borovik, in preparation)), I did not expect *being in control* becoming a recurrent theme in childhood testimonies of my colleagues. I also did not anticipate the depth of emotions involved.

When one asks a question to a mathematician, one has to be prepared to get a whole theory as an answer. One of the first stories came from

[2] Only to read childrens' books,
only to love childish things ...
(O. Mandelstam, translated by A. S. Kline)

Leo Harrington[3] and instantly crystallized the "being in control" theme. But LH's story is best told in his own words:

> I have three stories that I think of as related. My stories may not be of the kind you want, since they involve no mathematics; but for me they involve some very primitive meta-mathematics, namely: who is in control of the meta-mathematics.
>
> This is my mother's memory, not mine. When I was three my mother pushed my baby cart and me to a store. At the time prices were given by little plastic numbers below the item. When we got home my clothes revealed lots of plastic numbers. My mother made me return them.
>
> When I was nine in third grade I came home from school with a card full of numbers which I had been told to memorize by tomorrow. I sat on my bed crying; the only time I ever cried over an assignment. The card was the multiplication table.
>
> When I was around twelve, in grade school, a magazine article said how many (as a decimal number) babies were born in the world each second. The teacher asked for someone to calculate how many babies were born each year. I volunteered and went to the blackboard. When I had found how many babies were born each day, the number had a .5 at the end. The teacher said to forget the .5. I erased the .5.

Every teacher and parent knows that children can be very sensitive to issues of control. Not surprisingly, the memories that are described by mathematicians as their first memories of mathematics are frequently memories of attempting to control the world.

The following is from Victor Maltcev[4]:

> When I was 5, I was once playing with toys early in the morning while all the rest were sleeping. When my mother saw me alone, she thought I would like some company and asked my elder brother to go to me. He came and immediately started messing around soldiers. I shouted at him and asked to put everything were it used to be. He put them back but I said he did not. He asked why but I could not find any explanation for the feeling that the probability of that is 1.

And the following is from Jakub Gismatullin[5]:

> The wallpaper near my crib has been put up in a very carelessly way. I remember when I was at around 4 or 5 years old, I got up every morning in a bad mood because of this. Every morning I was trying to move some parts of the wallpaper in my mind. However, after some time (2–3 year) I got used to see inexact pattern. Some time [later] I even imagined the whole plane covered by the wallpaper, however not in a perfect way, but in a way that looks around my crib. That is, I think I have found a certain shape on a wallpaper near my crib, that can be used to tessellate the whole plane (of course this shape contains some inaccuracies).
>
> In our flat in Poland we decided not to have a wallpaper, but just a plain wall. My personal reason to make this choice was not to irritate my 2-years old daughter.

[3] LH is male, American, professor of mathematical logic in an American university.

[4] VM is male, Ukranian, a PhD student in a British university.

[5] JG is male, has a PhD in mathematics, is a professional research mathematician. At the time of described episode, his family was speaking Tatar, Russian and Polish, but mostly Russian.

This is a testimony from SC[6]:

> As a very young lad walking home from school in the Euclidean grid of streets that are the suburbs of Chicago, I thought about avoiding sidewalk cracks: "Step on a crack, break you back." Somehow I knew, or had been told by an older brother that lines were infinite. I reasoned that I didn't know if sidewalk cracks were perfectly regular, and the cracks running north to south from a block to the east might extend to my path. It was Chicago and from the point of view of a child, virtually infinite, so there clearly was no way to avoid sidewalk cracks. I missed an opportunity to become obsessive compulsive.

Yes, SC missed a chance. In their extreme cases, mathematical obsessions could become clinical—but I omit discussion of medical details (in particular, the role of autism spectrum conditions) from the paper—more details can be found in my forthcoming book *Shadows of the Truth* (Borovik, in preparation). Here I will only record that Simon Baron-Cohen (famous for his suggestion that mathematicians should avoid marriages between them because of a higher risk of producing autistic offspring) emphasizes the following:

> People with autism not only notice such small details and sometimes can retrieve this information in an exact manner, but they also love to predict and control the world. ***(Baron-Cohen, 2003, p. 139)***

Children seek control because, left unsupported, they start to feel dangers. This is a testimony from Pierre Arnoux[7]:

> When I was 10 years old, I remember very clearly the feeling I had when I first learnt the idea of a variable. The best comparison is that I felt I walked on very thin ice, which could break at any moment, and I only felt safe when I arrived at the solution.

Olivier Gerard[8]:

> I concur with Pierre Arnoux with a similar image.
>
> When I started solving equations in junior high school, I had the feeling that the variable was a very precious thing, that could become ugly if badly handled and following the rules such as "passing numbers from one side to the other and inverting signs" carefully was like walking softly or stepping slowly when moving a large stacks of things or books in one's hands. If something was done too roughly the things ended on the floor, some of them bruised or broken.

Please notice that that awareness of danger described by Pierre Arnoux and Olivier Gerard is not the same as paralyzing fear; it is a stimulus for being alert.

[6] SC is male, American, professor of mathematics.

[7] PA is male, French, a professor of mathematics in a French university.

[8] OG is male, French, a mathematical researcher and computer scientist in the private industry.

The quest for control makes children to seek help from adults; usually they can easily communicate to their parents and teachers their questions about the real world. In case of the ideal world of mathematics, finding the common language is much more difficult, and children feel bitterly disappointed by their inability to get help, and even more frustrated when they see the outright ignorance of adults.

Lawrence Braden[9]:

> When I was twelve years old, Mr. P, my math (maths) teacher, told the class that π equalled 22/7. Also that π equalled 3.1416. […]
>
> Excited, I went home with the grandiose notion of finding π to a hundred decimal places by the process of long division, and wondered if anyone had ever done that sort of thing before. Well, of course, I kept getting the wrong answer! Not 3.1416 at all! I did it over four or five times, and was really disturbed. The book said that π equalled 3.1416 and the teacher said that π equaled 3.1416 so I logically came to the conclusion that I did not know how to do long division! A truly disturbing notion; I thought I was pretty good at it and here I couldn't even do this simple problem!
>
> "Oh", Mr. P said the next day. "I didn't mean that π was exactly equal to 22/7." It was at that point that I learned not to take everything a maths teacher (or any sort of teacher) said as hewn in stone. Years later I came upon Niven's truly beautiful and elementary proof of the irrationality of π, and Lindemann's proof of its transcendence.

The next testimony comes from Nicola Arcozzi[10]: his truly impressive mathematical discovery was completely ignored by her parents.

> I was about eight-nine years old (Italian third-fourth grade) and I was learning about continents. "Is Australia a continent or an island?" I asked my father. He answered it was both a continent and an island; an answer I found deeply unsatisfactory. I thought for a while about islands and what makes them different from continents, until – weeks later – I reached the conclusion that, by stretching and contracting, Eurasia could be an island of the Oceans as well as an island of the Como lake ("all its shores are on the Como lake").
>
> A sunny day right after rain I was walking with my mother, I pointed to a puddle and I said: "we are on the island of that puddle".
>
> She shrugged and replied "why do you always say such stupid things". (Only many years afterwards I learned that was part of something called topology).

Relations with peers can also be complicated, as explained by RW[11]:

> I became unpopular with my class mates and some teachers because sometimes, maths questions appeared ambiguous to me. When a question was hidden in the text and you had to think about how to translate it into a mathematical problem, I frequently came up with at least one solution different from what the teacher expected, referring to different ways of understanding what was written. I used to be very stubborn when a teacher would try to "sell me" that only one unique answer was correct.

[9] LB is male, American, a recently-retired math teacher of 40 years experience.

[10] NA is male, Italian, has a PhD in mathematics, teaches at a university.

[11] RW is female, German, professor of mathematics in a German university.

> Of course I was wrong sometimes and just misunderstood what was said, but sometimes I was right and some teachers made me respect them a lot by discussing my views in an open way (independent of whether I was right or wrong).

In later sections we shall discuss children's perception of infinity, and the following almost unbelievable story, from GCS,[12] is closely related to this theme.

> Age 6, a state primary school in a working class London area. I always enjoyed playing with numbers. A teacher tried to tell us that when you broke a 12 inch ruler into two pieces that were the same, each would be 6 inches long.
>
> I went to see her, because I couldn't see how you could break the ruler into equal pieces, because of the point at 6; it wouldn't know which piece to join.
>
> In adult notation [0,6] and [6,12] are not disjoint, but [0,6), [6,12] are not isometric. No matter how I tried to explain the problem, she didn't understand. It was a valuable lesson, because from that point on my expectations of schoolteachers were much reduced.

If we reconstruct this truly spooky episode in adult terms, we have to admit that a 6-year-old boy was concerned with the nature of *continuum*, one of the most fundamental questions of mathematics.

A philosophically inclined reader will immediately see a parallel with Plato's Allegory of the Cave: children see shadows of the truth and sometimes find themselves in a psychological trap because their teachers and other adults around them see neither truth nor its shadows.

But I wish to make that clear: I include in my book Borovik (in preparation) only those childhood stories where I can explain, in rigorous mathematical language, what the child had seen in "the Cave." I apply the same criterion to this paper.

CAVEATS AND DISCLAIMERS

This paper touches emotions so raw that "Caveats and disclaimers" becomes its focal section.

Even an innocent, at a first glance, observation (I put it in the abstract of this paper) that mastering mathematics brings a gift of personal empowerment, the precious

> realisation that you know and understand something that no-one else in the world knows or understands – and that you can prove that.

is a very emotive statement as it was reminded to me by one of my colleagues who commented on an earlier version of this paper:

> What a wonderful example of the macho, competitive, arrogant, dog-eat-dog world of pure mathematics!!!!!

[12] GCS is male, English, lecturer of mathematics in a British university.

I accept that pure mathematics can be indeed described as a dog-eat-dog world. I write about that in my book, *Mathematics under the Microscope*, (Borovik, 2009), Section 8.9, and plan to say more elsewhere. But this paper is about children. They do not eat anyone—in most cases, they do not even dare to bite the adult's hand that tortures them.

The stories told to me by my fellow mathematicians are so unusual, so unexpected, and occasionally so spooky that caveats and disclaimers are due.

The stories that my correspondents conveyed to me cannot be independently corroborated or authenticated; they are memories that my colleagues have chosen to remember.

The real-life material in my research is limited to stories that my fellow research mathematicians have chosen to tell me; they represent tiny but personally significant and highly emotional episodes from their childhood. So far my approach is justified by the warm welcome it found among my mathematician friends, and I am most grateful to them for their support. For some reason (and the reason deserves a study on its own) my colleagues know what I am talking about!

Also, my colleagues' testimonies are consistent with my own memories of my first encounters with mathematics, it joys, and its sorrows. I do not include my childhood stories in this paper, but some of them can be found in Borovik (in preparation).

I direct my inquiries to mathematicians for a simple but hard to explain reason: early in my work, I discovered that not only lay people, but also school teachers of mathematics and even many professional researchers in mathematics education, as a rule, cannot clearly retell their childhood stories in *mathematical terms*. It appears that only professional mathematicians/computer scientists/physicists possess an adequate language that allows them to describe in some depth their experiences of learning mathematics.

Other people simply express their emotion without giving any mathematical details. The following statements are taken from a Reddit site,[13] they are answers to the question

> Why do so many children (and adults) hate advanced math? Is it how it's taught, or what is taught?

As the reader will immediately see, this is all about control. And abuse. And violence. Just read the responses:

> cerebral_monkey: I can definitely say this is why I hated math as a kid. It was always just "memorize this equation" or "if you see a problem that looks like this, follow these steps exactly," without an explanation of how or why.

[13] https://www.reddit.com/r/IAmA/comments/4p4poa/im_helping_parents_and_teachers_who_do_calculus/ (Accessed 19 August 2016). I do not know the identity or background of the contributors.

meatmeatpotato: My experience with math was always put X here because the book said to, divide there because that just what you have to do. I never understood the larger picture.

myintellectisbored: I had a math teacher that managed to make me hate it. I still am not even sure how he did, but he did. Maybe it was his archaic methods or insisting that the methods I used were wrong. [...] I think the whole "do the math problem the teacher's way" is part of the problem with math instruction. There are so many different ways to solve a problem and still get to the correct solution. And, a lot of the time, the math techniques in school are the more difficult ways to do a problem.

lucafishysleep: There is a general stigma, especially among teenagers (both when I was a teenager, and teenagers I know now) about showing enthusiasm or aptitude for anything. Sticking your head above the parapet and saying: "This is who I am, this is what I can do, this is what I like doing" is 'not cool'.

Everybodygetslaid69: Teachers who insisted I couldn't/it was wrong to solve problems in my head were the bane of my school career.

devolve: I at 14 realized I could solve for x intuitively without really understanding what I did. Instead of getting help understanding my own process I got reprimanded for doing things too fast and not documenting every step.

Anthro_Fascist: I absolutely hate it when the book just told me to just "suck it up and accept this equation as fact".

[deleted entry] I specifically remember the feeling of defeat I had when I got a geometry question marked wrong in high school because I had derived the distance formula rather than memorized it and regurgitated it.

FourOfFiveDentists: It is the most demoralizing experience ever. It makes you angry, depressed and feel like a loser. My experience is that math teachers are beings of pure evil devoid of any emotion.

IHartRed: Put my head down and cried in 5th grade because I just couldn't get division. Teacher thought I was sleeping and walked over and sprayed me with a plant mister. Needless to say I got to lift my head with tears streaming down my face with the entire room staring at me. Never tried again.

I trust no further comments are needed.

Another point that I wish to clarify: I understand that I encroach onto the sacred grounds of developmental psychology. In contrast with the accepted methodology, I stick to individual case studies. Statistics is always instructive (and, as a mathematically educated person, I hope I have a reasonable grasp of statistics), but I would rather understand the intrinsic logic of individual personal stories. I find an ally in the neurologist Vilayanur Ramachandran (2005), who said the following about statistical analysis (pp. xi–xii):

> There is also a tension in the field of neurology between the 'single case study' approach, the intensive study of just one or two patients with a syndrome, and sifting through a large number of patients and doing a statistical analysis. The criticism is sometimes made that it's easy to be misled by single strange cases, but this is nonsense. Most of the syndromes in neurology that have stood the test of time [...] were initially discovered by a careful study of single case and I don't know of even one that was discovered by averaging results from a large sample.

For that reason, I feel that I have to make the following qualifying remarks:

- I am neither a philosopher nor a psychologist.

- This paper is not about philosophy of mathematics; it is about mathematics.
- This paper is not about psychology of mathematics; it is about mathematics.
- This paper is not about mathematics education; it is about mathematics.

As you will soon see, Antoine de Saint-Exupéry has the same level of authority to me as my mathematics education or psychology of mathematics colleagues. Is anything wrong with that?

TAMING MATHEMATICAL ENTITIES

When I told these stories to my wife Anna,[14] she instantly responded by telling me how she, aged 9, was using the Russian word приручить, *"to tame,"* to describe accommodation of new concepts that she learned at school: the concept had to become tame, obedient like a well-trained dog. Importantly, the word was her secret, and she never mentioned it to parents or teachers—I was the first person in her life to whom she revealed it.

Anna was not alone in her invention; here is a story from Yagmur Denizhan[15]:

> Although I obviously knew the word before, my real encounter with and comprehension of the concept of "taming" is connected with my reading The Little Prince. As far as I can figure out I must have been nearly 12 years old. Saint-Exupéry offered me a good framework for my potential critiques in face of the world of grown-ups that I was going to enter.
>
> I also must have embraced the concept "taming" so readily that it became part of my inner language. Some years ago a friend of mine told me of a scene from our university years:
>
> One day when he entered the canteen he saw me sitting at a table with notebooks spread in front of me but seemingly doing nothing. He asked me what I was doing and I said (though I do not remember having said it, it sounds very much like me) "I am taming the formulae". (Having heard this story I can recall the feeling. Most probably I must have been studying quantum physics.)

Please notice the appearance in this narrative of Antoine de Saint-Exupéry's book *The Little Prince*, with its famous description of *taming*:

> "Come and play with me," proposed the little prince. "I am so unhappy."
> "I cannot play with you," the fox said. "I am not tamed."
> "Ah! Please excuse me," said the little prince.
> But, after some thought, he added:

[14] AB is female, Russian, for many years taught mathematics at a university.

[15] YD is female, Turkish, professor of electrical and electronics engineering in a leading Turkish university.

"What does that mean – 'tame'?"

[...]

"It is an act too often neglected," said the fox. "It means to establish ties."

"'To establish ties'?"

"Just that," said the fox."To me, you are still nothing more than a little boy who is just like a hundred thousand other little boys. And I have no need of you. And you, on your part, have no need of me. To you, I am nothing more than a fox like a hundred thousand other foxes. But if you tame me, then we shall need each other. To me, you will be unique in all the world. To you, I shall be unique in all the world ..."

NOMINATION

To move beyond taming, I have to give a brief explanation of the role of *nomination*—that is, *naming*—in mathematics. To avoid fearful technicalities, I prefer to do that at a childish level, without stepping outside of elementary school arithmetic—trust me, it already contains the essence of mathematics.

One of the rare books on that particular topic, *Children's Mathematics* by Carruthers and Worthington (2006), documents, using dozens of children's drawings, spontaneous birth of mathematics in preschool children. The very first picture is taken from *Le Petit Prince*, Antoine de Saint-Exupéry's famous drawing of a boa constrictor after swallowing an elephant.

In his picture, little Antoine expressed his understanding of some fragment of the world. The picture worked for him because there were two names attached to the picture: *boa constrictor* and *elephant*. And he famously complained how difficult it was to explain the meaning of the picture to adults:

> Les grandes personnes ne comprennent jamais rien toutes seules, et c'est fatigant, pour les enfants, de toujours et toujours leur donner des explications.[16]

Right now Şükrü Yalçinkaya, my research collaborator for many years, and I are writing a hardcore mathematics paper (Borovik & Yalçinkaya, in preparation) where we manipulate with mathematical objects made of some abstract symmetries. We think of them as two-sided mirrors, with labels attached on the opposite sides: one of them says *"point"* and another one says *"line,"* and these labels serve, respectively, as points and lines of some abstract geometry. It is exactly the same kind of the boa constrictor/elephant kind of thinking—as a mathematician, I do not see the difference—and I wish to emphasize that this vision is fully shared by my coauthor.

[16] *Grown-ups never understand anything by themselves, and it is tiresome for children to be always and forever explaining things to them.*

I have already said that I systematically collect stories from my mathematician colleagues about challenges in their early learning of mathematics. Besides "being in control," another common thread in the stories is again the same as in *Naming Infinity*: children need *names* for the concepts, objects, and structures they meet in their first encounters with mathematics.

A testimony from Jűrgen Wolfart[17] is quite typical:

> Probably I was four years old when my mother still forced me to go to bed after lunch for a while and have a little sleep (children don't need this rest after lunch, but parents need children's sleep). Quite often, I couldn't sleep and made some calculations with small integers to entertain myself, and afterwards I presented the results to my mother. Soon, I did not restrict myself to addition ("und") and invented by myself other arithmetic operations – unfortunately I don't remember which, probably "minus" – but I invented also a name for it, of course not the usual one. I don't remember which name, but I remember that my mother reconstructed from my results what operation I had in mind and told me what I did in official terminology. So I forgot my own words for it, but I had a new toy for the siesta time.

Mathematics is a plethora of names, and even memorizing them all could already be a challenging intellectual task for a child. Not surprisingly, the following observation belongs to a poet; it is taken from *Cahiers* by Paul Valéry:

> Vu Estaumier, nommé Directeur de l'Ecole Supérieure des PTT. Me dit que, enfant, à 6 ans il avait appris à compter jusqu'à 6 – en 2 jours. Il comprit alors qu'il y avait 7, et ainsi de suite, et il prit peur qu'il fallût apprendre une infinité de noms. Cet infini l'épouvanta au point de refuser de continuer à apprendre les autres nombres.[18] ***(Valéry, 1974)***

Notice that a child was frightened not by infinity of numbers, but by infinity of *names*; he was afraid that the sequence of random words lacking any pattern or logic:

> un, deux, trois, quatre, cinq, six, …

would drag on and on forever. I would agree—it was a scary thought; the poor little child was not told that a two dozen of numerals would suffice, that the rest of the arithmetical universe could be built from a handful of simplest names. He was not reassured in time that mathematics can

[17] JW is a professor of mathematics.

[18] Seen Estaumier, appointed Head of the "Superior School for Postal Service and Telecommunications". Told me that, as a child, aged six he had learned how to count up to 6 – in two days. He then understood that there was 7, and so on, and was afraid there would be an infinite amount of names to learn. This infinity horrified him to the extent that he refused to keep learning the other numbers.

bring safety back by providing very economic means for a systematic production of the infinity of names. This is evidenced by Roy Stewart Roberts[19]:

> At some point [...] I had discovered that you can continue counting forever, using the usual representation of numbers if one ran out of names.

As soon as a child discovers that he or she can combine "hundred" and "thousand" to form "one hundred thousand," as a soon as a child gets control over the names for numbers, the counting becomes unstoppable.

A testimony from John R. Shackell[20]:

> I would have been three years old, getting towards four. My mother was confined for the birth of my sister and so I was being cared for by an aunt. I don't think she had an easy task.
>
> I would stand on my head on the sofa and read the page numbers from an encyclopaedia. I was very persistent. The conversation went approximately as follows:
>
> – "One thousand three hundred and twenty three, one thousand three hundred and twenty four."
>
> – "John, stop that counting."
>
> – "One thousand three hundred and twenty five, one thousand three hundred and twenty six."
>
> – "Oh John do stop that counting."
>
> – "One thousand three hundred and twenty seven. I wish you were one thousand three hundred and twenty seven."
>
> – "Well you wouldn't be so young yourself!"

It is worth mentioning that John Shackell is a professor of Symbolic Computing; the work of his life is the book *Symbolic Asymptotics* (Shackell, 2004); in lay terms, these words mean computing (moreover, computing automatically, on a computer) names for certain types of infinity. As we can see, as soon as a child has control over the names for numbers, control over the names for infinity also becomes possible—and can even turn into a professional occupation for life.

Another story comes from Theresia Eisenkölbl[21]:

> My brother and I had learned (presumably from our parents) how counting goes on and on without an end. We understood the construction but we were left with some doubt that you could really count to high numbers, so we decided to count up to a million by dividing the work and doing it in the obligatory nap time in kindergarten

[19] RSR tells about himself: "As an adult I obtained a PhD in mathematics [...], and now am retired if mathematicians ever retire." The episode took place before he went to school.

[20] JRS is male, a professor of mathematics.

[21] TE is female, German, has a PhD in Mathematics (and a Gold Medal of an International Mathematical Olympiad), teaches mathematics at a university. At the time of this episode she was 3 years old, her brother 5 years old.

in our heads. After a couple of days, we had to admit that it took too long, so we debated whether it was ok to count in steps of thousands or ten-thousands, now that we had counted to one thousand many times. We ended up being convinced that it is possible to count to a million but slightly unhappy that we could not really do so ourselves.

NAMES AS SPELLS

Nomination (that is, naming, giving a name to a thing) is an important but underestimated stage in development of a mathematical concept and in learning mathematics. I quote mathematician Semen Kutateladze (2005):

> Nomination is a principal ingredient of education and transfer of knowledge. Nomination differs from definition. The latter implies the description of something new with the already available notions. Nomination is the calling of something, which is the starting point of any definition. Of course, the frontiers between nomination and definition are misty and indefinite rather than rigid and clear-cut.

And here is another mathematician talking about this important, but underrated concept:

> Suppose that you want to teach the 'cat' concept to a very young child. Do you explain that a cat is a relatively small, primarily carnivorous mammal with retractible claws, a distinctive sonic output, etc.? I'll bet not. You probably show the kid a lot of different cats, saying 'kitty' each time, until it gets the idea. ***(Boas, 1981)***

And back to Kutateladze (2005):

> We are rarely aware of the fact that the secondary school arithmetic and geometry are the finest gems of the intellectual legacy of our forefathers. There is no literate who fails to recognize a triangle. However, just a few know an appropriate formal definition.

This is not just an accident: definitions of many fundamental objects of mathematics in the *Elements* are not definitions in our modern understanding of the word; they are *descriptions*.

For example, Euclid (or a later editor of his *Elements*) defines a straight line as

> a line that lies evenly with its points.

It makes sense to interpret this definition as meaning that a line is straight if it collapses in our view field to a point when we hold one end up to our eye.

We have to remember that most basic concepts of elementary mathematics are the result of nomination not supported by a formal definition: number, set, curve, figure, etc.

And we also have to remember that as soon as we start using names, we immediately encounter logical difficulties of varying degree of subtleness—especially if we attempt to give a name to a definite object. I should mention in passing the classical Bertrand Russell's analysis of propositions like

> 'The present King of France is bald'

and arguments like

> 'The most perfect Being has all perfections; existence is a perfection; therefore the most perfect Being exists'

(see Russell, 1905). Russell points out that the correct reading of the last phrase should be

> 'There is one and only one entity x which is most perfect; that one has all perfections; existence is a perfection; therefore that one exists.'

And he comments further that

> As a proof, this fails for want of a proof of the premiss 'there is one and only one entity x which is most perfect'.

According to Russell, a definite nomination (emphasized, as it frequently happens in English, by the use of the definite article "the") amounts to assertion of *existence* and *uniqueness* of the nominated object and has to be treated with care.

Perhaps, I would suggest introducing a name for an even more elementary didactic act: *pointing,* like pointing a finger at a thing before naming it.

A teacher dealing with a mathematically perceptive child should point to interesting mathematical objects; if a child is prepared to grasp the object and play with it, a name has to be introduced—and, in most cases, there is no need to rush ahead and introduce formal definitions.

Life of definitions in mathematics is full of intellectual adventures; the size of this article does not allow me to go into any details; I would like to mention only that some definitions eventually become *spells.* My old university friend reminded me that Semen Kutateladze in his lectures on functional analysis frequently used a peculiar phrase:

> "By reciting standard spells, we prove that ..."

The words "by reciting standard spells" mean here "by invoking the canonical conceptual framework" (and you may even wish to use the word "sacred" instead of "canonical"). Phrases like this can be used when a definition has overgrown itself and has become a metadefinition, a pointer to the whole new host of names, nominations, definitions.

SOME CONJECTURES

Everything in this section is strictly conjectural; I am not making any claims, only a few tentative suggestions triggered by rereading an old book by Russian historian and anthropologists Porshnev (1974).[22]

I propose that specific structures of the human brain that are responsible for religious feelings are archaic voice communication centers that in the prehistoric times were used to process voice signals as absolute commands, like a dog processes a command from its master, without separation of the signifier from the signified. Therefore activation of these centers in a modern human (say, in experiments with electrodes inserted in the brain) results in the subject perceiving the most common words as having "supervalue" as revelations. But so do dogs: the master's command is perceived by a dog as a revelation, one master's word instantly changes a friend into a foe.

I follow Porshnev and propose that a prayer is a communication of a person with his/her supervalue centers in the brain. It is easy to suggest that these supervalue centers, being older than the parts of the brain responsible for consciousness, are better connected with various other archaic parts of the brain, and, first of all, with those responsible for emotions, which explains the undisputed psychological value of a prayer:

В минуту жизни трудную,
Теснится ль в сердце грусть,
Одну молитву чудную
Твержу я наизусть …
(М. Ю. Лермонтов)[23]

When a definition is used as a spell (or a prayer), it invokes (interiorized earlier, on earlier occasions) mechanisms for feasibility filtering of raw mathematical statements and images produced by subconsciousness. In this phrase, I would even downgrade the word "statement": why not

[22] Porshnev's approach is parallel to, but very different from, that of Jaynes (1990).

[23] When my life is arduous,
If sadness freezes blood,
I say one prayer marvelous,
I learned it all by heart …
(M. Yu. Lermontov. Translated by Yevgeny Bonver.)

say "utterance"? But for a person, the perception of processing of a prayer (usually it is described in subtler words, something like "dissolving in a prayer") and of a ritualized definition could be very similar.

In my book Borovik (2009, pp. 138–140), I conjecture that mathematics has a facet systematically ignored in mathematics education—it is a language for communication with subconsciousness. Children, when learning mathematics, need command words for subconscious parts of their mind that actually do their mathematics for them.

I'll write about that in more detail elsewhere.

CHILDREN AND INFINITY

Infinity is a name that can be adopted and used by a mathematically perceptive child in a very natural way, the same way as a child absorbs the words of mother tongue; even more, a child may start inventing synonyms, because infinity might stand for something real in his mind's eye.

Here are two stories of the discovery of infinity; as you will see, *naming* is its crucial component; the second crucial component is child's *control* over his mental constructions.

First comes a testimony from DD[24]:

> When I was 4 [and 2/3] I first went to nursery school. One day a girl came in with a pencil which had a sort of calculator on the back: a series of five or six wheels with 0–9 on each, allowing simple addition, counting, etc. This was in 1943 in New York. Her calculator absolutely fascinated me, and I kept watching as, when the numbers got larger, there would be all 9s and then a new column on the left would pop up with a 1. I just got a feeling for how the whole system operated and it definitely made me feel really satisfied, though I did not know why or what I would do with this information. Also, no one of my friends seemed the least bit interested: I don't think I explained it very well. That weekend, on the Sunday I got up at just before 6AM and went into my parents' bedroom, quietly, as I was allowed to do, went over to the window and looked out down the empty street, at the far end of which was East River Drive as it was then called, bordering the East River. After a bit I started thinking about those wheels. It seemed to me that more important than the 9s were numbers like 100, 1010, 110, 111 then 1000, 1001, 1010 and so on, and I played out these in longer and longer columns in my head until I was absolutely clear how it worked, and I just knew that what I would now call the place-integer system fitted together in a completely satisfactory way. It was still early so I continued thinking about these numbers, and remembered that we used to argue over whether or not there was a largest number. We would make up peculiar names in these arguments (the boys in the nursery class, that is): so somebody would say that a zillion was largest, and someone else might say, no, a squillion was, and so on, nonsense on nonsense. But if these discussions meant anything, I thought, it should all clear itself up in the column pictures I now had in my mind. I then tried to picture

[24] DD is a mathematical physicist and works in a British university. He preferred not to give his full name.

> the 0/1 arrangement of the largest number, and was tickled at the thought that if I then cranked everything up by rolling the smallest wheel round and then seeing (in my mind's eye) the spreading effect it had, I would get an even larger number. Great. Then I got upset: I already had the largest number, according to nursery class arguments. So what was going on. I do not know where it came from, but I suddenly realized that there was no largest number, and I could say exactly why not: just roll on one more, or add 1 (I did know addition quite well by then.) Aha! So I woke up my Dad and excitedly told him that there was no largest number, I could show it, and recited what I had thought out. Poor fellow: it was the overtime season and he worked more than 8 hours a day six days a week – he was not impressed. I won't tell you what he said. Later that day my mother was pleased that her first born son had done something, but I don't believe that either of my parents, or even any of the others in the nursery class, ever really understood the point I was making, and certainly never got intense pleasure from thinking about numbers. Although this incident is filtered through my decades of doing mathematical physics, it remains clear in my mind and always has.

Leaving the world of mathematics and mathematicians, we may listen to a story from philosopher MM[25]:

> I started thinking about death and wanted to convince myself I would never die, instead of thinking about life after death … So I started thinking about an infinity in this way: first, I assumed that my entire life was only one dream in one night in another life where I am still the same person but could not fully realize that a full life goes on in each dream (an interesting point about personal identity, I guess). Now, that other life would be finite and have only a finite number of nights. So, I thought further that in each night there must be a finite number of dreams, encapsulating a finite number of lives. This was still short of infinity, so I started thinking that in each of these finitely many dreams of the finitely many nights, I would live a life that would in turn contain finitely many nights, which would contain finitely many dreams, and so on. I was not so sure that I was safe that way (i.e. that I would go on living forever), but I convinced myself that these were enough lives to live, so that even if the process would end, I would still have lived enough, and stopped thinking about it.

A reader of my blog, who signed his comment only as JT, remarked that little MM was safe because of Koenig's Lemma in Set Theory:

> Every infinite finitely branching tree has an infinite path (with no repeated vertices).

EDGE OF THE ABYSS

I have said before that children may feel the dangers of navigating on unknown mathematical terrain. However, when given security and protection, children prefer the blissful ignorance of dangers of the world; and

[25] MM is male, French, a professional philosopher with research interests in philosophy of mathematics. The episode took place at age 7 or 8.

the world of infinity is dead dangerous. Here is a story from Alexander Olshansky[26]:

> In 1955 I was 9 years old. My father, Yuri Nikolaevich Olshansky, a lieutenant colonel-engineer in Russian Air Force, was transferred to a large air base in Engels. Every Sunday on the sport grounds of the base there were some sport competitions. A relay race of
>
> 800 meters + 400 meters + 200 meters + 100 meters
>
> was quite popular; it was called Swedish relay. After two or three races I have come to an obvious conclusion that the team wins which has the strongest runner on the first leg (or on the first two legs) because this runner stays in the race for longer.
>
> But the question that I asked to my father was in the spirit of Zeno's paradoxes: if the race continues the same way,
>
> 50 meters + 25 meters + ...,
>
> will it be true that the runners will never reach the end of the 4-th circle (one circle is 400 meters)? (My father was retelling my question to his fellow officers; before World War II, he graduated from the Mathematics Department of Saratov University).

Little Sasha was walking on the edge of an abyss; being an educated mathematician, his father had a false sense of security because perhaps he believed that Zeno's "arrow" paradox (of which Swedish relay is an obvious version) is resolved in elementary calculus by summation of the geometric progression

$$800 + 400 + 200 + 100 + 50 + 25 + \ldots = 1600 = 4 \times 400.$$

This is true; the runners will indeed reach the end of the fourth circle, and fairly quickly.

But if you think that Zeno's paradox ends here, you are wrong; be prepared to face one of its most vicious forms.[27] Indeed, the real trouble starts after the successful finish of the race:

> where is the baton?

Indeed, the whole point of the relay is that each runner passes the baton to the runner on the next leg. After the race is over, each runner can honestly claim that he is no longer in possession of the baton because he passed it to the next runner.

> I repeat: can you explain where is the baton?

In adult terms, this is a classical conundrum of potential and actual infinity—but, as we see, the problem can be formulated in terms accessible

[26] AO holds professorships in mathematics in Moscow and the United States. Some of his famous results in group theory can be described as a subtle and paradoxical interplay of finite and infinite.

[27] See Silagadze (2005) for a more detailed discussion of various versions of Zeno's paradox.

to a child. In the ideal world, a teacher of mathematics should follow the dictum from J.D. Salinger's *Catcher in the Rye* and gently guide a child through dangers of mathematics:

> I keep picturing all these little kids playing some game in this big field of rye and all. Thousands of little kids, and nobody's around – nobody big, I mean – except me. And I'm standing on the edge of some crazy cliff. What I have to do, I have to catch everybody if they start to go over the cliff – I mean if they're running and they don't look where they're going. I have to come out from somewhere and catch them. That's all I'd do all day. I'd just be the catcher in the rye and all.

CONCLUSIONS

In learning and doing mathematics, a person may experience a range of states of mind and emotions:

> frustration, suspicion, hunch, trust/mistrust, confidence, interest, ardour, chase, "going for the kill", concentration, focus, serenity, inspiration, epiphany, joy, ecstasy,

to name a few. This paper focused on a particular one, the feeling of "being in control"—mostly because it is not frequently discussed in the literature.

By ignoring the issue of control in mathematics, mainstream mathematics education ignores the point, the purpose, and the reason for the existence of mathematics. After all, an apocryphal saying describes mathematics as

> the language of contracts with Nature which Nature accepts as binding.

As you can see, it is all about control. In our computer age, I would add to that two more points.

> Mathematics is the language of orders to computers.
> Mathematics is also the art of finding precise and reproducible solutions to problems that cannot be solved by, or entrusted to, computers.

Blind trust in computers is the ultimate form of the loss of control. But we should not underestimate the intellectual courage of children: children come to this world to be its masters. We simply have to try not turning them into slaves of computer technology.

We also should not underestimate the power of a child's view of the world. I am not prepared to accept the wisdom of the following words:

> When I was a child, I spake as a child, I understood as a child, I thought as a child: but when I became a man, I put away childish things. (1 Corinthians 13:11)

Indeed I stand for my former (or inner?) child. In my stance, I find support in words of Michael Gromov[28] (for those not in the know: he is a really famous mathematician):

> My personal evaluation of myself is that as a child till 8–9, I was intellectually better off than at 14. At 14–15 I became interested in math.
>
> It took me about 20 years to regain my 7 year old child perceptiveness.

Acknowledgments

The list of people who helped me in many ways in my work on the project reflected in this paper is becoming almost as long as the paper itself; I refer the reader to *Introduction* of my book *Shadows of the Truth* (Borovik, in preparation).

But, in addition, I thank Adrien Deloro for help with translations from French and Peter McBurney, Roman Kossak, Robert Kotiuga, Maria Droujkova, and Galina Sinkevich for useful comments on the previous versions of this paper.

References

Baron-Cohen, S. (2003). *The essential difference. The truth about the male and female brain.* New York, NY: Basic Book.

Boas, R. P. (1981). Can we make mathematics intelligible? *The American Mathematical Monthly, 88*, 727–731.

Borovik, A. V. (2009). *Mathematics under the microscope. Notes on cognitive aspects of mathematical practice.* Providence, RI: American Mathematical Society.

Borovik, A. V. *Shadows of the truth: Metamathematics of elementary mathematics* (in preparation).

Borovik, A., & Yalçinkaya, Ş. *Structural recognition of black box groups encrypting PGL(3)* (in preparation).

Carruthers, E., & Worthington, M. (2006). *Children's mathematics: Making marks, making meaning*. New York, NY: Sage Publications.

Graham, L., & Kantor, J.-M. (2009). *Naming infinity: A true story of religious mysticism and mathematical creativity.* Cambridge, MA: Harvard University Press.

Jaynes, J. (1990). The origin of consciousness in the breakdown of the bicameral mind. In 1st Mariner Book ed. (Ed.), Boston, MA: Houghton-Mifflin [Originally published 1976].

Kutateladze, S. S. (2005). *Nomination and definition.* http://www.math.nsc.ru/LBRT/g2/english/ssk/nomination_e.html.

Porshnev, B. F. (1974). *On the beginning of human history (problems of paleopsychology).* Moscow: Mysl' [in Russian].

Ramachandran, V. (2005). *The emerging mind*. New York, NY: Profile Books.

Russell, B. (1905). On denoting. *Mind, 14*, 479–493.

Shackell, J. R. (2004). *Symbolic asymptotics (algorithms and computation in mathematics).* New York, NY: Springer.

Silagadze, Z. K. (2005). Zeno meets modern science. *Acta Physica Polonica B, 36*(10), 2887–2929.

Sinkevich, G. (2012). *Рецензия на книгу Лорена Грэхема и Жана-Мишеля Кантора«Имена бесконечности. Правдивая история о религиозном мистицизме и математическом творчестве»*, Вопросы истории естествознания и техники №2, 196–204.

Valéry, P. (1974). *Cahiers*. Paris: Gallimard.

[28] MG is male, Russian, a professor at the Institut des Hautes Études Scientifiques, Paris, and the Courant Institute of Mathematical Sciences, New York. He is the recipient of the Abel Prize (2009).

CHAPTER

4

Epistemic States of Convincement. A Conceptualization from the Practice of Mathematicians and Neurobiology

Mirela Rigo-Lemini, Benjamín Martínez-Navarro

The Center for Research and Advanced Studies of Mexico's National Poly-technical Institute, Mexico City, Mexico

Presentation of the Problem

There are many studies in the mathematics education field (Balacheff, 2000; De Villiers, 1990, 2010; Duval, 1999; Fischbein, 1982; Foster, 2016; Harel & Sowder, 2007; Inglis, Mejia-Ramos & Simpson, 2007) that investigate, within several scenarios, empirical phenomena concerning convincement, conviction, certainty, presumption, and doubt that various groups—such as professors and students with different levels of education—experience regarding mathematics facts; some examples are rules, principles, laws, operational processes, or solutions to school tasks. Notwithstanding, these works often fail to precisely define the meaning of these convincement-associated terms, which renders such terms polysemic, and further complicate or even prevent agreement and consensus from being reached among experts. Consequently, this hinders academic progress on this research topic.

This chapter investigates precisely what those internal states associated with convincement are. The proposal, motivated by the grounded theory (Corbin & Strauss, 2015), has been expressed in two parts; so the document is divided accordingly.

http://dx.doi.org/10.1016/B978-0-12-802218-4.00004-2

In the first part, empirical data gathered from the professional practice of mathematics is complemented and confronted by conceptual elements taken from various disciplines such as philosophy and sociology; from which the undeniable presence of states of convincement in the mathematical practice follow and relationships are inferred between said states, on the one hand; and beliefs, groundings, and motives are on the other, as well as some specific attributes pertaining to said states of convincement, like the various types of convincement. The second part employs a model proposed by Damasio to describe emotions and feelings in order to interpret the epistemic states themselves as certain types of emotions and/or feelings, which then allows for a deepening of the analysis to account for, although prematurely, convincement-generating phenomena in the framework of a deductive-type proof. The last part of the document shows that significant implications for mathematics learning follow from the study of epistemic states, some of which are presented here.

STATES OF CONVINCEMENT IN THE PROFESSIONAL PRACTICE OF MATHEMATICS

The first part is focused on investigating whether convincement is present in the practice of mathematicians. Should this be the case, the goal then is to identify where and how the states of convincement occur in this practice—by unveiling its role and finding out if the expert community is conscious of its presence in their activities and of the significance of this presence.

Within the context of the research, wich results are partially presented here, empirical work has been undertaken with mathematics teachers and students in classroom settings; however, this document only reports the results of the analysis using empirical data that has been taken from the professional practice of mathematics. As we shall see throughout the first section, mathematicians, with their keen attitude, continuously carry out metacognitive activities from which they gain insight about the components of their work or the processes being used; specifically, this self-reflection has allowed them to realize and remain cautious of the role that convincement and other similar states have in their habitual discipline practice, which they have also seen fit to make explicit. This represents an important benefit for research on the subject—which has the tendency to be evasive and elusive—because it is very difficult to find similar introspections, with similar levels of detail and depth, in other communities that practice mathematics, such as with professors and students in a classroom setting.

Convincement in the Endeavors of Mathematicians

René Thom explains that "any proof is rigorous if it wins acceptance by all readers who are adequately educated and prepared to understand it"

(Thom, 1986, p. 72), and Brousseau considers that "(To establish a theorem) requires an adherence, a personal conviction, an internalization, that may not be received by mere definition" (Brousseau, 1997, p. 15). Volmink, on the other hand, recognizes that "we bother to prove theorems to convince people (and ourselves)... we may regard a proof as an argument that is sufficient to convince a reasonable skeptic" (quote in De Villiers, 1990, p. 17).

This work aims to study the states of/for gaining acceptance (1986) referred to by René Thom, the adhesion or personal conviction states discussed by Brousseau, or the states of convincement that Volmink describes. In summary, the goal is to discern the meaning and role that professional mathematicians, and other participants in the practice of mathematics, confer on these internal states.

Professionals in this field have expressed the results of their introspective reflections regarding the role of convincement in mathematical proofs. Balacheff, for instance, suggests that "We distinguish three types of situations that call for validation processes, that is the production of proofs. Such situations may be characterized by the function of proofs in each of those situations: proof to decide, proof to convince and proof to know" (Balacheff, 1987, p. 155); Harel and Sowder (1998, p. 241), on the other hand, hold that a proof includes two subprocesses, namely to ascertain and to persuade; Hanna and Jahnke (1996) assert that the importance of a proof is in correspondence with the greater or lesser degree of certainty that it produces and the amount of insight it grants in terms of the truth of the proven claim; and De Millo, Lipton, and Perlis (1986, p. 268) "recognize that convincingness is an essential feature of mathematical proof."

Clearly, convincement is significantly present in the stages of mathematical discovery. History shows that mathematicians experience a certain type of conviction during experimentation activities, which De Villiers (2010) calls experimental. This experimental conviction precedes and provides a motive for proof. As a sample, the author refers to the remark of Neubrand (1989) about the proof of the Bieberbach conjecture (1916, now the De Branges theorem, 1984):

> As in many other cases, in this example mathematicians first started with the consideration of special cases, restricted cases, etc., in order to convince themselves of the possibility of the validity of the conjecture.

De Villiers (2010, p. 28) reinforces this position, regarding the presence of experimental convincement in mathematical work, turning to the self-reflection that Paul Halmos shares about his own practice as a mathematician:

> The mathematician at work ... arranges and rearranges his ideas, and he becomes convinced of their truth long before he can write down a logical proof. The conviction is not likely to come early—it usually comes after many attempts, many failures, many discouragements, many false starts

Convincement, then, precedes and guides proof. It also serves as a criterion to distinguish a proof from other types of reasoning; that is, it serves as a limiting criterion: a justification needs to be convincing to be considered a proof. De Millo, Lipton, and Perlis (1986, p. 267) state that: "proofs—in the practice of mathematics—must be convincing; otherwise, they wouldn't be recognized [as such]."

What does a proof require for it to be convincing?

From the internal perspective of mathematics and the rationality of last century—a rationality that is affiliated with the underpinnings of Euclid's Elements and that was formalized by Hilbert's proof theory—Tymoczko (1986, p. 247) suggests that a proof is convincing when it can be surveyable (i.e., that the proof is perspicuous and checked by hand by a rational agent) and formalized (i.e., it satisfies certain conditions of formal theories).

From a sociological vantage point, on the other hand, "…formal logic is rarely employed in proving theorems, the real truth of the matter is that *a proof is just a convincing argument, as judged by competent judges*" (Hersh, 1993, p. 389) or, similarly, it is necessary that the proof go through a social process of multiple triangulations that generally take place over (very) long periods of time: When a mathematician considers that he has discovered a proof for a theorem, he verbally communicates it to his students and peers: In its first incarnation, a proof is a spoken message… If they find it tolerably interesting and believable, he writes it up … and distributes it; after it has circulated in draft for a while, if it still seems plausible, he does a polished version and submits it for publication. If the referees also find it attractive and convincing, it gets published. If enough members of that larger audience believe it and like it and if the results are not questioned, then the proof is paraphrased and the theorem takes different versions, … each case reinforcing the belief… And a theorem that is believed then begins to be used. It may take the form of a lemma in other proofs and if it does not lead to contradictions… one is more inclined to believe it (De Millo, Lipton & Perlis, 1986, pp. 272–273).

If during this time … "alternative proofs are invented, if it has applications and generalizations and is analogous to known results in related areas, then it comes to be regarded as 'rock bottom'" (Davis and Hersh, quote in De Villiers, 1990, p. 214); or in the words of De Millo and collaborators, the proof becomes (more) credible and convincing and the truthfulness of the theorem is accepted (in a more definitive way).

Despite its obvious importance—according to De Millo et al—this social aspect of a proof tends to be considered philosophically irrelevant, because it is argued that many beliefs or ideas may become convincing under the pressure of propaganda or prejudice. The authors claim that conviction in mathematics is obtained by a highly evolved set of processes, with very little arbitrariness about them. These general characteristics of mathematical practice allow for bad proofs to be discarded and useful, attractive, and memorable theorems to be chosen (De Millo et al., p. 268).

Historical evidence shows that convincement regarding mathematical facts is experienced in a sensible and conscious manner among mathematicians; it is present in the heuristic stage of proof; it acts as a criterion for separating proof from reasoning that fails to meet the standards; and verifying that a proof is convincing (whether from the internal perspective of mathematics or from the sociological perspective) assumes an expert or a collective expert trained in the practice of mathematics and in decidedly distinguishing its criteria for rigor.

This first exploratory and open approach to the professional practice of mathematics has been prompted by sensible questions (Corbin & Strauss, p. 92) related to the possible presence of convincement in the practice of mathematicians, the role it plays in the mathematical endeavor, and their awareness of these roles. Early results, recently submitted, have in turn lead to a great diversity of topics for research: What are these internal states that mathematicians call convincement, adhesion, or security? What are these states of convincement attached to? How are they formed during the development of a proof? What are its main attributes? Are there several types of these internal states? How are these internal states expressed? And there are many other questions. To outline some initial responses to these questions we turn to the literature and define terms and, in a back and forth process, pair the terms with historical data so they may gain empirical validity, clarity, and they may be broadened and enriched. This work is shown in the following section, explaining the first concepts that are defined in the research framework and presenting some of the historical cases that suggested and endorsed them.

The daily endeavor of mathematicians, described by De Millo et al., clearly shows how convincement is the result of social interactions occurring in the framework of socially introduced and constructed cultural values. The authors of this work recognize the social and cultural nature of states of convincement; however, in this work they focus their attention on an intrasubjective dimension of the phenomena being studied that, among much else, allows for an initial understanding of what these states of convincement consist of, how they are gestated internally by the subject, and what their relationships are with cognitive and affective processes; this finally yields an answer outline to some of the foregoing questions.

The First Theoretical Constructs: Beliefs, Epistemic Schemes, Epistemic States

Beliefs

Most experts, despite their disagreements on the topic, suggest that for a person to believe s/he must have a representation of the belief, and s/he must have associated said representation with a true truth value

(Gilbert, 1991; Villoro, 2002). From this point of view, the representation of an object is a necessary condition for belief; it is not possible to believe without first grasping the believed object or objective situation, without understanding, conceiving, or coding in some representation register in which one believes. However, it is not possible to believe a proposition that is supposed to be false (Wittgenstein, 1988); beliefs make up the truth baggage of people, beliefs are part of their reality (according to how they grasp it) and are a guide to interaction with their internal and external world and to regulate their decisions and actions (Villoro, 2002).

Beliefs (and knowledge) are powers, in the Aristotelian sense (Villoro, 2002), that are conserved unconsciously and may be updated in consciousness under certain conditions. So humans hold beliefs unconsciously and without realizing it (Thompson, 1992)—probably in the same way as small children and some animals—although older humans also have conscious beliefs and awareness of said consciousness.

People assign truth values to their beliefs, although they tend not to choose only true or false. They assign different probabilities of truth in the (0,1] continuum to their beliefs (Rigo, 2013; Villoro, 2002).[1] This phenomenon may be observed in the professional practice of mathematics. Mathematicians assign the maximum probability of truth (of 1) to certain beliefs, such as in the example of the Poincaré conjecture; the community of specialists assigned it maximum credibility after Perelman (in 2008) provided a proof of its truth and experts rushed to corroborate, to provide alternative approaches for the proof, and to find various applications. There are other beliefs to which mathematicians grant very high probabilities, but less than one; these include conjectures that, albeit lacking proof, have computational evidence that strongly supports their truth yet is not sufficient for the community to grant full certainty. The Collatz conjecture, for instance, (consider the following operation on an arbitrary positive number: if the number is even, divide it by two, and if the number is odd, triple it and add one. The Collatz conjecture asserts that, starting from any positive number m, repeated iteration of the operations eventually produces the value 1) has been verified for a very large amount of numbers (1.64×10^{11} 64-bit numbers per second) using multi-coprocessing systems (Ito & Nakano, 2011). Another example is the ternary Goldbach conjecture (every odd number can be written as a sum of three primes), which has been experimentally verified up to 4×10^{14} (Richstein, 2000).

In general, the process of assigning a probability of truth to a belief takes place through significantly varied practices of supporting them, as was shown in the foregoing paragraph and as shall be seen

[1]The probability of truth may never be 0, since as explained above, it makes no sense to think that someone will believe a proposition they consider false.

below. Of particular interest are examples that show mathematicians not always relying on proof—unrestricted rigor criterion—to assign relatively high probabilities of truth to mathematical facts. The most salient are cases where notable members of the community resort to extramathematical considerations to back mathematical statements.

Epistemic Schemes for providing grounds

In the vast majority of cases, especially those related to the perceptual system, a belief or significant idea is accepted almost simultaneously to its representation or grasping (and its rejection or justified confirmation occurs subsequently, if at all). This is what was meant by William James, who suggested the existence of spontaneous credibility, at least in relation to objects that are not contradicted; he asserted that, in those cases, "objects are believed *ipso facto*; the belief is caused by the current stimulus" (quote in Villoro, 2002, p. 80).

Nevertheless, the way the perceptual system processes beliefs, in terms of which the credibility of objects arises spontaneously and in the absence of reasons, does not always prevail in human beings. There is a varied and well substantiated body of evidence suggesting that people almost always have reasons supporting their beliefs, and though they are not always explicit, they may eventually be so.

So, if a person believes because from her perspective the reasons are sufficient, and if adequately encouraged to do so, that person may make explicit and explain the "reasons" justifying a belief (Krummheuer, 1995; Villoro, 2002); that is, people do not usually believe gratuitously and always they will find a cause, justification, or grounds for their beliefs if they seek one with determination.

In some cases, the supports are sufficient and objective from an external perspective; that is, they make consistent and relevant reference to the contents of the belief. Some examples include reasoning using deductive justification or experimental and empirical arguments; or supporting a belief with well-educated intuition; or adopting a belief because it is consistent with others; or basing a belief of a proposition because of a lack of evidence against it. Cases of objective reasoning abound in the professional practice of mathematics. Indeed, one of the paradigmatic features of their work is the way they justify their results, through proof, which is motivated by reason of the discipline, structured in a deductive manner and expressed through several different levels of formalization. Mathematicians also turn to experimental empirical work and, during the last few decades, to work based on computer software, not only in discovering plausible conjectures but also as an alternative to proof in supporting results; this leads to an alternate rationality, to that which drove mathematics based on formal deductive proof up to the end of the twentieth century. A widely-cited example in the literature of this new experimental computer assisted rationality is that of the two "proofs" of the four color "theorem"; the first was published by Appel and Hanken in 1976, and the second was published by Robertson, Sanders, Symour, and Thomas in1997.

It is not uncommon for a listener to deem the grounds that lead a person to give credibility to a statement as limited, subjective, inadmissible or, definitely, as extrarational. Some of the ways of backing beliefs include repetition, customs, and habits; authority; the biographical background of a person (I believe it because I was taught it and I wouldn't be taught something false); practical reasons (I believe it because it is easier); religious or aesthetic emotion; a profound personal experience; social consensus; or faith in a person, loyalty, to previous belief, or even a gut feeling (the gut has its reasons unbeknownst to reason) (Cf. Villoro, 2002). In principle, one might think of these forms of backing as foreign to the mathematical endeavor. Indeed, they are not; some practices, like that of Saccheri, provide for extrarational reasons in their work. Lobachevski (1974, p. 2) states that mathematicians throughout history resorted to different means to support their belief of Postulate V of *The Elements,* where Euclid states that given a straight line 1 and a point P that is not on 1, there exists exactly one line parallel to 1 that goes through P. Saccheri, for instance, decided to establish the veracity of the Postulate by using a double *reducto ad absurdum*: negating the existence of the parallel line (no line parallel to 1 crosses P) or negating uniqueness (more than one line parallel to 1 cross P). The negation of the existence led to a contradiction. Saccheri deduced several theorems from the second option, which from his standpoint were odd; although free of contradiction, the seeming lack of believability of the consequences—in the sense that they were contrary to the ontological epistemological belief by which Euclidean geometry is the only possible model of space—were sufficient for him to reject this second possibility, from which he derived the truth of Postulate V as the only possibility. Saccheri's behavior and that of other mathematicians that tried to prove Postulate V in a similar manner shows how loyalty to ontological or metaphysical beliefs assumed to be uncontestable—that is, abiding by extramathematical considerations—may disrupt mathematical reasoning that in this case abide by and respond to sources foreign to the discipline.

Mechanisms—both implicit and explicit, conscious and unconscious, deliberate or involuntary, and rational or irrational—commonly used by people to support their beliefs are called "epistemic schemes for providing grounds" (or only epistemic schemes) (Rigo, 2013) in this document. It is a case of behavior patterns that are invariant under similar conditions. We use the notion of epistemic schemes rather than commonly used terms, such as "justification," "argument," "reason," or "reasoning" because the latter alludes to practices that are deliberate, explicit, and subject to a certain strict logic, conditions that are not always satisfied by epistemic schemes.

People do not only believe because they have reasons they deem as sufficient to believe. Concomitantly, people also believe because they have a motive to do so, sometimes a strong or vitally important motive to believe. According to Villoro (2002, p. 103), a motive is anything that moves or induces a person to act in a certain way in order to reach an end. Motives may be reduced to needs, such as volitions, interests, intentions, and desires, not only conscious ones, but also unconscious and uncontainable instincts that take over the actions of a person. Among others, these motives are in the background of a person's impulse to believe, because beliefs perform useful and beneficial psychological functions, not only for safety and comfort or to reduce anxiety or stress, but also even in integrating one's personality and sense of self. The needs and motives are also in the background of what leads people to deem reasons as adequate for supporting what is believed; motives may explain, for instance, why people adopt and accept backings that seem irrational as "reasons," and it can account for the benefits they seem to obtain from holding these reasons and from the needs they satisfy.

These considerations have also been derived from history. Take Cantor's attempts to prove the continuum hypothesis[2] as an example; in addition to the insights based on some initial intuition, Cantor believed the proposition because proving it would be the "most beautiful result in set theory" and "could mean the successful conclusion of his works on the subject" (Dauben, 1980). Truth of the hypothesis would have likely given Cantor great aesthetic pleasure; also—and he certainly needed this due to the strong pressure exerted by his detractors—he would have gained great satisfaction as a professional, reinforcing his relevance in the mathematics community and his identity as an expert in the field. The continuum hypothesis is perhaps merely one small piece of all the work Cantor produced in the development of set theory. As Dauben (1980) asserts:

> Perhaps more than most branches of modern mathematics, set theory bears the special stamp of its originator's interests and personality. Thus the historical development of Cantorian set theory demonstrates how the abstract objectivity so often ascribed to scientific theory may be influenced by the character and interests of those who contribute most to its development ***(p. 181)***

Motives may cause beliefs to be adopted, which may be of a mathematical or a general nature. However, motives may also lead to acceptance of the justifications on which the beliefs are supported, because both the beliefs and justifications address and respond to the motives of people and fulfill a certain psychological function by satisfying a specific need. This does not exclude people from, concomitantly, supporting their beliefs with

[2]The continuum hypothesis asserts that there are no sets that have a cardinality strictly comprised between that of the set of natural numbers and the set of real numbers.

reasons and justifying and validating their reasons with other broader and more general reasons.

Epistemic States of Convincement

This document recognizes—as was done by Newman—that there are no degrees of belief: one either believes or does not; but there are different probabilities of truth associated with beliefs (Villoro, 2002), as was already mentioned.

This document also admits that, as shown before, convincement can be experienced as conviction, certainty, need, security, trust, or doubt linked with beliefs about mathematical content, and concomitantly, with the epistemic schemes upon which they are supported and the motives that lead a person to believe.

These internal states that are based on beliefs and/or their backings, in the line of research in which partial results are presented here, are referred to as "epistemic states of convincement."

Epistemic states do not always accompany beliefs because, similarly to beliefs, they are indeed potential states or latent states; however, these states are updated and expressed under the right conditions (personal, social, cultural), accompanying almost all types of beliefs, including, of course, beliefs about mathematical content. The foregoing is a point of contention among some researchers (e.g., Villoro, 2002) who claim that only vitally important beliefs, such as religious, political, or ideological beliefs (mathematical beliefs are among those excluded) may be accompanied by feelings of security and trust. This document concedes that, given a favorable context, different epistemic states may be experienced regarding the Pythagorean Theorem, the existence of the limit of a sequence, or the Riemann hypothesis.

Different Types of Epistemic States of Convincement

There are different types of epistemic states of convincement. On the one hand, there are those in which complete convincement may be experienced. In the latter case, two states can be seen: those associated with what is evident, when faced with a proposition that does not require any reasons to support its truthfulness (such as tautologies[3] or perceptual propositions), or those related to certainty, where beliefs require reasons to validate them. This is very likely the difference that mathematicians noticed when trying to prove Euclid's Postulate V, about which they were certain albeit lacked evidence. This was possibly because in contrast to the first four postulates, V establishes a commitment with infinity (Kline, 2009). A complete state of convincement may also be experienced regarding the

[3]Tautological propositions are those that are true by their mere structure, such as pv-p. An example of a structurally contradictory proposition which is therefore always false is p∧−p.

falsity of absurd propositions, which are evidently impossible (as is the case of contradictory propositions. See foregoing note).

On the other hand, epistemic states of partial convincement can also be identified. These include the state of presumption, in which people have some reasons supporting the truth of the believed proposition, but they are not definitive; doubt, in which reasons suggest, but not conclusively, the falsity of the proposition in question; and the state of "radical doubt" that is experienced when faced with indeterminacy[4] that occurs when one is not in a position to decide on the truth or falsity of an idea—that is, "there is no reason to agree or to disagree" (Bárcia, 1883).

The various types of epistemic states described above may be illustrated with the security or insecurity experiences that the contemporary expert community experiences regarding the so called "millennium problems." For instance, they seem to experience presumption regarding the Riemann hypothesis (which claims that nontrivial zeros of a complex function s have real part equal to ½, where the complex function s is defined on the half-plane $\Re(s)>1$ by the absolutely convergent series: $\zeta(s)=\sum_{n=1}^{\infty}\frac{1}{n^s}$). Bombieri's remarks (2006) suggests that mathematicians may possibly experience a state of presumption regarding this hypothesis, stating that "Notwithstanding some skepticism voiced in the past, based perhaps more on the number of failed attempts to a proof rather than on solid heuristics, it is fair to say that today there is quite a bit of evidence in its favor" (p. 112). The epistemic state insinuated by the remarks of mathematicians surrounding the unsolved problem known as P versus NP is different. The P versus NP problem is to determine whether every language accepted by some nondeterministic algorithm in nonpolynomial time (NP) is also accepted by some (deterministic) algorithm in polynomial time (P), given in the form of the following question: Is P=NP? According to Cook (2006), "Most complexity theorists believe that P≠NP. Perhaps this can be partly explained, by the potentially stunning consequences of P=NP mentioned above [devastating consequences for cryptography of P=NP], but there are better reasons. We explain these by considering the two possibilities in turn: P=NP and P≠NP…" (p. 98). The reasons given by mathematicians against P=NP suggest some doubt regarding the truth of said conjecture. This epistemic state seems very different from that which specialists experience regarding the

[4]The indeterminate value is introduced by some authors, such as Duval(1999), as another truth value in addition to true and false. From the foregoing considerations, in this study it is introduced as a degree of radical doubt. This sense differs from the mathematical notion of an undecidable proposition; in these cases, it is impossible to formally assign a truth value to these propositions—that is, it is formally proven that neither its truth nor its falsity may be proven. An example is Cantor's Continuum hypothesis. That is, there are reasons to decide that a proposition is not true but also not false. In the case of radical doubt there are no reasons to decide on whether a proposition is true or false.

problem of deciding on the possible existence (or inexistence) of smooth and physically reasonable solutions to the Navier–Stokes equations (Fefferman, 2006). Currently—according to Fefferman—"There are many fascinating problems and conjectures about the behavior of solutions of the Euler and Navier–Stokes equations...Since we don't even know whether these solutions exist, our understanding is at a very primitive level. Standard methods from PDE appear inadequate to settle the problem. Instead, we probably need some deep, new ideas."(p. 66). Fefferman's comments on the work done by mathematicians regarding the possible existence of solutions to the Navier–Stokes equations suggest a state of radical doubt by contemporary experts regarding this mathematical problem of existence.

In summary, this text differentiates between epistemic states of complete convincement (of evidence and certainty, as well as absurdity) and epistemic states of partial convincement (presumption, doubt and radical doubt). Their characterization is based on the type of reasons considered sufficient and necessary for accepting or rejecting the truth of a proposition. Table 1 summarizes this information.

The authors of the text moreover accept that security and confidence[5] in the truth of mathematical statements are epistemic states. They have,

TABLE 1 Subconcepts, Properties, and Dimensions of the Concept of Epistemic States of Convincement

	Types of epistemic states of convincement	**Degree of convincement**	**Reasons involved**
Complete Convincement	Evidence	Completely convinced of the truth	Does not require reasons to verify the truth.
	Certainty	Completely convinced of the truth	Requires reasons to verify the truth.
	Evidence of what is false.	Completely convinced of the falsity	Does not require reasons to verify the falsity.
Partial Convincement	Presumption	Relatively convinced of the truth	There are nondefinitive reasons
	Doubt	Relatively convinced of the falsity	There are nondefinitive reasons
	Radical Doubt	Indeterminacy regarding truth or falsity	Lack of reasons to determine truth or falsity

[5] One can hold that security or disagreement in beliefs or ideas stems from a certain type of a broader security (emotion or feeling) that includes, inter alia, physical security, social security, or economic security. For the purposes of this text, security and confidence are deemed as synonymous.

nevertheless, not been included in Table 1 because, in view of the conceptual elements presently at hand, the relationships of those states with the states included in the table are not clear. The authors have every intention of clarifying such relationships on the basis of Damasio's theoretical framework concerning emotions and feelings (see about the Damasio's Testimony section).

Relationships Among the Notions Introduced

Epistemic states are linguistically expressed through the verbs of propositional attitude (believe, know, be convinced, have doubt, amongst many others). These denote the attitude (or state) of a subject to a proposition and are formulated by sentences of the form "S *v* that *p*" (*v*: propositional attitude verb). For instance, Pedro is convinced that the Rule of Three can help solve missing value problems. In other words, epistemic states are experienced with respect to propositions believed. They are a matter of a complex experience involving the belief itself, but as already mentioned, the reasons and motives the subject has for supporting the proposition are also involved. Consider the certainty regarding the belief that "Euclidean geometry is the only possible model of the physical space." This may arise from the repetition of an experience grouping the various reasons that back the idea (e.g., physical applications and day-to-day life; internal coherence), which is also impacted by the motives for believing—that is, the needs covered by that belief (coherence with philosophical and metaphysical values; belonging to the group of pairs; self-assertion) as well as the motives for accepting these reasons as sufficient (coherence, agreement with standards of rigor, confidence). If this relationship between certainty and belief is also endorsed at community levels over long periods of time, then the level of confidence increases.

Fig. 1 presents the information regarding the possible relationships among beliefs, reasons, and motives, notions that have been introduced in the theoretical framework developed thus far.

The empirical study undertaken on historical data, which complements and validates some of the conceptual elements taken from philosophy and psychology, have yielded significant results in the form of answers to some of the initial research questions: it shows that epistemic states are essentially linked to beliefs, reasons (epistemic schemes), and motives and that there are various types of epistemic states.

These results require further clarification, of a very different ilk, but clarifications that seem to not be postponable: to elucidate the nature of the epistemic states of convincement and analyze the relationships these states have with the affective and cognitive domains. This problem is the focus of the second part of this document, which uses the neurobiological model of feelings and emotions proposed by Damasio.

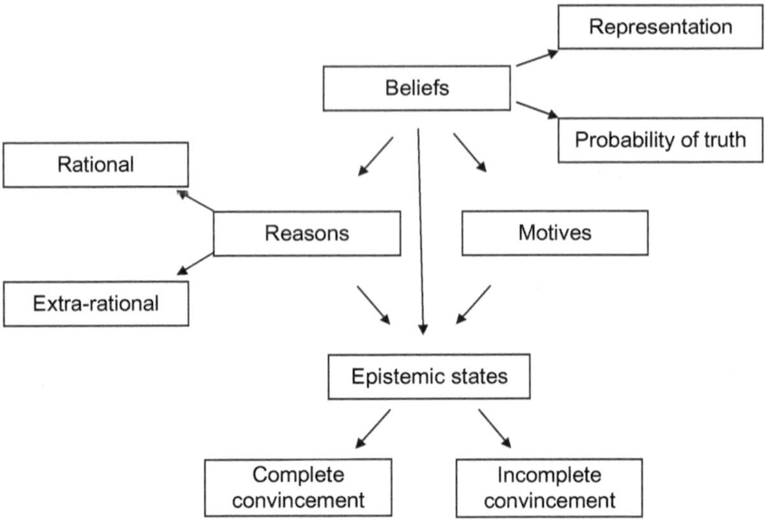

FIG. 1 Beliefs associated with backings, motives, and epistemic states.

EPISTEMIC STATES AS EMOTIONS AND FEELINGS

Work performed on empirical data taken from the professional practice of mathematics (and from classroom mathematics practice, which is omitted here) suggests that epistemic states are internal states experienced by mathematicians in their professional practice and that hold a certain relationship with their cognitive world, as well as with their affective world. To elaborate on these topics, this part of the chapter turns to the work of Damasio (1994, 2003), a pioneer in the neurobiological description of emotions and feelings. The first section extensively details his ideas (otherwise the information that follows is incomprehensible) and in the second part we apply the Damasio model to the case of epistemic states. The conclusion is that these states are a type of emotion and feeling and that, as such, they conserve and confirm the properties and attributes based upon which they were characterized in the first part.

The Neurobiology of emotions and Feelings in the Work of Damasio. Presentation of some of the ideas

From a systematic and comprehensive perspective, and under the scope of cognitive science, Damasio undertakes a neurobiological examination of emotions and feelings and their relationship with mind and body.

Taking an integral point of view of the human being, Damasio reinterprets clinical cases that suggested that the loss of emotional knowledge—that is, the weakening of a memory that registers the relationship between certain decisions or actions with specific emotions—prevents reasoning and decision making in sociocultural contexts. This shows that emotions may eventually disturb reasoning, as is often thought to be the case. But it also suggests very different and surprising phenomena: emotions are equally indispensable for reasoning, for learning success and failure, and for proper decision making in the cultural framework of social interaction with others. In other words, Damasio provides a neurophysiological demonstration of what Vygotsky suggested many decades ago, that cultural functions underlie our reasoning and decision making (Immordino-Yang & Damasio, 2011).

Damasio's experimental research found that brain systems responsible for emotions generally participate in the treatment of cognitive processes and those related to social behavior. He claims that the reasoning system was developed as an extension of the automatic machinery of emotion: the evolutionarily ancient brain—the lower level so to speak (the brain stem, hypothalamus, amygdala)—regulates the bodily functions needed for survival, and it regulates the processing of emotions and feelings, while it also plays a role in the development and activity of evolutionarily modern brain structures—the upper level—responsible for managing cognition and social behavior. Thus, structures in the upper level are embodied in the evolutionarily ancient structures and, therefore, this is not a case of two functionally separate mechanisms.

When the organism, composed of the body and brain, interacts with the environment, it does not merely generate external responses, known as behavior. It elicits internal responses, some of which take the form of images (visual, auditory, olfactory, somatosensory, etc.). Damasio postulates that the mind (cognition or cognitive processes; Damasio, 1994, p. 112) consists precisely of the capacity to internally represent images and order them by way of a process we call thought. Damasio explains what occurs during mental phenomena at a neurobiological level. He suggests that biological changes take place in neural circuits during the processes that occur in learning, which then take on the form of a neural representation and become images, which, as already mentioned, may be eventually manipulated by thought. Although representations are stored in memory, latently and expectedly as dispositional representations that trigger image generation, people actually think about and reason from images. In fact, we share our image-based concept of the world with other humans and some animals. It is by way of images that objective knowledge required for reasoning and decision making reaches the mind (Damasio, 1994, p. 116).

Damasio suggests that an emotion in itself, such as sadness, shame, or sympathy, is a complex set of chemical and neural responses generating a distinctive pattern. Responses are produced by the normal brain when it detects an emotionally competent stimulus (ECS)—that is, the object or event, whether it's real or a mental remembrance, that triggers emotion (Damasio, 2003, p. 55–56). The immediate result of these answers is a temporary change in the state of the body itself and the state of brain structures mapping the body and holding thought. The ultimate result of the responses, directly or indirectly, is to place the organism in circumstances that are favorable to survival and wellbeing. Indeed, within a few hundred thousandths of a second, the emotional cascade transforms our body, the rhythm of our mind, and the object of our thoughts (Damasio, 2003).

Despite the brain being prepared by evolution to respond to certain ECS with specific action repertoires, the list of ECS is not confined to the records prescribed by evolution. It includes many others that are learned throughout one's life experience. That is, although humans share clearly similar biological mechanisms of emotion, the circumstance under which certain stimuli have become emotionally relevant vary among individuals, their education, and their cultural milieu. Although there are genetic mandates that govern emotions biologically, culture and society also play a role in their development, execution, and mode of expression.

This circumstance leads Damasio (1994), p. 157) to suggest the existence of primary and secondary emotions. In primary emotions, certain environmental stimuli excite, by way of an inflexible and innately established mechanism, specific instructions for bodily reactions. Secondary emotions vary in the degree and intensity of said preestablished instructions, which are modulated by personal experience.

The machinery of secondary emotions is started as reflections or considerations regarding images of objects or events that are either taking place or being evoked or remembered and, from their cognitive appraisal (that may be preconscious or conscious), they act as emotionally competent stimuli; and this takes place in the frontal cortex of the brain. At the subconscious level, the prefrontal cortex automatically responds to certain images, which activate dispositional representations that hold knowledge accumulated by a person through prior experience in which the type of event relates to certain emotions (in a long list of sorts). This information is then sent to the limbic system, a region which itself sends signals to the body (the gut, motor system—skeletal muscles that unleash certain facial expressions and body posture, and other endocrine signals) and also sends signals to the brain itself (neurotransmitter nuclei in the brain stem), which significantly impacts the style of mental processing and the efficiency of cognitive and thought processes, leading to additional mental changes. Altogether, the above preforms an emotional body state (Damasio, 1994), from which certain actions are taken (e.g., run away from a fearful situation) and certain ideas and plans are brought to consciousness, ideas and plans that are in harmony with the general signal of the emotion (sadness, e.g., slowing down thought and may lead to insisting on the situation that aroused it).

The next step in the emotional process occurs when a feeling arises.

Emotional feelings arise from the perception of everything that occurs during emotion, both in body and mind, as well as the changes that take place: on the one hand, there is the subjective perception of the object (i.e., the ECS); on the other there is a perception of the particular state of what is occurring in our body; and finally there is the perception of an altered state of cognitive resources and a particular style of mental processing. These perceptions cause a change in the cognitive process and generate thoughts that concur, in terms of the topic, with the emotion being felt. The identity of the person is mobilized throughout this process.

While emotions take on an important role in "basic" survival, feelings play a role in a much finer and sophisticated type of survival. The reason for this is that while emotions offer a vast field of emotional experience in body and mind, feelings allow for their registration in memory and for learning from experiences, providing more personal and less automated responses in the future, or responses that are freer of the "tyranny" of the routine emotional machinery.

In contrast to emotions, feelings last much longer and, in some cases, the physical sensations are not as clear. Some examples of feelings are love, hate, trust, and gratitude. Feelings based on secondary emotions occur when more subtle branches of the cognitive state are connected to more subtle variations of the emotional state of the body, allowing for a finer and more detailed experience of certain emotions (e.g., they allow us to feel varying degrees of remorse or shame).

Epistemic States Interpreted as Certain Types of Emotions and Feelings

Evidence from empirical research undertaken throughout the history of mathematics (or with students) has allowed us to presume that epistemic states of convincement, doubt, or presumption are internal states experienced by people and have the following distinctive features:

Epistemic states are associated with beliefs; beliefs act as a stimulus or trigger that activates and encourages these states. Given that beliefs tacitly or implicitly carry their truth—people can't believe what is false, as Wittgenstein said—we can say that beliefs and their probability of truth drive epistemic states and enliven and favor their presence.

Epistemic states depend on the reasons that back the belief, and they also depend on a person's motives for believing. That is, they are sensitive or susceptible to reasons and motives. Additionally, epistemic states are acquired in the cultural environments where the person is trained to feel certain epistemic states when faced with certain beliefs and reasons, and they are generated, driven, and consolidated as a result of social interactions.

Epistemic states, similarly, are firmly expressed through certain behavior patterns and body expressions, patterns that tend to refer to those states exclusively.

The empirical research cited presumes, as previously mentioned, that epistemic states are internal states (characterized by the set of traits just described). However, further research is required to unravel the specific states involved.

In order to continue to delve deeper into the research, this text argues that, according to Damasio's neurobiological model, people's experiences of internal states of doubt, certainty, or presumption—the epistemic states—may be conceptualized as a certain type of emotion and feeling and that, as such, they depend on beliefs and their truth, they are sensitive to reasons and motives, and they have certain patterns of expression. The latter conceptualization facilitates delving into the analysis and moving some of the explanations forward concerning convincement-related phenomena that have to date been left behind in research, particularly in mathematics education research.

The methodological strategy used to support the argument is as follows: first, based on a meaningful Damasio quotation, one can see that Damasio offers the possibility of conceiving epistemic states, such as security or agreement, as certain type of emotions and feelings accompanying beliefs and their backings. Further, evidence is provided to indicate that interpretation of epistemic states as emotions and feelings is consistent with the Damasio model (nonsense and dissonance apparently do not follow from this interpretation). Moreover, seen from this standpoint, epistemic states do not only maintain their distinctive traits, but that standpoint enables delving further into and, albeit incipiently, accounting for such traits. For these purposes, we initially propose possible neurobiological route that follows the configuration of a state of certainty, if conceived as emotion or feeling, associated with the truth of a belief of mathematical content, and, subsequently, offer a possible explanation for how the certainty is generated and established in the framework of a deductive argument of mathematical content—an important phenomenon related to convincement in the field of mathematics and the teaching of mathematics.

Damasio's Testimony

Damasio facilitates and allows for the conception of epistemic states as certain types of emotions and feelings. In one of his works, he holds that

> Even when we somewhat misuse the notion of feeling—as in "I feel I am right about this" or "I feel I cannot agree with you"—we are referring, at least vaguely, to the feeling that accompanies the idea of believing a certain fact or endorsing a certain view. This is because **believing and endorsing** *cause* a certain emotion to happen... Through either innate design or by learning, we react to most, perhaps all, objects with emotions, however weak, and subsequent feelings, however feeble ***(Damasio, 2003, p. 93) (italics appear as in the original; bold lettering was added).***

This passage is very revealing in terms of this research, because Damasio explicitly asserts that security about being right ("I feel that I am right about this") and disagreement are types of emotions and feelings, but they are also special emotions that stem from beliefs and backings. Taking security or disagreement in beliefs or ideas as "generic" epistemic states, epistemic states (like those included in Table 1) can be seen as variations of the secondary emotion of security in beliefs or ideas. That is to say, it is a matter of states of varying degrees of security, which are also distinguished amongst themselves by the reasons that support them; for instance, evidence can be conceived as a state of total and unquestionable security with respect to the veracity of a statement that does not require reasons to justify it; however, presumption is a state of security in the truth of an assertion that can be low, fair, or high and that requires reasons that justify it (see Table 1). Epistemic states as feelings are associated with and act on the secondary emotion of security in beliefs or ideas (or to epistemic states seen as secondary emotions), that is to say, they refer to the conscious perception or realization of what happens during that emotion, both in the body and mind, as well as the changes that said emotion produces there.

But if this is indeed the case, what possible neurobiological route is taken to mobilize said epistemic states? Below is an outline of what that route could ultimately be.

Possible Neurobiological Route Associated with Activating the Emotion of Certainty

In order to set these ideas in the context of mathematics and its didactics, let us suppose that the epistemic state in question is the certainty that a mathematician might feel regarding a tautology, for instance, pv-p (read as "p or not p," with p being any proposition; an instance of this might be "this number is prime or this number is not prime"). Let us suppose also, for purposes of complementing this idea, that the mathematician has experience in handling this type of proposition.

The emotional mechanism starts as the conscious deliberations the mathematician makes regarding the tautological belief, deliberations that are expressed as mental images organized in thought processes. In this case, thoughts may turn to certain attributes and properties of tautological propositions or to their truth.

Processing of said images generates an unconscious response that occurs automatically and involuntarily in the prefrontal cortex networks. The response stems from (dispositional representations that it contains) knowledge the subject has accumulated from experiences pairing certain categories of situations (related to those images) with specific types of emotions. As previously mentioned, people store this knowledge as relatively unequivocal "long lists" (short lists are used for primary emotions) that act as assessment criteria. In each of the above cited cases, the knowledge that the mathematician will likely mobilize relates to the connections he has established between the category of tautological propositions and secondary emotions of total and unquestionable security in the truth of the proposition.

Prefrontal assessment criteria act as the trigger that unconsciously, automatically and involuntarily unleashes signals to the amygdala and anterior cingulate. This has a twofold effect: on the somatic level the effects are given through a series of changes in the body that occur according to some pattern, causing an "emotional body state" through the transfer of signals to the gut and the motor system (skeletal muscles triggering specific facial expressions and body posture), and by sending other endocrine signals. Another repercussion takes place at the mental level, which has a significant impact on the style and efficiency of cognitive processes, which constitute a parallel route for the emotional response. Bodily and cognitive processing changes may be anywhere from barely perceptible to clearly notable and distinguishable. Continuing with the case of the mathematician, mobilization of his long list will automatically declare propagation of a body state in step with certain patterns that characterize the "security" emotion that he will must likely show through a body in a state of relaxation and of body language expressing openness and confidence. Activation of the long list will also modify the contents of his thoughts and the cognitive mode of processing his ideas: being faced with such unquestionable evidence may even lead him to scarce mental proliferation concerning the topic.

The above places the organism in the circumstances needed to display certain behaviors that in principle will tend towards corporal or social wellbeing of the person experiencing belief. Our mathematician will be in the position to perform actions consistent with his emotions: he will use emphatic language that may reveal a level of commitment to the truth of the proposition, a firm, suggestive, clear tone of voice; he will perhaps make eye contact with others and may act with determination and consistently.[6] Should consideration of the truth of the tautology be no great challenge to the mathematician, he may simply conclude the activity, experiencing a secondary emotion of very basic and immediate evidence.

The Possible Neurobiological Path of a Feeling of Certainty

Let the reader now consider a different scenario. Suppose the mathematician wants to convince a student of the formal truth of tautologies. This might cause him to more closely reflect in greater detail upon the connections of the evaluation that he automatically activated (i.e., tautologies trigger unquestionable security) and urge him to delve deeper into and become

[6] Cognitive and bodily states, as well as the actions a person takes as a result of experiencing an epistemic state, are described in greater detail as "Criteria for Identifying Epistemic States" in Martínez and Rigo (2014).

aware of the emotions and associated body and mental changes triggered by said evaluation. The circumstance might even lead to him, either unconsciously or deliberately, entertaining thoughts about the logicoformal truthfulness of this type of proposition, *inter alia*. Given this self-reflection and his emotion-related thoughts on the subject, not only will the mathematician be in a position to engage didactic tools to convince his students, he will indeed to experience a **feeling** of **evidence** regarding the tautology—referenced to the emotion of security albeit different from it due to the fact that it is conscious and the integral realization of lived experience.

Damasio holds that a feeling depends on two basic processes: i) becoming aware of emotions, that is, the realization by an individual that they have certain "tendencies, customs, habits" (that correspond to evaluative criteria or the long list) that cause them to respond to certain emotionally competent stimuli with certain emotions and ii) a certain style of cognitive processing, accompanied by certain thoughts consistent with the emotion, of which the subject is conscious. Certainty as a feeling associated with the secondary emotion of security in the mathematician would consist of his reflections about the connections he established between tautological propositions and their evoked states of unquestionable security, as well as of other thoughts related to the topic of tautologies or those connections.

Applying Damasio's model to epistemic states, that is certainty interpreted as a certain type of secondary emotion related to unquestionable security and as a derived feeling, is consistent with the findings of empirical evidence but also reveals other much subtler and more complex phenomena under the surface. It suggests, among much else, profound and unfathomable connections between epistemic states and the beliefs that give rise to them. It also reveals the role of thought in gestating and developing that state of certainty, which may have been expected, but nonetheless reveals the role of the body in the entire process, providing evidence of the existence of a narrative through which epistemic states are actually expressed and communicated to others, just as what actually occurs with many other phenomena in the affective world (and as was assumed in the research). Finally, the application of epistemic states to Damasio's model shows how a person may experience certainty in the absence of all realization (thinking about certainty strictly as an emotion) or, on the contrary, how he may display sharp reflections regarding his experience of that state, to transform it into an intense feeling.

The Dynamic and Complex Process of Constituting Certainty Associated with a Deductive Argument of Mathematical Content

We would now encourage the reader to consider an even more complex circumstance, associated with an important phenomenon for mathematics and its teaching, related to conforming certainty about the truth of the conclusion of a deductive argument, specifically of the theory of even numbers.

Some clarifications are in order prior to getting started. First, deductive arguments, as opposed to any other type of argumentative pieces (such as those of a factual nature, in which truth is established when facing concrete and finite empirical facts), allow for universal affirmations that refer to infinite sets of objects. Because mathematics is interested in precisely establishing the veracity (or deducibility, in case of formal uninterpreted theories) of this type of universal proposition of infinite reference, the importance of this type of artifact of demonstration is clear for mathematics and for mathematics education. On the other hand, it should be clarified that when students of basic and intermediate levels of mathematics develop deductive arguments, they tend to focus on the validity of the rules of inference that they apply (almost always implicitly), while they also consider the truth of the propositions that arise. The latter is due to the fact that these types of deductive reasoning tend to be part of already interpreted theories, which refer to specific objects that take part in propositions that have already been assigned a truth value (such as the theory of even numbers, with reference to which we make the argument analyzed below). This differs from the case of formal theories (not interpreted), which have formulas that lack meaning and have not yet been associated truth values. It therefore makes sense to consider validity rather than truth when developing such arguments. It is worth noting that truth is a property of propositions while validity is a property of arguments.

Damasio, once again, gives rise to a possible explanation of the phenomena at hand—constituting certainty in the framework of proof. To him (especially in his work *Searching for Spinoza*), emotions and feelings are changing, alterable, and temporary phenomena that also act as links in chain reactions, where one thing leads to the next, continuing to produce a domino effect by which an event my possibly drive several further events (by magnifying or buffering them) until some circumstance puts an end to the process. The author gives shape to a similar idea in the following passage:

> As thoughts normally causative of emotions appear in the mind, they cause emotions, which give rise to feelings, which conjure up other thoughts that are thematically related and likely to amplify the emotional state. The thoughts that are conjured up may even function as independent triggers for additional emotions and thus potentiate the ongoing affective state. More emotion gives rise to more feeling, and the cycle continues until distraction or reason put an end to it ***(Damasio, 2003, p. 70)***

Emotions and feelings, however—the neurobiologist clarifies—don't operate as linear and mechanical cycles, but rather as an shell of multiple parallel branches and alternate bifurcations, all interconnected, juxtaposing, and weaving into very complex nets. In Damasio's own words:

> For the purposes of providing a manageable description of the processes of emotion and feeling, I have simplified them to fit into a single chain of events beginning

> with a single stimulus and terminating with the establishment of the substrates of the feeling related to the stimulus. In reality, as might be expected, the process spreads laterally into parallel chains of events and amplifies itself. This is because the presence of the initial emotionally competent stimulus often leads to the recall of other related stimuli that are also emotionally competent… Relative to that initial stimulus, the continuation and intensity of the emotional state is thus at the mercy of the ongoing cognitive process. The contents of the mind either provide further triggers for the emotional reactions or remove those triggers, and the consequence is either the sustaining or even amplification of the emotion, or else its abatement ***(Damasio, 2003, p.p. 64–65)***

The dynamic and interdependent perspective that Damasio assumes in his work on emotions and feelings suggests that building the state of certainty, considered as an emotion and a feeling and in the context of a deductive argument, may take several parallel paths. Below is an attempt at identifying such paths.

For purposes of consolidation, consider the proof of a simple theorem of the theory of even numbers: the sum of two even numbers is even, offered by a student enthusiast (hypothetical) in first year of pursuing a mathematics bachelor degree. The second column of Table 2 includes the steps of the proof, the third shows the justification for validly deducing those steps (or the propositions that allow said deduction, such as axioms, definitions, or theorems—what Duval (1999) calls "third sentence"), and the last column has possible logical connections among the propositions involved in the various steps established by the student writing the proof—probably unconsciously and implicitly. As we shall see below, the rules of inference that are applied in each step are instances of a third sentence in the generic case of p and q (so they have not been made explicit in Table 2), except in step 10, where instead of this rule there is a universal generalization.

In this proof, the student started by assuming that p and q are even numbers (P1). Then the student applied the definition of an even number (third sentence), assuming the proposition is true. This is a definition (universal) that refers to **any** even number:

$$(r)P(r) \Leftrightarrow \exists m \in N \ni r = 2m, \text{read as: for any integer, it is even, if and only if there is an N such that } r = 2m.$$

This definition was immediately applied to two generic numbers, p and q; that is, the "instantiation" rule of inference was applied to said definition of even number to construct a proposition in terms of p and q:

$$P(p) \wedge P(q) \Leftrightarrow \exists(m, n \in N) \ni (p = 2m \wedge q = 2n),$$

which is read as the following: p and q are even numbers, if and only if there are integers m and n, such that $p = 2m$ & $q = 2n$. From which it follows that $P1 \Rightarrow P2$ and then P2.

TABLE 2 Possible Paths in the Construction of Certainty in a Deductive Proof

	Steps in the proof	Third sentence	Logical connections
C	The sum of two even integers is even	Conjecture High presumption, result of the somatic marker	
P1	Suppose p and q are any even numbers	Supposition (suggested by C)	
P2	There exists an m in N such that p = 2m; There exists an n in N such that q = 2n;	By definition of an even number (applied to P1)	P1 ⇒ P2
P3	Consider p + q	Supposition based on properties of the sum operation in N (suggested by C)	
P4	p + q = 2m + 2n	Properties of equality and sums (applied to P2 and P3)	P2 ⇒ P4
P5	2m + 2n = 2(m + n)	Distributive law of multiplication with respect to the sum (applied to P4)	P4 ⇒ P5 [P1 ⇒ P5]
P6	2(m + n) = p + q	By the transitive property of equality (applied to P4 and P5)	(P4∧P5) ⇒ P6 [P1 ⇒ P6]
P7	2(m + n) is even	by definition of an even number (applied to P6)	P7
P8	p + q is even	By the properties of equality (applied to P6 and P7)	(P6∧P7) ⇒ P8 [P1 ⇒ P8]
P9	If p and q are even, then p + q is even	Tautology Of P2, P4, P5, P6, P7, and P8 (see last column, rows P5, P6, and P7, below in brackets)	P1 ⇒ P8
P10	(p)(q) (If p and q are even, then p + q is even).	Universal Generalization[a] (applied to P9)	(p)(q) (P1 ⇒ P8)

[a]*It is possible to define the truth of general sentences without taking on the complexities of mathematical logic. We can say that (x)F(x) is a truthful proposition if and only if all of the sentences Fa, Fb, Fc, … Ft are true, i.e., where each object of the universe of discourse referred to by the universal quantifier is present in the list. It is plausible to suppose that this is the case above, as even numbers taken at the start are arbitrarily chosen, not distinguished from others, so anything that applies to them applies to all others (Mates, 1979, p. 81).*

Now, if the general definition is true, then the specific assertion or the instantiation—that is, P1 ⇒ P2 and P2 is also true, and the student surely considered that to be the case even if implicitly. We should clarify that said logical consideration is based on the following: if a rule of logical inference (such as instantiation) is validly applied to a true assertion, another true assertion will always stem from it; that is, no false propositions can be derived from this true proposition and under this valid rule of inference. In general, we can say that the valid application of the rules of inference in a deductive argument "inherit" or "convey" the truth of the premises (provided they are true) to the conclusion or other propositions derived from said premises. It is possible that the student lacked this explicit knowledge of formal logic but that, in his proof and deductive reasoning experiences, he developed sufficient intuition to enable him to work on these considerations.

And in these cases, just how was security woven in? Specifically, how did the student feel secure about the truth of P1⇒P2 and P2? It would seem that the security had two sources: the unrestricted confidence the student experienced from the truth of the universal definition of even numbers (that, as was mentioned, served as a third sentence, from which P1⇒P2 was derived and then P2) and from the firm sense of security offered to him by validly applying a rule of logical inference, which entails the property of "conveyance of truth." Thus, the general proposition of mathematical content (the third sentence) and a valid rule of inference acted as emotionally competent stimuli that concurred in order to unleash an epistemic state of certainty in the student, a state that was then the product of a combination of harmonized securities. It is clear that for this to occur, it was necessary for the student to have a long list containing evaluation criteria that would allow him to associate total security with theorems of the theory of even numbers and the unquestionable certainty of the rules of logical inference and the truth of the propositions that derive from them.

The above described mechanism was likely to take place throughout all steps of the proof, except for the last one. In step 10, rather than instantiation, the student applied a rule of logical inference called "universal generalization," which states that if an arbitrary element of a set possesses a certain property, then so do all other elements of the set. It is necessary to apply this rule of inference because what the proof seeks is precisely to show the conjecture for all even numbers, and P9 is, strictly speaking, an assertion that applies only to a generic pair of even numbers and not to all of them.[7] This may be clearer if we turn to a symbolic expression of the propositions. P9 is represented as

$$P(p) \wedge P(q) \Rightarrow P(p+q),$$

[7]In fact, what we see in the last column are not formulas for the propositional calculus; they are formulas of predicate calculus (relative to objects possessing certain qualities that may be predicated on them); for instance, P1 can be represented using symbolic language as $P(p) \wedge P(q)$, where "P" denotes the predicate "is an even number," so P1 means that the objects p and q are even numbers; P1⇒P2 is expressed symbolically as $P(p) \wedge P(q) \Rightarrow \exists (m, n \in N) \ni (p=2m \wedge q=2n)$ and in natural language as the following: if p and q are even numbers then there exist integers m and n such that $p=2m$ and $q=2n$.

which means that if p and q are any even numbers then their sum is even (with P denoting the predicate "even number"), while P10 can be expressed symbolically as

$$(p)(q)(P(p) \wedge P(q) \Rightarrow P(p+q)),$$

meaning that, for all pairs of objects, if they are even numbers, then their sum is also an even number. In less formal mathematical proofs tend to omit the last step, which explicitly generalizes to all objects in the universe, because is an obvious step in proofs such as that of the example, in which the starting point is generic objects.

Now, as the student progresses through the proof, his own self-reflections come into play; this coming into consciousness gives rise to the feelings that triggered various thoughts. Some of these short-range concerns allowed him to plan the next step (e.g., consider what scenario or third sentence to include) in order to then make the decision that he assessed to be the best, possibly driving emotions of security (or doubt, eventually). The student, directed by those feelings and with the conjecture in mind, also imagined ideas with a larger scope, allowing him, for instance, to "keep a logical tally" of what he had done and what was still required—a balance that he could prepare by turning to his intuitions and formal and informal knowledge of logic and that he organized implicitly in logical connections similar (although expressed in natural language) to those appearing (in square brackets) in the last column of Table 2. For instance, in step 5 he may have noticed that if P1 $\Rightarrow$ P2, P2 $\Rightarrow$ P4 and P4 $\Rightarrow$ P5, then P1 $\Rightarrow$ P5 and that if (P6 & P7) $\Rightarrow$ P8 then P6 $\Rightarrow$ P8 (as also appears in P6). Perhaps he validly concluded that P1 $\Rightarrow$ P8 from all these implications. Since each of those connections displayed a positive balance, it possibly contributed to the student's sense of security, an emotion that somehow became part of the pool of epistemic states that he forged throughout the proof.

The claim, as such, is that the student built secondary emotions of security throughout the proof process and then gradually experienced a solid feeling of certainty, with which he concluded the proof, traveling along three paths in synchronicity. One of the paths fed him with the security of truth of the third sentences; on a second, his security came from the rules of inference he applied at each step; and on a third path, he felt the emotions of security and confidence that came from the feelings that arose at each step of the process, feelings that were basically associated with the "logical tally" and the derived logical connections, which noted the balance between what had been done and what was still pending.

Therefore, as Damasio suggests for emotions and feelings, epistemic states do not seem to be definitive, nor are they a point of arrival but rather a point of departure that evokes new experiences. Additionally, the constitution processes of these epistemic states seem to "spread into parallel chains of events that are amplified." The above means two things: first, that these constitution processes are unfolded into multiple avenues, each of which contains experiences that are concomitant to those unfolding in the others; and second, that a constant transformation of the epistemic states occurs within each of these avenues: those that take the form of emotions give rise to other epistemic states, now as feelings, which in turn generate thoughts that once again serve as emotionally competent stimuli. As such, as already mentioned, it is very possible that the state of certainty reached by the student performing the proof was the result of a stockpiling of various epistemic states that he continually juxtaposed and combined. However, the final result was the not arithmetic sum of those aspects, but something far more complex because in the trajectory of building each component was affected by the preceding component. As Damasio says, "the juxtapositions of nature generate complexity, ... combining the right parts produces more than their mere sum" (Damasio, 2003, p. 164).[8]

These considerations reveal other aspects of the complexity of the phenomenon at hand. We can determine that epistemic states are actually located somewhere between cognition and emotion (as was originally presumed), but they are also integrated in a very sophisticated canvas, the components of which constantly speak to each other and produce a joint configuration in a step-by-step process.

Clearly it all started with establishing the conjecture to which the student associated a high degree of presumption. This likely brought on a feeling of great interest in the student regarding the proof because it represented a challenge that made sense to accept. The challenge then operated as another emotionally competent stimulus that unleashed, through a fourth path and in parallel to the other emotions and feelings above described, emotions of interest and certain security, playing an equally important role in incentivizing the necessary cognitive work (memorization, attention, and an accelerated rate of thought) that would lead to proving, with a great degree of certainty, the result that was sought.

[8]In terms of what was stated above, there are many unanswered questions, such as what happens if the proof is produced by a person that is not an expert in deductive proof. In that case, possibly no security follows from previous connections made by this nonexpert person (related to the topic of the proof). For example, that person may not realize that the definition of an even number applies to all numbers and then that person may associate $P1 \Rightarrow P2$ (or to some other derived proposition) only as a presumption; or that person may lack the logical intuition that would allow her to discover the validity of the instantiations or generalizations rendering those states of security, or allow her to map the logical connections that are sequentially made during the proof, providing a rich and complete picture to foresee and plan the remaining ones.

The Heuristic Work: Epistemic States as Somatic Markers

A natural question comes to mind from the above: How important is it in proof development to be able to associate a certain degree of presumption to a conjecture?

As was shown in part one, De Villiers (2010, 1990) asserts that mathematicians experience a certain type of experimental conviction during their heuristic work and that prior to establishing a proof, a person must be reasonably convinced of the truth of a result. Otherwise, the person might be tempted to mobilize her efforts to finding a counter example rather than a proof. Further, De Villiers (2010, p. 208) considers that in real mathematics research, while personal conviction generally depends on the existence of logical proof (even if not rigorous), it also depends on the security that was experienced during the experimentation stage.

But by what mechanism does that association arise between a certain level of presumption and a conjecture?

It makes sense to offer an initial explanation of this phenomenon using the Damasio model. He holds that there is a possible presence of a biological mechanism, which he calls "the somatic marker," responsible for undertaking an automatic preselection from an array of possibilities, from which a person must choose at a given point in time. This means that under these circumstances a person may not need to resort to reasoning in order to choose from among the field of possible options. The somatic marker "automatically chooses for them."

Damasio suggests that the somatic markers are present in works of mathematical content, specifically in the heuristic stages; Poincaré empowers him to inquire the following:

> In fact, what is mathematical creation? It does not consist in making new combinations with mathematical entities already known. Anyone could do that, but the combinations so made would be infinite in number and most of them absolutely without interest. To create consists precisely in not making useless combinations and in making those which are useful and which are only a small minority ***(of trained experts)***
>
> The mathematical facts worthy of being studied are those which, by their analogy with other facts, are capable of leading us to the knowledge of a mathematical law
>
> The sterile combinations do not even present themselves to the mind of the inventor. Never in the field of his consciousness do combinations appear that are not really useful ***(Poincaré, quoted in Damasio, 1994, p. p. 188–189)***

The hypothesis of the somatic marker is based on learning through previous experience: as we grow, says Damasio,

> experience has made our brain solidly pair a triggering stimulus (emotionally competent) with the most advantageous response. The "strategy" for selecting the response consists of activating the strong connection between the stimulus and response, such that when put into practice the response seems automatic and quick, without any effort or deliberation. ***(Damasio, 1994, p. 198)***

So, prior to any reasoning regarding the solution to the problem, something important happens: "When the bad outcome connected with a given response option comes into mind, however fleetingly, you experience an unpleasant gut feeling" (Damasio, 1994, 173). Consequently, the result is discarded mechanically and without further consideration. It is not that somatic markers deliberate on our behalf.

> They assist the deliberation by highlighting some options (either dangerous or favorable) and eliminating them rapidly from subsequent consideration. You may think of it as a system for automated qualification of predictions, which acts, whether you want it or not, to evaluate the extremely diverse scenarios of the anticipated future before you ***(Damasio, 1994, p. 174)***

We should clarify, however, that in most cases somatic markers are not enough to make a decision; the subsequent and complementary process of reasoning is then brought into play and on which basis a final decision is reached.

For Damasio, somatic markers are a special case of feelings generated from secondary emotions. These emotions and feelings have been connected, through learning, to future results predictable under specific scenarios (Damasio, 1994, p. 205).

The epistemic states as feelings and emotions act, under certain conditions, like somatic markers. A paradigmatic case is shown in the discovery work undertaken by mathematicians, work that requires parsimonious and efficacious decision making. Epistemic states as a certain type of somatic marker associate, for instance, a certain level of security with a conjecture; that security may follow from the consistency of the conjecture with other mathematical facts or from an inductive reasoning or even from an analogy to a well-known theorem. This gives us the possibility of deciding how to proceed accordingly: to work hard in finding a formal proof, perform experimental work, find a counterexample or simply discard it. Part of the training of mathematics professionals consists of building a long evaluative list that allows for decision making by efficiently discarding "sterile combinations," as Poincaré put it. Epistemic states as somatic markers also drive the focus of attention and functional memory toward the option chosen, whether in developing a deductive proof or verifying experimentally by way of software, for example, with the understanding that the "somatic marker also operates as an amplifier of continued attention and functional memory" (Damasio, 1994, p. 232).

A Remark for Learning

The universality of emotional expressions advances automation of the emotional program of action: activation of the mechanism per se is automatic, both in triggering primary and secondary emotions, because it is part of the instructions that are written in the genome. However, as we have seen, activation of the short and long lists connecting certain types of situations with certain emotions is also mechanical, since the limbic system acts quickly and automatically in a preconscious manner without leaving any room to consider whether the evaluation is appropriate or not.

Nevertheless, Damasio holds that it is possible to delete automation and the tyrannical unconsciousness of the emotional machinery (Damasio, 2003, p. 64) by paying attention to the relationship between certain objectives and certain emotions and by making intentional efforts to control our emotions (Damasio, 2003, p. 64). He describes it as follows:

> As social beings, our emotions (may be) triggered only after an evaluative, voluntary, non-automatic mental process. ...[That is] The reaction to that broad range of stimuli and situations can be filtered by an **interposed mindful evaluation**. And because of the thoughtful, evaluative filtering process, there is room for variation in the extent and intensity of preset emotional patterns ***(Damasio, 1994, p. 130) (bold font added)***

Elsewhere in his work, he delves further into the idea:

> Emotions provide a natural means for the brain and mind to evaluate the environment within and around the organism, and respond accordingly and adaptively. Indeed, in many circumstances, we actually evaluate consciously the objects that cause emotions, in the proper sense of the term "evaluate". We process not only the presence of an object but its relation to others and its connection to the past. In those circumstances the apparatus of emotions naturally evaluates, and **the apparatus of the conscious mind thinkingly coevaluates**. We even can modulate our emotional response. In effect, one of the key purposes of our educational development is to interpose a nonautomatic evaluative step between causative objects and emotional responses. We attempt by doing so to shape our natural emotional responses and bring them in line with the requirements of a given culture. ***(Damasio, 2003, p. 54) (bold font added).***

In preceding quotes, Damasio obliquely entertains the possibility that, parallel to the "natural" preconscious evaluation, a "second evaluation" based on rational consciousness-raising processes may occur; this "second conscious evaluation" that occurs throughout the emotional response allows for creative actions to be generated in tune with social and cultural demands that prevent behaviors that reactively and automatically respond to unconscious and involuntary stimuli. In that sense, Paul Ekman speaks of a "reflexive evaluation," registered in the emotional process, by which consciousness of the initial evaluation is taken and reevaluated, achieving a change in the emotional path (Cf. Kavindu, 2013). Thus, the raising of consciousness, especially in the context of emotion and feelings currently felt, may lead to a change in emotional responses and an ultimate rewriting of the long list.

Another resource present in the work of Damasio and one that avoids reactive conduct is simply getting around the triggering stimulus:

> We can decide which objects and situations we allow in our environment and on which objects and situations we lavish time and attention. We can, for example, decide not to watch commercial television, and advocate its eternal banishment from the households of intelligent citizens. By controlling our interaction with objects that cause emotions we are in effect exerting some control over the life process and leading the organism into greater or lesser harmony, ... We can control our exposure to the stimuli that bring on the reactions. We can learn over a lifetime to engage modulating "brakes" on those reactions. We can simply use sheer willpower and just say no. ***(Damasio, 2003, pp. 52)***

Emotions and feelings, and long and short lists on which their activation is based, are not fatally and inevitably undetachable.

Similarly, epistemic states of convincement, which are seen as emotions and feelings, are not unmovable. Stimuli that produce convincement or doubt are never definitive. That is, the connections between beliefs and backings on the one hand and epistemic states on the other are not fixed; these connections that are based on people's evaluative long lists are retrainable and may be redefined in certain conditions. We observe this continually in everyday life, where propaganda, especially electronic, plays a role of convincement—for instance, through repetition until exhaustion. Directing our sights to the realm of the didactics of mathematics, we observe the same thing in mathematics schools. We subject students to a redeveloping of their long lists so they learn to be convinced by backings that establish criteria of rigor for current mathematics. In fact, it is not at all infrequent that education centers subject students to retraining their epistemic states by strictly turning to authority or coercive practices through evaluation.

Damasio seems to suggest that a key point in the redevelopment of the long list is experiencing a second conscious evaluation through perception, both of the connections that have been established between beliefs and epistemic states and of the connections between backings and epistemic states. In the above example on deductive proof, one possible strategy of retraining the epistemic states could consist of subjecting the student to continuous experiences, accompanied by a teacher's timely and careful guidance, so that the student can reflect and begin to perceive the security provided by mathematics assertions—the security activated within them by rules of inference and promoted by other logical connections—to then eventually reinforce or redirect such connections. And they then contrast this with the type of backing that they were previously convinced by, both in and out of the mathematics classroom.

Another route for retraining epistemic states of convincement that could stem from the work of Damasio consists of directly avoiding situations that act as stimuli for certain emotions. In the case of epistemic states in a mathematics classroom setting this would be translated, for instance, into avoiding certain backings—such as those based on authority, on customs, or on practical reasons—that frequently act as emotionally competent stimuli that drive emotions of security concerning mathematical facts, sometimes irrefutably. If this type of backing cannot be avoided, one safe path would be to raise awareness related to the role they play vis-à-vis our emotional state. This facilitates rebroaching the substantive elements of Damasio's suggestion, based on the conscious and lucid realization of genesis and development of affective phenomena.

Finally, it is worth reiterating that the ideas presented here are preliminary in nature, that they may be the bases for directing future exploratory research, and they definitely require verification in concrete situations through in-depth empirical work. The authors are presently already in the process of developing that work.

Final Considerations. Relevance of the Study for the Didactics of Mathematics

The study on the practice of professional mathematicians offers evidence, among much else, of the value of epistemic states in their endeavor: a heuristic engine, guiding the work of proof and serving as a criteria for defining what a proof is. These studies also show the level of consciousness the mathematician has of the value and the role these internal states play in their discipline practice, in addition of the studies also reveal the need for mathematicians to be well trained in being faithfully convinced by current rigor criteria. This represents a salient revelation regarding the area of mathematics learning, because it shows the scope that epistemic states have for the classroom and school practices, likewise providing standards for distinguishing the roles these states may play in driving the activity as a guide and in the final stages of solving assignments, specifically in verifying results.

The studies included in the first part of this document also clarify the existing relationship between epistemic states on the one hand, and the beliefs, reasons, and motives on the other. Likewise, they attest to the presence of various types of epistemic states that mathematicians may experience. At the mathematics teaching and research level, this information outlines the domain for exploring the phenomena of convincement, and it allows for learning expectations to be planned according to the type of epistemic state students are expected to experience when faced with certain beliefs and backings.

On the other hand, the interpretation of epistemic states of convincement as certain types of emotions and feelings refreshes the perspective and information generated in the first part of the document and helps confirm and deepen its details. With this new viewpoint, it is possible to foresee that epistemic states actually lie halfway between the cognitive and affective domains and make it possible to clarify, albeit superficially and among much else, the complex processes occur during the generation of those states in the framework of deductive reasoning. Similarly and notably, it offers evidence that epistemic states are trainable and shapeable, in any sociocultural context but specifically in formal education settings, and it also suggests how this may eventually happen, clearly showing that consciousness participates in the process of epistemic state enculturation, as it happens for mathematicians mainly.

This work has provided several clues indicating that convincement is not only reduced to personal matters: historical data related to the endeavor of professional mathematicians show that convincement is an individual phenomena that is fed and ordered according to principles and guidelines related to rigorousness, cultural standards that occur in collective environments. Neurobiology, on the other hand, shows that cultural values assimilated by the subject in the framework of social interactions definitely influence constitution of the subject's secondary emotions and development of his long list. If we conceive of epistemic states as a certain type of emotion and feeling, then this means that those states are culturally conditioned. Notwithstanding the sociocultural nature of convincement and other analog states, this document focuses on providing initial explanations concerning some aspects of epistemic states in an intrasubjective dimension, since it is not possible to encompass the phenomenon of convincement in its entire complexity.

Evidence has been provided here of the importance of convincement in building mathematics, both in professional and classroom settings. It is also highly relevant because, as Tymozcko says, convincement is key in understanding mathematics as a human activity (1986, p. 247).

References

Balacheff, N. (1987). Processus de preuve et situations de validation [proof processes and situations of validation]. *Educational Studies in Mathematics, 84*(1), 147–176.

Balacheff, N. (2000). Procesos de prueba en los alumnos de matemáticas [proof processes among mathematics students]. *Bogotá: una empresa docente*. Andes: Universidad de los.

Bárcia, R. D. (1883). *Primer diccionario general etimológico de la lengua española* [First Spanish Etimological Dictionary]. Madrid: Establecimiento Tipográfico de Álvarez Hermanos.

Bombieri, E. (2006). The Riemann hypothesis. In J. Carlson, A. Jaffe, & A. Wiles (Eds.), *The millennium prize problems* (pp. 107–128). Massachusetts: Clay Mathematics Institute.

Brousseau, G. (1997). *Theory of didactical situations in mathematics: Didactique des mathématiques [mathematics learning], 1970–1990* [N. Balacheff, M. Cooper, R. Sutherland, & V. Warfield, Eds. & Trans.]. Dordrecht: Kluwer.

Cook, S. (2006). The P versus NP problem. In J. Carlson, A. Jaffe, & A. Wiles (Eds.), *The millennium prize problems* (pp. 87–106). Massachusetts: Clay Mathematics Institute.

Corbin, J., & Strauss, A. (2015). *Basics of qualitative research: Techniques and procedures for developing grounded theory* (4nd ed.). California: Sage Publications, Inc.

Damasio, A. (1994). *Descartes' error: Emotion, reason and the human brain*. New York, USA: Avon Books.

Damasio, A. (2003). *Looking for Spinoza: Joy, sorrow and the feeling brain*. London, United Kingdom: William Heinemann.

Dauben, J. W. (1980). The development of Cantorian set theory. In Grattan-Guinness (Ed.), *From the calculus to set theory, 1630–1910. An introductory history* (pp. 181–219). London: Duckworth.

De Millo, R., Lipton, R., & Perlis, A. (1986). Social processes and proofs of theorems and programs. In T. Tymoczko (Ed.), *New directions in the philosophy of mathematics* (pp. 267–287). USA: Birkhäuser Boston, Inc.

De Villiers, M. (1990). The role and function of proof in mathematics. *Pythagoras*, *24*, 17–24.

De Villiers, M. (2010). Experimentation and proof in mathematics. In G. Hanna, H. Niels, & H. Pulte (Eds.), *Explanation and proof in mathematics: Philosophical and educational perspectives* (pp. 205–221). New York: Springer.

Duval, R. (1999). *Semiosis y pensamiento humano. Registros semióticos y aprendizajes intelectuales [Semiosis and human thought. Semiotic registers and intellectual learnings]*. CA: Universidad del Valle.

Fefferman, Charles L. (2006). Existence and smoothness of the Navier-Stokes equation. In J. Carlson, A. Jaffe, & A. Wiles (Eds.), *The millennium prize problems* (pp. 57–70). Massachusetts, USA: Clay Mathematics Institute.

Fischbein, E. (1982). Intuition and proof. *For the learning of mathematics*, *3*(2), 9–24.

Foster, C. (2016). Confidence and competence with mathematical procedures. *Educational Studies in Mathematics*, *91*(2), 271–288.

Gilbert, D. T. (1991). How mental systems believe. *American Psychologist*, *46*(2), 107–119.

Hanna, G., & Jahnke, H. N. (1996). Proof and proving. In A. J. Bishop, K. Clements, C. Keitel, J. Kilpatrick, & C. Laborde (Eds.), *International handbook of mathematics education* (pp. 877–908). Dordrecht: Kluwer.

Harel, G., & Sowder, L. (1998). Students' proof schemes: Results from exploratory studies. In A. Schonfeld, J. Kaput, & E. Dubinsky (Eds.), *Research in collegiate mathematics education III* (pp. 234–282): American Mathematical Society. Issues in Mathematics Education, Vol. 7.

Harel, G., & Sowder, L. (2007). Toward comprehensive perspective on the learning and teaching of proof. In F. K. Lester Jr., (Ed.), *Second handbook of research on mathematics teaching and learning* (pp. 805–842). Charlotte: National Council of Teachers of Mathematics.

Hersh, R. (1993). Proving is convincing and explaining. *Educational Studies in Mathematics*, *24*, 389–399.

Immordino-Yang, M. H., & Damasio, A. (2011). We feel, therefore we learn: The relevance of affective and social neuroscience to education. *Learning Landscapes*, *5*(1), 115–132.

Inglis, M., Mejia-Ramos, J. P., & Simpson, A. (2007). Modelling mathematical argumentation: The importance of qualification. *Educational Studies in Mathematics*, *66*(1), 3–21.

Ito, Y., & Nakano, K. (2011). Efficient exhaustive verification of the collatz conjecture using DSP blocks of Xilinx FPGAs. *International Journal of Networking and Computing*, *1*(1), 49–62.

Kavindu. (2013). *Mindfulness*. Barcelona: Editorial Kairós.

Kline, M. (2009). *Matemáticas para los estudiantes de humanidades* [Mathematics for Students of Humanities]. México: FCE.

Krummheuer, G. (1995). The ethnography of argumentation. In P. Cobb & H. Bauersfeld (Eds.), *The emergence of mathematical meaning: Interaction in classroom cultures* (pp. 229–270). Hillsdale: Lawrence Erlbaum Associates.

Martínez, B., & Rigo, M. (2014). Mathematical certainties in history and distance education. In P. Liljedahl, S. Oesterle, C. Nicol, & D. Allan (Eds.), In *Proceedings of the joint meeting of PME 38 and PME-NA 36: Vol. 4* (pp. 177–184). Vancouver, Canada: PME.

Mates, B. (1979). *Lógica matemática elemental [Elementary Mathematical Logic]*. Madrid: Editorial Tecnós.

Neubrand, M. (1989). Remarks on the acceptance of proofs: The case of some recently tackled major theorems. *For the Learning of Mathematics*, *9*(3), 2–6.

Richstein, J. (2000). Verifying the Goldbach conjecture up to 4×10^{14}. *Mathematics of Computation, 70*(236), 1745–1749.

Rigo, M. (2013). Epistemic schemes and epistemic states. A study of mathematics convincement in elementary school classes. *Educational Studies in Mathematics, 84*(1), 71–91.

Thom, R. (1986). "Modern" mathematics: An educational and philosophic error? In T. Tymoczko (Ed.), *New directions in the philosophy of mathematics* (pp. 67–78). USA: Birkhäuser Boston, Inc.

Thompson, A. (1992). Teachers' beliefs and conceptions: A synthesis of the research. In D. A. Grouws (Ed.), *Handbook of research on mathematics teaching and learning* (pp. 127–146). New York: Macmillan.

Tymoczko, T. (1986). The four-color problem and its philosophical significance. In T. Tymoczko (Ed.), *New directions in the philosophy of mathematics* (pp. 243–266). Boston: Birkhäuser.

Villoro, L. (2002). *Creer, saber, conocer [belief, personal and propositional knowledge]* (14ªth ed.). México: Siglo veintiuno editores.

Wittgenstein, L. (1988). *Sobre la Certeza [On Certainty]*. Barcelona: Editorial Gedisa.

CHAPTER

5

The Impact of Anxiety and Working Memory on Algebraic Reasoning

Kelly Trezise, Robert A. Reeve

University of Melbourne, Melbourne, VIC, Australia

INTRODUCTION

Poor mathematics achievement has long been associated with mathematics anxiety (MA) and impaired cognitive abilities (Betz, 1978; Geary, Brown, & Samaranayake, 1991; Morris, Kellaway, & Smith, 1978; Raghubar, Barnes, & Hecht, 2010; Suinn, Edie, Nicoletti, & Spinelli, 1972; Suinn & Richardson, 1971). However, understanding the respective effects of MA and cognitive abilities on math performance is complex due to relationships between MA and cognition. MA has also been associated with cognitive abilities, causing difficulty in developing an understanding of their respective effects on math abilities. Poor cognitive abilities (e.g., WM) are associated with MA (Maloney, Ansari, & Fugelsang, 2011; Maloney, Risko, Ansari, & Fugelsang, 2010; Mammarella et al., 2015). And MA is thought to affect cognitive abilities (e.g., WM), which in turn impairs math problem-solving ability (Ashcraft & Kirk, 2001; Faust, Ashcraft, & Fleck, 1996; Hopko et al., 2003; Kellogg, Hopko, & Ashcraft, 1999; Miller & Bichsel, 2004; Raghubar et al., 2010). We suggest the inferences that can be made about the relationship between cognitive abilities, MA, and math problem solving from previous work are limited in at least two respects. First, somewhat surprisingly, we know little about the ongoing reciprocal relationships between cognition and MA in math over time. Second, it is often assumed that MA and cognition are general traits, rather than specific states. We suggest that treating MA and WM as states (rather than trait-like) and investigating the interaction between cognition and emotion longitudinally would provide a more complete understanding of the relationships between WM and MA that would benefit researchers and practitioners alike.

http://dx.doi.org/10.1016/B978-0-12-802218-4.00005-4

The chapter is divided into three sections. First, we begin by reviewing research that has investigated the impact of MA on math problem solving. Most previous studies have investigated arithmetic anxiety, rather than anxiety associated with complex math (e.g., algebra). This focus is surprising since increases in math anxiety commonly occur in early adolescence (i.e., at a time when students encounter complex math reasoning). Second, we discuss the nature of interactive relationships between math anxiety and cognitive abilities within the context of a specific math problem-solving domain (algebra). In this section we discuss our own research that has examined algebraic anxiety-WM patterns in students across time and how these patterns predict algebraic problem-solving abilities (Trezise & Reeve, 2014a, 2014b, 2016). In particular, we show how individual differences in WM and MA can affect each other and change over a short period of time. Although this research comes from a cognitive psychology perspective, its findings have relevance for educational and professional contexts. In the fourth section, we discuss the implications of our research for assessment and intervention practices. We conclude by discussing the ways in which a better understanding of individual differences in the relationships between MA and WM may foster new approaches for reducing the impact of poor cognitive abilities and/or high math anxiety on math problem-solving abilities.

Mathematics Anxiety

MA is usually described in terms of feelings of tension, anxiety, or stress that occurs in situations that involve math problems (Richardson & Suinn, 1972). Research has demonstrated complex relationships between MA, WM, and math problem solving. High MA is thought to impair problem solving ability by reducing available WM capacity (Ashcraft & Kirk, 2001). WM is thought to provide a workspace for mathematical problem solving (DeStefano & LeFevre, 2004; LeFevre, DeStefano, Coleman, & Shanahan, 2005; Raghubar et al., 2010). For example, in arithmetic, WM is thought to support the encoding and maintenance of problems, calculation, and generation of responses (DeStefano & LeFevre, 2004). Poor math cognitive abilities are also thought to lead to increases in MA: poor cognitive abilities may impair math learning and performance, which may lead to individuals' MA increasing (Ma & Xu, 2004; Maloney et al., 2010, 2011; Rubinsten & Tannock, 2010). Although MA is claimed to affect WM, and WM is thought to affect MA, few studies have explored the interaction between them. In this section, we review previous MA and WM research and outline how findings that suggest evidence of a mutual relationship between WM and MA, and we also discuss implications for math problem solving.

Math Anxiety and Math Problem Solving

Ashcraft and Faust (1994) examined how different MA levels affect math problem solving by investigating whether MA is related to the cognitive processes underlying mental arithmetic. Ashcraft and Faust examined mental arithmetic problem-solving speed and accuracy in college students divided into four math anxiety groups (low to very high). Participants solved simple arithmetic (e.g., single-digit addition and multiplication) and complex arithmetic (e.g., double-digit addition and mixed arithmetic) problems. Ashcraft and Faust found that accuracy was lower and response speed was higher for the high anxiety groups. Moreover, the highest anxiety group was likely to sacrifice accuracy for rapid performance in complex arithmetic problem solving; in other words, they displayed very fast response times and low accuracy. This pattern of findings is important because it demonstrates that math anxiety is associated with impaired mathematical processing, and this relationship is affected, in turn, by task demands. In short, high MA and high task demands are likely to have a negative impact on mathematical processing, and the effect of MA is likely to be smaller when task demands are low. These findings suggest that individual differences in MA may interact with cognitive task demands to affect math problem-solving abilities.

Similarly, Beilock and Carr (2005) examined the moderating effects of WM capacity on the anxiety-problem-solving relationship by examining WM, arithmetic problem solving, and different pressure conditions. The effect of anxiety was by varying "pressure" conditions. Low pressure was described as practice, and high pressure was induced by rewards and peer pressure. Math was assessed by a modular arithmetic task in which appraisal was made about the correctness of a complex arithmetic statement. Low pressure and high WM capacity was associated with better performance, compared to low WM. However, in high pressure situations, the performance of high WM individuals was impaired so their advantage over individuals with low WM disappeared. These findings showed that individuals' WM capacity interacts with anxiety and task demands to affect math performance. However, the research also showed high WM individuals were more impaired by anxiety/pressure, compared to low WM individuals.

Together, Ashcraft and Faust's (1994) and Beilock and Carr's (2005) work show a complex relationship between MA, WM, and math problem solving. Individual differences in MA interact with task demands to affect math problem solving (Ashcraft & Faust, 1994), and individual differences in WM interact with anxiety to affect math problem solving (Beilock & Carr, 2005). Moreover, both sets of findings show that these relationships vary with math type (e.g., single addition or complex arithmetic) and/or math task demands (math complexity).

Influences Between Math Anxiety and WM

While the evidence shows that WM can modulate the effect of MA on math problem solving, individual differences in MA and WM may also interact to affect WM and MA. In other words, WM may affect MA and MA may affect WM, in addition to their respective effects on math problem solving. Ashcraft and Kirk (2001), for example, examined the relationship between MA and WM and how these factors interact to impact arithmetic problem solving. College students completed an MA questionnaire and a mental arithmetic task under varied WM demands. Addition accuracy was lower when WM demands were high, and this "cost" increased in individuals with higher MA. That is, as task demands increased, problem solving was more affected for students with high MA than low MA. These findings show that increased task demands with high MA lead to greater WM demands and greater performance impairments. Ashcraft and Kirk's work shows that high math-anxious individuals have lower WM. Ashcraft and Kirk's findings are particularly important for two reasons. First, they demonstrate that high MA is associated with lower WM. Second, their findings suggest that MA and WM demands interact to affect math problem solving. Overall, the study showed that MA level is associated with differences in WM capacity, and performance impairments induced by high WM demands are greater in individuals with high MA compared to those with low MA.

While Ashcraft and Kirk (2001) demonstrated that MA can reduce WM capacity, low WM capacity may also impact MA. Maloney et al. (2010, 2011) suggest that impaired cognitive functions that underlie math abilities (e.g., WM) may increase MA. Children with math learning disabilities show anxious reaction to math words (Rubinsten & Tannock, 2010). In a longitudinal study examining MA and math achievement throughout high school, Ma and Xu (2004) found that low math achievement predicted subsequent increases in MA. It would appear that these math cognition-MA relationships are age-independent.

Maloney et al. (2010, 2011) examined the association between MA and core numerical abilities in adults. Core numerical abilities, such as enumeration and processing of numerical magnitude, are thought to underlie the development of math ability (see Maloney et al., 2011; Reeve, Reynolds, Humberstone, & Butterworth, 2012). Maloney et al. (2010, 2011) found high math-anxious groups are slower at counting and numerical magnitude processing than low math-anxious groups. These studies suggest that impaired cognitive functions that underlie math problem solving may also increase MA. Combined, MA-WM research suggests there a may be a mutual relationship between MA and WM: Ashcraft and Kirk (2001) have shown individual differences in MA affect WM and findings from Ma and Xu (2004), Maloney et al. (2010, 2011), and Rubinsten and Tannock (2010) suggest that individual differences in WM may affect MA.

Summary

Given the fact that individual differences in MA and WM affect each other, as well as math problem solving, several implications follow. First, MA and WM may interact additively to effect math problem solving. Second, individual differences in WM may interact with individual differences in MA and effect changes to WM and/or MA, resulting in decreases/increases to WM and/or increases/decreases in MA. In other words, reciprocal relationships between WM and MA may result in individuals' WM/MA changing over time. Third, it is possible that changes to WM and/or MA over a short period of time may impact math problem solving. Finally, given that MA and WM interact with task demands to affect problem solving, it is also possible that MA-WM relationships vary with different math contexts/demands.

It is possible that MA and WM may interact within an individual and (1) affect WM/MA, which results in changes to WM-MA relationships over time, and (2) they may affect math problem solving. Moreover, MA-WM relationships may interact with the task demands to respond differentially in different math contexts. Specifically, iterative and reciprocal relationships between WM and MA relationships may change over time. If WM/MA relationships do change over time, these changes may directly affect math problem-solving abilities.

Algebra

Currently, most research into MA and WM relationships and math have been examined in the context of arithmetic, or generalized math assessments. This research shows that WM-math and MA-math relationships vary between different forms of arithmetic (Ashcraft & Faust, 1994; Raghubar et al., 2010). And students report complex math, such as long division and fractions, as more stressful than simpler math, such as mental arithmetic (Chinn, 2009). It is therefore possible that MA-WM-problem-solving patterns will also differ in reaction to different types of math (e.g., algebra).

One math domain strongly associated with math anxiety is algebra. Ng (2012) found high school students specifically identified algebra as a source of math anxiety. And among preservice teachers, algebra was most often identified as the type of math causing anxiety (Uusimaki & Nason, 2004). Moreover, sharp declines in math achievement coincide with the introduction of algebra courses (National Mathematics Advisory Panel, 2008). Many students find the transition from arithmetic to algebra troublesome (Knuth, Alibali, McNeil, Weinberg, & Stephens, 2005). Given achievement in algebra predicts the likelihood of continuing to study STEM subjects (Adelman, 1999; Carnevale & Desrochers, 2003; Evan, Gray, & Olchefske, 2006; Gandal, Kraman, Pelzman, & McGiffert Slover, 2006; McGregor,

1994; National Mathematics Advisory Panel, 2008), it is important to understand the factors that impair algebra problem-solving abilities.

However, little research has examined the cognitive and math problem-solving consequences of anxiety in an algebraic context. Given that the WM-math and MA-math relationship varies between different forms of arithmetic (Ashcraft & Faust, 1994; Raghubar et al., 2010), it is likely these variations would be larger between arithmetic and algebra. Algebra is hypothesized to require WM resources because it requires maintenance of math expressions and retrieval of algorithms and math facts from long-term memory in WM (Tolar, Lederberg, & Fletcher, 2009).

EXAMINATION OF MA, WM, AND ALGEBRA RELATIONSHIPS IN THREE STUDIES

We examined the interactive nature of MA, WM, and algebraic problem solving in adolescent students (Trezise & Reeve, 2014a, 2014b, 2016). We aimed to examine (1) different relationships of MA and WM, (2) how WM and MA interact to affect each other over time, and (3) the implications of these interactions on algebraic problem solving. In order to address these issues, we adopted research methodologies that would allow for the examination of individual differences and changes in MA and WM over time.

Rationale for the Research Agenda

The interactions between WM, MA, and math problem solving are currently unclear. Although traditionally MA is thought to impair math performance, recent research suggests that poor math ability may increase MA. Thus, it is possible that MA impairs math processing by reducing WM capacity, as suggested by Ashcraft (2002), and/or that poor math processing results in high MA, as proposed by Maloney et al. (2010, 2011). Alternatively, mutual effects between MA and WM may be also evident.

Methodological Issues Limiting Current Understanding

Methodological approaches investigating WM-MA interactions limit our ability to characterize the relationship in at least three ways:

Assumption of trait WM and MA limits characterization of direction of influence. Though evidence suggests that WM and MA may be domain-specific and dynamic over time, WM and MA are assessed in ways that assume both trait and domain general properties. Previous research has not examined the state nature of MA and WM involved in math problem solving. Cross-sectional studies show a relationship between MA and WM and an additive effect of WM and MA on math problem solving; but these studies are limited in the ability to characterize the direction of influence. It is possible that there is only a single direction of influence between WM and MA. Alternatively, there may be mutual influences between WM and MA, which may result in changes in WM and/or MA over time.

Analytic approaches to examining interactive relationships between WM and MA. Statistical analyses used to examine WM-MA relationships are inappropriate for testing possible interacting relationships (Bergman, Magnusson & El-Khouri, 2003; MacCallum, Zhang, Preacher, & Rucker, 2002). Traditional analytic methods used to examine WM-MA relationships include linear regressions, which are limited in their ability to accurately approximate interactive relationships (Bergman et al., 2003), and dichotomizing variables, which can result in a distorted picture of relationships between variables (Altman & Royston, 2006; Irwin & Mcclelland, 2003; MacCallum et al., 2002; Preacher, Rucker, MacCallum, & Nicewander, 2005; Ruscio & Ruscio, 2008).

WM and MA outside of arithmetic. The relationship among math, WM, and anxiety has largely been explored in the arithmetic domain. Arithmetic is introduced early in education and by adolescence is familiar and well-learned. Raghubar et al. (2010) suggested that WM-math relationships can alter with the math domain, strategy-use, and math skill development. MA can increase with math problem domain and complexity (Chinn, 2009; Trezise & Reeve, under review), and the effects of MA can change with different forms of arithmetic (Ashcraft & Faust, 1994). Consequently, it is possible that for adolescents and adults, patterns of WM-anxiety relationships in arithmetic may differ from WM-anxiety relationships in more complex math, such as algebra.

These issues suggest a need to (1) employ "longitudinal" studies to investigate WM-anxiety relationships, (2) use statistical analyses suitable for testing interacting factors and longitudinal relationships, and (3) examine these relationships outside of an arithmetic domain.

Measurement of Algebraic Problem Solving, Math Anxiety, and WM

Algebra is an ideal domain to examine WM-anxiety relationships. Math problem solving, such as in algebra, is WM demanding, and math abilities are related to individual differences in WM capacity (Geary, Hoard, Byrd-Craven, Nugent, & Numtee, 2007; Raghubar et al., 2010; Tolar et al., 2009). And algebra elicits anxiety in many individuals (Ng, 2012; Uusimaki & Nason, 2004).

Algebraic problem-solving abilities are often assessed in using algebraic word problems (e.g., Dooren, Verschaffel, & Onghena, 2002; Lee, Ng, & Ng, 2009), which require students to both correctly decode as well as solve the problem. While algebraic word problem solving is thought to involve WM, it is also related to individual differences in literacy ability, which is thought to affect students' ability to decode algebraic problems (Lee et al., 2009). To minimize the impact of literacy ability, we used algebraic equations, rather than word problems, to assess algebraic problem-solving ability. Compared to numerical equations, solving symbolic algebra equations requires more abstract thinking (Susac, Bubic, Vrbanc, & Planinic, 2014). Students transition from concrete to abstract thinking in early adolescence and, consequently, show greater difficulty in solving symbolic algebra equations than numerical equations (Küchemann, 1981; Susac et al., 2014).

We examined linear algebraic equation problems suitable for 14-year-olds. Specifically, we manipulated problem-solving difficulty using the equivalence sign, symbolized as "=." Studies of conceptual difficulties with algebra have identified the equivalence sign as a stumbling block in the acquisition of algebra (Alibali, Knuth, Hattikudur, McNeil, & Stephens, 2007; Küchemann, 1981). We manipulated the number of terms on each side of the equivalence sign to vary problem difficulty (Humberstone & Reeve, 2008; Küchemann, 1981).

MA is traditionally assessed with psychological scales designed to assess trait math anxiety. Most indices of math anxiety tend to be based on self-reported difficulties experienced in mathematical situations (Capraro, Capraro, & Henson, 2001). Few studies have assessed the anxiety of students as they complete math tasks. The items on math anxiety scales tend to only include arithmetic referents, and they do not refer to more complex problems (e.g., algebraic problems) (Capraro et al., 2001). Conceptually, math anxiety scales appear to be based on the assumption that they assess a domain-general trait which remains constant over time and context. The research focused on math worry, since it has been identified as the cognitive component of state anxiety (Eysenck, Derakshan, Santos, & Calvo, 2007; Liebert & Morris, 1967), is thought to affect WM (Hayes, Hirsch, & Mathews, 2008; Owens, Stevenson, Hadwin, & Norgate, 2012),[1] and the term "worry" has been used in previous research to assess MA in the context of attitudes and emotions in math and other academic domains (e.g., Dowker, Bennett, & Smith, 2012; Punaro & Reeve, 2012).

We examined "worry" in the context of algebraic problems (Trezise & Reeve, 2014a, 2014b, 2016). Students were shown pairs of algebra equations (e.g., $x+2=5$; $2x+4=10$) and judged whether the value of x in the two equations had the same value (i.e., equivalent) or different values (see Fig. 1). Immediately after each judgment, students rated the worry experienced while making a judgment. A faces anxiety scale (Bieri, Reeve,

[1]The distinction between worry and anxiety has long been made (Morris, Davis, & Hutchings, 1981) and reiterated in recent papers (Derakshan & Eysenck, 2009; Eysenck et al., 2007).

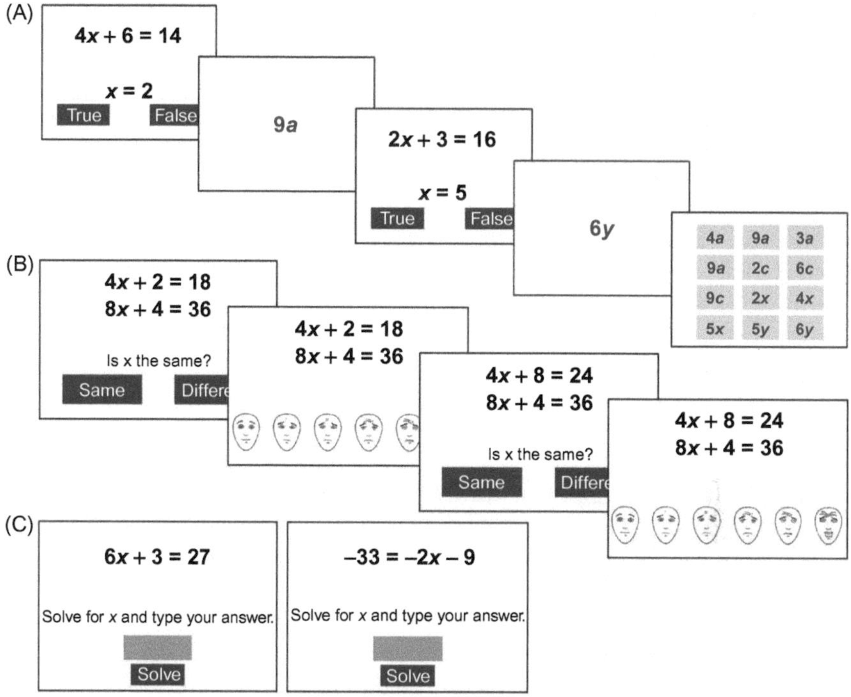

FIG. 1 Example stimuli for the (A) WM, (B) worry, and (C) problem-solving tasks. A two-trial sequence on the algebraic WM task is depicted: two trials each containing a processing (correct/incorrect) and memory (algebraic term) phase, and memory terms on a recognition screen. Algebraic worry task involves a judgment about the equivalence of two equations, followed by a worry rating; easy *(left)* and difficult *(right)* trials are shown. Easy and hard *(left to right)* problems for the algebraic problem-solving task are shown. From Trezise, K., & Reeve, R. A. (2014b). Working memory, worry, and algebraic ability. *Journal of Experimental Child Psychology, 121C*, 120–136. doi:10.1016/j.jecp.2013.12.001.

Champion, Addicoat, & Ziegler, 1990), comprising six faces of increasing anxiety was used to assess worry: students selected the face that corresponded to their level of worry. Assessing worry after each algebraic equivalence pair enabled us to measure possible changes in worry over time and in response to problem difficulty.

To examine the role of WM in WM-MA relationships, we assessed WM in a domain relevant to algebraic problem solving for two reasons (Trezise & Reeve, 2014a, 2014b, 2016). First, cognitive tasks, such as WM tasks, can act as a distractor from individuals' emotions, when the cognitive task is in a different stimulus domain separate from emotional tasks domains (e.g., faces and digits) (Van Dillen & Koole, 2007, 2009). Given we aimed to characterize WM-MA relationships in algebraic problem solving, it was important to avoid the WM task distracting from MA by assessing a WM task in the same domain as the worry and problem-solving tasks (i.e., algebraic WM).

A second reason for using algebraic stimuli in the WM task is that WM is thought to involve domain-specific knowledge. In all WM theories, an individual's skills and knowledge affect WM functioning in a specific domain. These factors have been conceptualized in different theories as domain-specific stores (Engle, 2010; Miyake & Shah, 1999), an episodic buffer (Baddeley, 2000), and long-term memory activated to WM (Oberauer, 2009). Within arithmetic problem solving, domain-specific relationships differ with type of problem (e.g., addition or multiplication), presentation of problem (e.g., vertical or horizontal), and skill level (see Raghubar et al., 2010). Little research has focused on algebra: most of the relationships between WM and math have been examined in the context of arithmetic. While some WM tasks use digit and arithmetic calculation stimuli (e.g., digit span, or calculation span tasks) and may provide a useful index of arithmetic WM, few WM tasks use algebra-relevant stimuli, such as alphanumeric symbols or algebraic processing.

We assessed WM using a dual span task. Dual-span tasks are considered good measures of WM: the interplay between storage and attention required in dual-span tasks is thought to vary as a function of domain-specific knowledge (Conway et al., 2005). The algebraic WM task was designed to assess the ability to process and remember alphanumeric symbols. Other dual-span tasks used in previous research, such as the language span task (Daneman & Carpenter, 1980) and the computation span task (Ashcraft & Kirk, 2001), have been designed to evaluate knowledge of the specific domain being assessed. The algebraic WM task examined the ability to both remember and process alphanumeric symbols (see Fig. 1).

EXAMINATION OF WM, WORRY, AND ALGEBRA RELATIONSHIPS IN THREE STUDIES

The aim of our research agenda was to characterize different patterns of WM-MA relationships in algebra and their relationships with algebraic problem solving. Each of the studies (Trezise & Reeve, 2014a, 2014b, 2016) examines different aspects of the WM-MA relationships. The first (Trezise & Reeve, 2014a) examined individual differences in patterns of WM-MA relationships at a single time point and the implications of these patterns of WM/MA for algebraic problem-solving abilities. The second study (Trezise & Reeve, 2016) investigated the direction of influence between WM and MA: it assessed whether WM influences MA, MA influences WM, or WM and MA mutually interact. It also examined if WM and/or MA change over a short period of time, the predictors of change, and whether WM and/or MA predicted algebraic problem-solving ability. In the final study (Trezise & Reeve, 2014b), inter- and intraindividual differences were characterized to identify patterns of WM-MA likely to remain stable/change over time as well as the consequences of WM-worry patterns for algebraic problem solving. Our overall goal was to characterize the interactions between algebraic WM and worry and the implications of these relationships for algebraic problem-solving abilities.

Study 1: Profiles of Algebraic WM-Worry Relationships

In the first study (Trezise & Reeve, 2014a) we examined the nature of WM-worry relationships at a single time point and their implications for math problem solving. Female high school students completed domain-relevant worry and WM tasks, an algebraic problem-solving task, and general cognitive measures (basic response time, calculation ability, and nonverbal IQ).

To examine individual differences in WM-worry relationships, students' worry levels and WM capacity were characterized using latent profile analysis (LPA). LPA is a classification method that identifies discrete classes of individuals that share similar response patterns to a set of continuous variables (Collins & Lanza, 2010). LPA seeks to identify the smallest number of distinct groups of similar individuals that best represent the patterns in the data; groups are represented by a categorical latent variable (Bergman & Magnusson, 1997; Collins & Lanza, 2010; von Eye & Bergman, 2003). It does not require some of the assumptions of many other forms of analysis (e.g., traditional modeling), such as normal distribution and homogeneity, and is therefore less prone to statistical biases in the data (Magidson & Vermunt, 2003). The number and characteristics of the profiles are not predetermined but are identified after the analyses to determine the best fitting model (Fletcher, Marks, & Hine, 2012; Pastor, Barron, Miller, & Davis, 2007). LPA assumes that members of one subgroup share a pattern of response probabilities that distinguishes them from other groups (Collins & Lanza, 2010).

Four patterns of WM-worry relationships were identified: high WM-low worry, moderate WM-low worry, moderate WM-high worry, and low WM-high worry. We examined how WM-worry relationships predict algebraic problem solving. Accuracy on the algebraic problem-solving task was compared across the four WM-worry profiles identified in the LPA. Profile membership predicted algebraic problem-solving accuracy (see Fig. 2): high WM with low worry was associated with accurate problem solving, moderate WM (with low or high worry) was associated with moderate problem-solving accuracy, and the (small) low WM-high worry subgroup showed inaccurate algebraic problem solving.

This study shows there are different relationships between WM and worry. Moreover, WM-worry subgroup membership predicted algebraic problem-solving accuracy. High WM with low worry was associated

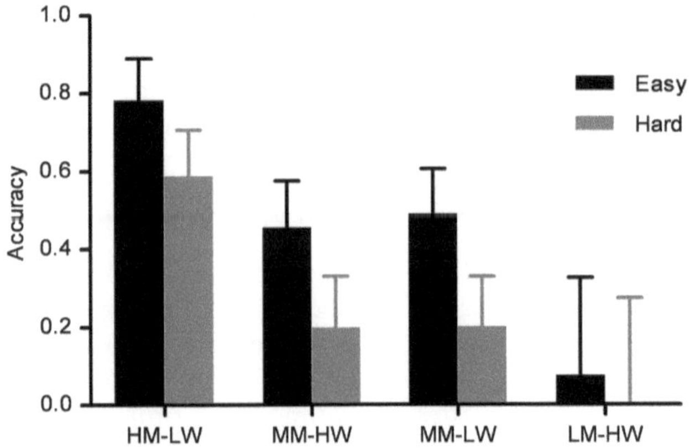

FIG. 2 Mean algebraic problem-solving accuracy, as a function of group membership (error bars = 2 × SE). From Trezise, K., & Reeve, R. A. (2014a). Cognition-emotion interactions: Patterns of change and implications for math problem solving. *Frontiers in Psychology, 5*(July), 1–15. doi:10.3389/fpsyg.2014.00840.

with more accurate problem solving, and low WM with high worry was associated with inaccurate problem solving. The associations between subgroups and problem solving suggest a relationship between WM and algebraic problem solving, which supports findings that WM is important for math problem-solving ability (DeStefano & LeFevre, 2004; LeFevre et al., 2005; Raghubar et al., 2010).

Study 2: Mutual Influences Between WM and Worry

The relationships between WM and anxiety/worry were established in study 1 (Trezise & Reeve, 2014a), but it was unclear from the findings whether WM affects worry, and/or vice versa, and how each of these contribute to algebraic problem-solving abilities. In a second study, we examined the direction of influence between WM and worry: whether WM influences worry, worry influences WM, or WM and worry are mutually interacting (Trezise & Reeve, 2016). If a reciprocal relationship between WM and anxiety does exist, then WM and/or anxiety may change over time, and we might expect that individuals' initial WM and/or anxiety levels may affect the nature of these changes. Therefore, the study examines whether WM and/or worry (1) change over time, (2) are predictors of change, and (3) predict algebraic problem solving ability. Finally, we examined how WM and worry, respectively, affect algebraic problem-solving abilities. In other words, our aim was to characterize how WM and worry affect each other and change within individuals over time, and to specify what these changes mean for individuals' algebraic problem-solving ability.

Using the algebraic WM and worry task, students were assessed several times over a single day as they prepared for an algebraic problem-solving task at the end of the day. To examine the direction of influence between WM and worry, and potential changes over time, we conducted analyses using latent change score models (also referred to as latent difference score models—Grimm, An, McArdle, Zonderman, & Resnick, 2012; McArdle, 2009). The analyses examine individuals' WM and worry over time: change in WM and worry between adjacent time points is identified, along with the predictors of these changes. The model allows for the characterization of the influences between WM and worry and the change in WM and worry over time. Models with different predictors of change were compared. This allowed us to identify the best fitting model, which was interpreted to be the best representation of the dynamics between WM and worry.

The model identified as the best fitting model showed that worry negatively predicted changes in WM, WM capacity negatively predicted changes in worry, and decreases in WM predicted increases in worry (see Table 1). These findings suggest that high worry led to decreases in WM, supporting Ashcraft and Kirk's (2001) findings that MA reduces WM. The findings also show that low WM or decreases to WM predicts an increase in worry, which supports the developmental hypothesis that poor math cognitive abilities predict increases in MA (Maloney et al., 2010, 2011). Together, our findings suggest mutual influences between WM and worry, which result in high WM and low worry predicting stability over time, and low WM and/or high worry predicting changes over time. Finally, the model indicates that both high WM and low worry predict

TABLE 1 Parameter Estimates for Model Fit to WM, Worry, and Problem-Solving Data

	WM		Worry	
Fixed Effects	**PE**	**SE**	**PE**	**SE**
Mean intercept	0.53*	0.02	0.29*	0.02
Mean slope	0.01	0.10	0.52*	0.16
WM level	0.06	0.22	**−0.36***	**0.14**
Worry level	**−0.30***	**0.07**	−0.62	0.51
WM change	−0.27	0.24	**−0.65***	**0.26**
Final score on problem solving	**0.79***	**0.14**	**−0.45***	**0.16**

*$P < .05$.

PE, parameter estimate; *SE*, standard error of the parameter estimate. Bold text indicates the statistically significant dynamic parameters that describe the interplay between WM and worry, and problem solving.

Adapted table from Trezise, K., & Reeve, R. A. (2016). Worry and working memory influence each other iteratively over time. Cognition and Emotion, 30(2), 353–368. doi:10.1080/02699931.2014.1002755.

accurate algebraic problem solving, which supports math cognition theories of the role of WM in supporting math problem solving (DeStefano & LeFevre, 2004; LeFevre et al., 2005; Raghubar et al., 2010) and MA theories that math anxiety impairs math problem solving (Ashcraft & Kirk, 2001; Beilock & Carr, 2005). Overall, the findings of our second study suggest that WM and worry influence each other and the nature of influence changes over time; also, these influences can lead to changes in WM and/or worry over time.

Study 3: Profiles and Changes in WM-Worry Relationships

In a third study, we (Trezise & Reeve, 2014b) examined WM-worry relationships and patterns of stability/change in these relationships and their implications for algebraic problem solving. We sought to combine WM-worry relationship profiles, similar to study 1, with changes over time, similar to study 2, to describe stability/changes in WM-worry relationships over time and their implications for algebraic problem-solving abilities.

We assessed WM and worry twice in a single day: first at the beginning of a math class and then at the end of a second math class later in the day. After the second WM and worry assessment, students completed an algebraic problem-solving task; both accuracy and response times in the algebraic problem-solving task were assessed. We chose the design in order to determine whether different patterns of WM-worry relationships remain stable or change over a short period of time and whether these patterns are associated with differences in algebraic problem solving. Of interest is whether particular WM-worry relationships are more/less stable than others and, in particular, whether initial differences in the WM-worry relationship differ over time.

We used a three-step latent transition analysis (LTA) to provide an integrative model of WM-worry stability/change relationships and their implications for algebraic problem solving (Trezise & Reeve, 2014b). LTA is a longitudinal version of LPA (used in Study 1). It examines patterns of WM-worry subgroups at an initial time point and the changes in subgroup membership over time, conditional on initial subgroup membership (Vermunt & Magidson, 2013). In this study, we examined WM-worry relationships by entering WM capacity and worry levels into the LTA. Initial WM-worry relationships were described by identifying initial subgroup membership. Changes in WM-worry relationships were described by the changes in subgroup membership between time 1 and time 2. In the third step of the analysis, we examined how WM-worry subgroup membership predicted algebraic problem-solving performance.

We identified six subgroups of WM-worry relationships (see Fig. 3): three WM (high, moderate, and low) × two worry (low and high). At the first time point, most students belonged to high WM (with low or high worry) or the

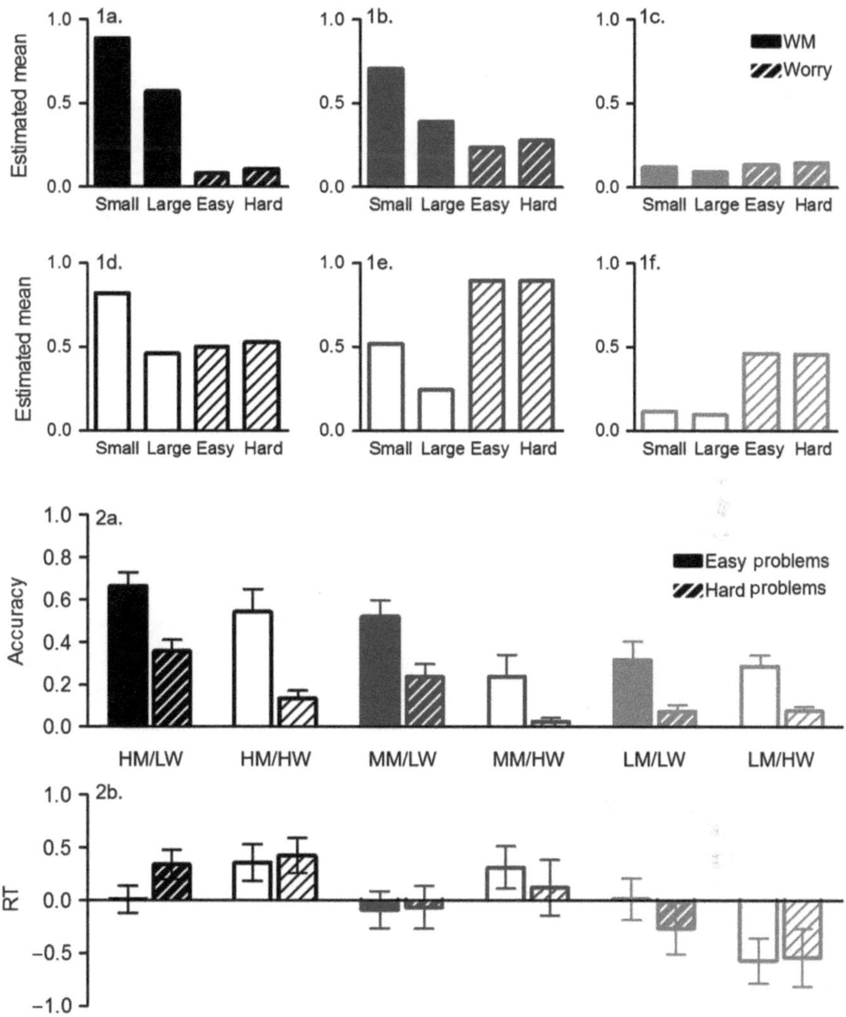

FIG. 3 Subgroups from the latent transition analysis and step 3 analysis outcomes for time 2 subgroup predicting problem solving. Mean scores on WM small (2 and 3), WM large (4 and 5) sets, and worry for easy and hard judgments are shown for (1a) high WM/low worry, (1b) moderate WM/low worry, (1c) low WM/low worry, (1d) high WM/high worry, (1e) moderate WM/high worry, and (1f) low WM/high worry subgroups. Three-step analysis outcomes for time 2 subgroup predicting problem solving (2a) accuracy and (2b) speed for easy and hard problems. For (2a), higher bars represent higher accuracy. Speed regression coefficients for easy and hard problems are shown in (2b); bars above the x-axis indicate slower responses and bars below the x-axis indicate faster responses. Error bars represent standard error. Subgroups are HM/LW (high WM/low worry), HM/HW (high WM/high worry), MM/LW (moderate WM/low worry), MM/HW (moderate WM/high worry), LM/LW (low WM/low worry), and LM/HW (low WM/high worry). Modified from Trezise, K., & Reeve, R. A. (2014a). Cognition-emotion interactions: Patterns of change and implications for math problem solving. *Frontiers in Psychology*, 5(July), 1–15. doi:10.3389/fpsyg.2014.00840.

moderate WM-low worry subgroups. Students in the high WM-low worry subgroup were likely to remain in that subgroup over time. However, we found that students in the high WM-high worry and moderate WM-low worry subgroups were likely either to remain in the same subgroup or transition to a lower WM subgroup. In particular, students in the high WM-high worry subgroup transitioned to the low WM-high worry subgroup, indicating high worry with high WM was associated with decreases in WM, supporting our previous research (Trezise & Reeve, 2016) and the findings of Ashcraft and Kirk (2001) and Beilock and Carr (2005). Students in the moderate WM-low worry subgroup were likely to transition to both the low WM-high worry and low WM-low worry subgroups, indicating that moderate WM was associated with decreases in WM over time and possibly worry increases. The two low WM subgroups each had a small *n* at the initial time point, and students moved into these subgroups over time. Moreover, the WM capacity for these subgroups was very low; which suggests that very low WM is likely to emerge over time from high cognitive demands. The moderate WM-high worry subgroup had a small *n* at both time points, but the subgroup was noticeably characterized by very high worry, supporting Trezise and Reeve (2014a).

Finally, we examined the relationship between WM-worry subgroups and algebraic problem solving (see Fig. 3). The WM-worry subgroups predicted algebraic problem-solving abilities: high WM-low worry predicted accurate problem solving; high WM with high worry predicted accurate problem solving for easy problems, but accuracy was lower for hard problems. Moderate WM with low worry predicted accurate problem solving for easy problems with lower accuracy for hard problems. Low problem-solving accuracy was predicted by membership to the moderate WM with high worry subgroup, the low WM with low worry subgroup, or the low WM with high worry subgroup.

The difference in problem-solving accuracy across different WM levels supports the notion that WM capacity is important for algebraic (and mathematical) problem-solving abilities (DeStefano & LeFevre, 2004; Geary et al., 2007; LeFevre et al., 2005; Raghubar et al., 2010). We also examined the relationship between subgroup and problem-solving speed. Low worry predicted average to fast problem solving. High worry was found to interact with WM to predict problem-solving speed: for high worry with high WM, problem solving was slow; for high worry with moderate WM, problem solving was slow for easy problems and moderate for hard problem solving; and for high worry with low WM, problem solving was very fast.

Summary of Findings

Three general findings are of interest. First, relationships between WM, worry, and algebraic problem solving were nonlinear, and they show patterns of individual differences. Second, individuals' WM and/or worry change over short time-frames as a result of mutual influences. Of particular interest, low WM capacity led to increases in individuals' worry, and worry decreased WM capacity. The mutual influences and resulting changes led to increases in variance between individuals over time. Third, WM capacity and worry interact to affect math problem-solving speed and accuracy. Over the three studies, the relationship between WM and worry is shown to be reciprocal, heterogeneous, and changing over time.

DISCUSSION

WM-Worry Interactions and Algebraic Problem Solving

The three studies show higher WM is associated with more accurate algebraic problem solving. In Studies 1 and 3, the subgroups with high WM were associated with accurate algebraic problem solving, and low WM s associated with inaccurate problem solving. In Study 2, the bivariate model showed a positive relationship between WM capacity and algebraic problem-solving accuracy. The findings support previous research that has examined the relationship between WM and math, which has found that individual differences in WM capacity predict math ability (e.g., Barnes et al., 2014; Geary et al., 2007; Swanson, 2006). The studies within the research extend the findings of these WM-arithmetic relationships to algebraic problem solving (DeStefano & LeFevre, 2004; Hecht, 2002; Imbo & LeFevre, 2010; Imbo & Vandierendonck, 2007a, 2007b; Metcalfe, Ashkenazi, Rosenberg-Lee, & Menon, 2013; Passolunghi & Mammarella, 2010; Xenidou-Dervou, van Lieshout, & van der Schoot, 2014).

Our studies also show a mutual relationship between algebraic anxiety and WM that predict changes in algebraic anxiety and/or WM over time. Students with low WM were likely to increase their algebraic anxiety over time, and students with high algebraic anxiety were likely to show reductions in WM capacity over time. Together, the mutual effects of algebraic anxiety and WM indicate WM and algebraic anxiety are likely to remain stable over time when students' initial anxiety is low and WM is high. Conversely, the students most likely to change are those who are initially disadvantaged: students with both high algebraic anxiety and low WM capacity were likely to further decline throughout the day, with WM reducing and anxiety levels increasing. In other words, the most disadvantaged students got worse over time.

The changes in students' algebraic anxiety and WM also interact to affect algebraic problem solving. Students with high WM were likely to show accurate problem solving, although if they also have high anxiety, problem solving is likely to be slower and less accurate with harder problems. Conversely, students with lower WM are likely to be inaccurate algebraic problem solvers. Crucially, students who have high anxiety in addition to lower WM are likely to show very fast, inaccurate problem solving, which suggests these students are "switching off" and avoiding algebra.

Insights Into Math Anxiety

The interaction between worry and WM to affect algebraic problem solving supports MA claims that anxiety affects math problem solving (Ashcraft & Faust, 1994; Ashcraft & Kirk, 2001; Beilock & Carr, 2005; Richardson & Suinn, 1972). Moreover, the findings support suggestions that math anxiety can change over time, and increases in MA are associated with poor math cognitive abilities (Maloney et al., 2010, 2011). In other words, individuals with poor WM/cognitive abilities are likely to show increases in anxiety over time. However, some individuals with high WM capacity also show high worry/anxiety, which affects their math performance. This might suggest that poor math cognition abilities may not lead to MA, contrasting with the development theory of MA (Maloney et al., 2010, 2011). However, given that math learning disorders, such as dyscalculia and math learning disabilities, are associated with high MA, it is likely that poor math cognitive abilities and/or poor math performance do contribute to MA (Lebens, Graff, & Mayer, 2011; Rubinsten & Tannock, 2010). More likely, the findings suggest that there are several causes of anxiety in math. Various factors have been proposed to contribute to the development of MA, including stereotype threat, poor math achievement, teacher's negative math attitudes, and poor cognitive abilities (Beilock, Rydell, & McConnell, 2007; Bleeker & Jacobs, 2004; Brien, Crandall, & O'Brien, 2003; Lebens et al., 2011; Lyons & Beilock, 2012; Maloney et al., 2011; Maloney & Beilock, 2012). It is possible that all or some of these factors interact to affect MA. Further research is required to characterize how different factors affect MA, and their relationships with math problem solving.

Interactions Between WM, Math Anxiety, and Math

Algebraic worry was found to interact with WM to affect algebraic problem-solving accuracy and response times. These findings support Beilock and Carr (2005) in showing that individuals' anxiety/worry interacts with WM to affect math problem solving. The interaction between WM and worry suggests that the effects of MA on math problem solving may vary between individuals. Moreover, under conditions of high worry, low WM, and high task demands, individuals show very fast and inaccurate algebraic problem solving. These findings support those of Ashcraft and Faust (1994), who found that a very high math-anxious group showed fast but inaccurate problem solving for demanding arithmetic problems.

Together, these studies show that the effect of MA and task demands on math problem solving depends on WM: high worry with high WM was associated with large response times and a nonsignificant increase with harder problems, indicating effortful processing; however, high worry with low WM was associated with very fast, inaccurate problem solving, which was not affected by problem difficulty, described by Ashcraft and Faust (1994) as avoidance.

Overall, the three studies suggest that examination of worry and task demands alone provides little insight into how worry/anxiety will affect math processing: the relationship is oversimplified. Rather, the results show a more complex relationship in which the effect of anxiety cannot accurately be characterized without accounting considerations of individuals' cognitive abilities. Overall, the findings suggest that *math performance is affected by interactions between anxiety and WM*, rather than anxiety influencing WM to affect math performance.

Implications

We see four implications of our findings. First, they provide insight into the responsive nature of math anxiety and cognitive functioning. Second, the research demonstrates the importance of understanding students' affect (e.g., anxiety) as well as their cognitive abilities in understanding and improving math achievement. Third, they provide insight into how students' ability to perform throughout a class or test may change over time. Our findings (Trezise & Reeve, 2014b, 2016) suggest that for high math-anxious or low WM capacity students, the ability to problem solve may decline as the problem-solving difficulty increases. Fourth, the findings provide some insight into how cognition and MA can be targeted to improve students' math problem solving.

Our findings show anxiety can change relatively quickly, which suggests that ways of managing anxiety during math problem solving (and math tests) need to be highlighted. Traditionally, treatments for math anxiety tend to be long-term interventions, such as reducing stereotype threat (Maloney & Beilock, 2012). Because our work shows changes within a short period of time, it underscores that interventions can directly address MA and/or WM within math classes. Moreover, our findings suggest that other factors (e.g., motivation) may interact with WM or MA to predict changes in WM/MA over time. Further work is required to understand the motivation linked to variation in WM, affect, and academic achievement (Brose, Schmiedek, Lövdén, & Lindenberger, 2012; Dweck, 1986) in relationship to WM, MA, and algebraic problem solving. Understanding predictors of change and the nature of variability in WM and worry may contribute to development of interventions that target WM and MA in order to improve math problem-solving abilities.

While WM and worry were shown to change over time, which in turn affected problem-solving ability, problem solving was only assessed once in each study. Further research is required to better understand how math problem-solving abilities change over time (e.g., throughout a math class). Moreover, while accuracy and speed were important predictors of math problem solving, it would also be useful to examine possible changes in strategy use. WM capacity has been associated with arithmetic strategy use (Imbo & Vandierendonck, 2007a, 2007b; Raghubar et al., 2010). Understanding how math problem-solving accuracy, strategies, and speed change over time may help to design educational approaches aimed at ameliorating possible negative effects of high anxiety or low WM.

Conclusion

Overall, the research demonstrated cumulative effects between WM and worry, which affected algebraic problem-solving abilities and resulted in changes in WM and worry over time. Subgroups of varying WM-worry relationships were identified, indicating heterogeneity in WM-MA interactions. Examination of influences between WM and worry over time revealed a mutual, iterative relationship. The mutual influences suggest WM and MA can change over a short period of time, and interindividual variability increases over time, such that the most disadvantaged individuals are likely to become more disadvantaged over time. Finally, WM-worry relationships were shown to change over time: high WM combined with low worry was associated with stability, whereas higher worry or lower WM was associated with declines in WM and possible increases in worry over time. These findings indicate that WM-MA relationships can change over time, and initial WM-MA relationships predict subsequent changes. Examination of how WM-worry relationships predicted algebraic problem solving revealed that WM and worry interact to predict speed and accuracy of problem solving. This supports research on the importance of WM in math problem solving (Raghubar et al., 2010) and the consequences of MA (Ashcraft & Faust, 1994; Ashcraft & Kirk, 2001; Beilock & Carr, 2005). The findings highlight that when examining the effect of anxiety or emotions on WM/cognition, overly simple may occur if the mutual influences between WM/cognition and anxiety/emotion is not accounted for (and vice versa).

References

Adelman, C. 1999. *Answers in the tool box: Academic intensity, attendance patterns, and bachelor's degree attainment*. Washington, DC: Office of Educational Research and Improvement, U.S. Department of Education. http://dx.doi.org/10.1002/yd.20075.

Alibali, M. W., Knuth, E. J., Hattikudur, S., McNeil, N. M., & Stephens, A. C. (2007). A longitudinal examination of middle school students' understanding of the equal sign and equivalent equations. *Thinking and Learning*, 9(3), 221–247. http://dx.doi.org/10.1080/10986060701360902.

Altman, D. G., & Royston, P. (2006). The cost of dichotomising continuous variables. *BMJ, 332*, 1080. http://dx.doi.org/10.1136/bmj.332.7549.1080.

Ashcraft, M. H. (2002). Math anxiety: Personal, educational, and cognitive consequences. *Current Directions in Psychological Science, 11*(5), 4. http://dx.doi.org/10.1111/1467-8721.00196.

Ashcraft, M. H., & Faust, M. W. (1994). Mathematics anxiety and mental arithmetic performance: An exploratory investigation. *Cognition & Emotion, 8*(2), 97–125. http://dx.doi.org/10.1080/02699939408408931.

Ashcraft, M. H., & Kirk, E. P. (2001). The relationships among working memory, math anxiety, and performance. *Journal of Experimental Psychology, General, 130*(2), 224–237. http://dx.doi.org/10.1037//0096-3445.130.2.224.

Baddeley, A. D. (2000). The episodic buffer: a new component of working memory? *Trends in Cognitive Sciences, 4*(11), 7. http://dx.doi.org/10.1016/S1364-6613(00)01538-2.

Barnes, M. A., Raghubar, K. P., English, L., Williams, J. M., Taylor, H., & Landry, S. (2014). Longitudinal mediators of achievement in mathematics and reading in typical and atypical development. *Journal of Experimental Child Psychology, 119*, 1–16. http://dx.doi.org/10.1016/j.jecp.2013.09.006.

Beilock, S. L., & Carr, T. H. (2005). When high-powered people fail: Working memory and "choking under pressure" in math. *Psychological Science, 16*(2), 101–105. http://dx.doi.org/10.1111/j.0956-7976.2005.00789.x.

Beilock, S. L., Rydell, R. J., & McConnell, A. R. (2007). Stereotype threat and working memory: Mechanisms, alleviation, and spillover. *Journal of Experimental Psychology, General, 136*(2), 256–276.

Bergman, L. R., & Magnusson, D. (1997). A person-oriented approach in research on developmental psychopathology. *Development and Psychopathology, 9*(2), 291–319. http://dx.doi.org/10.1017/S095457949700206X.

Bergman, L. R., Magnusson, D., & El-Khouri, B. M. (2003). Studying individual development in an interindividual context: A person-oriented approach. *Paths through life: Vol. 4*. Mahwah, NJ: Lawrence Erlbaum Associates Publishers.

Betz, N. E. (1978). Prevalence, distribution, and correlates of math anxiety in college students. *Journal of Counseling Psychology, 25*(5), 441–448. http://dx.doi.org/10.1037/0022-0167.25.5.441.

Bieri, D., Reeve, R. A., Champion, G. D., Addicoat, L., & Ziegler, J. B. (1990). The Faces Pain Scale for the self-assessment of the severity of pain experienced by children: Development, initial validation, and preliminary investigation for ratio scale properties. *Pain, 41*(2), 139–150. http://dx.doi.org/10.1016/0304-3959(90)90018-9.

Bleeker, M. M., & Jacobs, J. E. (2004). Achievement in math and science: Do mothers' beliefs matter 12 years later? *Journal of Educational Psychology, 96*(1), 97–109. http://dx.doi.org/10.1037/0022-0663.96.1.97.

Brien, L. T. O., Crandall, C. S., & O'Brien, L. T. (2003). Stereotype threat and arousal: Effects on women's math performance. *Personality and Social Psychology Bulletin, 29*(6), 782–789. http://dx.doi.org/10.1177/0146167203252810.

Brose, A., Schmiedek, F., Lövdén, M., & Lindenberger, U. (2012). Daily variability in working memory is coupled with negative affect: The role of attention and motivation. *Emotion (Washington, DC), 12*(3), 605–617. http://dx.doi.org/10.1037/a0024436.

Capraro, M. M., Capraro, R. M., & Henson, R. K. (2001). Measurement error of scores on the Mathematics anxiety rating scale across studies. *Educational and Psychological Measurement, 61*(3), 373–386. http://dx.doi.org/10.1177/00131640121971266.

Carnevale, A. P., & Desrochers, D. M. (2003). *Standards for what? The economic roots of K–16 reform*. Washington, DC: Educational Testing Service.

Chinn, S. (2009). Mathematics anxiety in secondary students in England. *Dyslexia, 15*(1), 61–68. http://dx.doi.org/10.1002/dys.381.

Collins, L. M., & Lanza, S. T. (2010). *Latent class and latent transition analysis: With applications in the social, behavioral, and health sciences*. Hoboken, NJ: Wiley and Sons. http://dx.doi.org/10.1002/9780470567333.

Conway, A. R. A., Kane, M., Bunting, M., Hambrick, D. Z., Wilhelm, O., & Engle, R. W. (2005). Working memory span tasks: A methodological review and user's guide. *Psychonomic Bulletin & Review*, *12*(5), 769. http://dx.doi.org/10.3758/BF03196772.

Daneman, M., & Carpenter, P. (1980). Individual differences in working memory and reading. *Journal of Verbal Learning and Verbal Behavior*, *19*, 450–466. http://dx.doi.org/10.1016/S0022-5371(80)90312-6.

Derakshan, N., & Eysenck, M. W. (2009). Anxiety, processing efficiency, and cognitive performance. *European Psychologist*, *14*(2), 168–176. http://dx.doi.org/10.1027/1016-9040.14.2.168.

DeStefano, D., & LeFevre, J. A. (2004). The role of working memory in mental arithmetic. *European Journal of Cognitive Psychology*, *16*(3), 353–386. http://dx.doi.org/10.1080/09541440244000328.

Dooren, W. Van, Verschaffel, L., & Onghena, P. (2002). The impact of preservice teachers' content knowledge on their evaluation of students' strategies for solving arithmetic and algebra word problems. *Journal for Research in Mathematics Education*, *33*(5), 319–351. http://dx.doi.org/10.2307/4149957.

Dowker, A., Bennett, K., & Smith, L. (2012). Attitudes to mathematics in primary school children. *Child Development Research*, *2012*, 1–8. http://dx.doi.org/10.1155/2012/124939.

Dweck, C. S. (1986). Motivational processes affecting learning. *American Psychologist*, *41*(10), 1040–1048. http://dx.doi.org/10.1037/0003-066x.41.10.1040.

Engle, R. W. (2010). Role of working memory capacity in cognitive control. *Current Anthropology*, *51*(S1), S17–S26. Retrieved from http://www.jstor.org/stable/10.1086/650572.

Evan, A., Gray, T., & Olchefske, J. (2006). *The gateway to student success in mathematics and science*. Washington, DC: American Institutes for Research.

Eysenck, M. W., Derakshan, N., Santos, R., & Calvo, M. G. (2007). Anxiety and cognitive performance: Attentional control theory. *Emotion*, *7*(2), 336–353. http://dx.doi.org/10.1037/1528-3542.7.2.336.

Faust, M. W., Ashcraft, M. H., & Fleck, D. E. (1996). Mathematics anxiety effects in simple and complex addition. *Mathematical Cognition*, *2*(1), 25–62. http://dx.doi.org/10.1080/135467996387534.

Fletcher, J. M., Marks, A. D. G., & Hine, D. W. (2012). Latent profile analysis of working memory capacity and thinking styles in adults and adolescents. *Journal of Research in Personality*, *46*(1), 40–48. http://dx.doi.org/10.1016/j.jrp.2011.11.003.

Gandal, M., Kraman, J., Pelzman, A., & McGiffert Slover, L. (2006). *Closing the expectations gap 2006: An annual 50-state progress report on the alignment of high school policies with the demands of college and work*. Washington, DC: Achieve, Inc.

Geary, D. C., Brown, S. C., & Samaranayake, V.a. (1991). Cognitive addition: A short longitudinal study of strategy choice and speed-of-processing differences in normal and mathematically disabled children. *Developmental Psychology*, *27*(5), 787–797. http://dx.doi.org/10.1037//0012-1649.27.5.787.

Geary, D. C., Hoard, M. K., Byrd-Craven, J., Nugent, L., & Numtee, C. (2007). Cognitive mechanisms underlying achievement deficits in children with mathematical learning disability. *Child Development*, *78*(4), 1343–1359. http://dx.doi.org/10.1111/j.1467-8624.2007.01069.x.

Grimm, K. J., An, Y., McArdle, J. J., Zonderman, A. B., & Resnick, S. M. (2012). Recent changes leading to subsequent changes: Extensions of multivariate latent difference score models. *Structural Equation Modeling: A Multidisciplinary Journal*, *19*(2), 268–292. http://dx.doi.org/10.1080/10705511.2012.659627.

Hayes, S., Hirsch, C., & Mathews, A. (2008). Restriction of working memory capacity during worry. *Journal of Abnormal Psychology*, *117*(3), 712–717. http://dx.doi.org/10.1037/a0012908.

Hecht, S. A. (2002). Counting on working memory in simple arithmetic when counting is used for problem solving. *Memory & Cognition*, *30*(3), 447–455. http://dx.doi.org/10.3758/BF03194945.

Hopko, D. R., McNeil, D. W., Lejuez, C. W., Ashcraft, M. H., Eifert, G. H., & Riel, J. (2003). The effects of anxious responding on mental arithmetic and lexical decision task performance. *Journal of Anxiety Disorders, 17*(6), 647–665. http://dx.doi.org/10.1016/s0887-6185(02)00240-2.

Humberstone, J., & Reeve, R. A. (2008). Profiles of algebraic competence. *Learning and Instruction, 18*(4), 354–367. http://dx.doi.org/10.1016/j.learninstruc.2007.07.002.

Imbo, I., & LeFevre, J.-A. (2010). The role of phonological and visual working memory in complex arithmetic for Chinese- and Canadian-educated adults. *Memory & Cognition, 38*(2), 176–185. http://dx.doi.org/10.3758/MC.38.2.176.

Imbo, I., & Vandierendonck, A. (2007a). Do multiplication and division strategies rely on executive and phonological working memory resources? *Memory & Cognition, 35*(7), 1759–1771. http://dx.doi.org/10.3758/BF03193508.

Imbo, I., & Vandierendonck, A. (2007b). The development of strategy use in elementary school children: Working memory and individual differences. *Journal of Experimental Child Psychology, 96*(4), 284–309. http://dx.doi.org/10.1016/j.jecp.2006.09.001.

Irwin, J. R., & Mcclelland, G. H. (2003). Negative consequences of dichotomizing continuous predictor variables. *Journal of Marketing Research, 40*(3), 366–371. http://dx.doi.org/10.1509/jmkr.40.3.366.19237.

Kellogg, J. S., Hopko, D. R., & Ashcraft, M. H. (1999). The effects of time pressure on arithmetic performance. *Journal of Anxiety Disorders, 13*(6), 591–600. http://dx.doi.org/10.1016/S0887-6185(99)00025-0.

Knuth, E. J., Alibali, M. W., McNeil, N. M., Weinberg, A., & Stephens, A. C. (2005). Middle school students' understanding of core algebraic concepts: Equivalence and variable. *ZDM, 37*(1), 68–76. http://dx.doi.org/10.1007/BF0255899.E.J.

Küchemann, D. (1981). Algebra. In K. M. Hart (Ed.), (Vols. 11–16). *Children's understanding of mathematics* (pp. 102–119). London: John Murray.

Lebens, M., Graff, M., & Mayer, P. (2011). The affective dimensions of mathematical difficulties in schoolchildren. *Education Research International, 2011*, 1–13. http://dx.doi.org/10.1155/2011/487072.

Lee, K., Ng, E. L., & Ng, S. F. (2009). The contributions of working memory and executive functioning to problem representation and solution generation in algebraic word problems. *Journal of Educational Psychology, 101*(2), 373–387. http://dx.doi.org/10.1037/a0013843.

LeFevre, J.-A., DeStefano, D., Coleman, B., & Shanahan, T. (2005). Mathematical cognition and working memory. In J. I. D. Campbell (Ed.), *Handbook of mathematical cognition* (pp. 361–378). New York, NY: Psychology Press.

Liebert, R. M., & Morris, L. W. (1967). Cognitive and emotional components of test anxiety: A distinction and some initial data. *Psychological Reports, 20*(3 PT.1), 975–978. http://dx.doi.org/10.2466/pr0.1967.20.3.975.

Lyons, I. M., & Beilock, S. L. (2012). Mathematics anxiety: Separating the math from the anxiety. *Cerebral Cortex, 22*(9), 2102–2110. http://dx.doi.org/10.1093/cercor/bhr289.

Ma, X., & Xu, J. (2004). The causal ordering of mathematics anxiety and mathematics achievement: A longitudinal panel analysis. *Journal of Adolescence, 27*(2), 165–179. http://dx.doi.org/10.1016/j.adolescence.2003.11.003.

MacCallum, R. C., Zhang, S., Preacher, K. J., & Rucker, D. D. (2002). On the practice of dichotomization of quantitative variables. *Psychological Methods, 7*(1), 19–40. http://dx.doi.org/10.1037//1082-989X.7.1.19.

Magidson, J., & Vermunt, J. K. (2003). A nontechnical introduction to latent class models. *DMA Research Council Journal*. Retrieved from http://www.statisticalinnovations.com/articles/lcmodels2.pdf.

Maloney, E. A., Ansari, D., & Fugelsang, J. A. (2011). The effect of mathematics anxiety on the processing of numerical magnitude. *The Quarterly Journal of Experimental Psychology, 64*(1), 10–16. http://dx.doi.org/10.1080/17470218.2010.533278.

Maloney, E. A., & Beilock, S. L. (2012). Math anxiety: Who has it, why it develops, and how to guard against it. *Trends in Cognitive Sciences*, *16*(8), 404–406. http://dx.doi.org/10.1016/j.tics.2012.06.008.

Maloney, E. A., Risko, E. F., Ansari, D., & Fugelsang, J. (2010). Mathematics anxiety affects counting but not subitizing during visual enumeration. *Cognition*, *114*(2), 293–297. http://dx.doi.org/10.1016/j.cognition.2009.09.013.

Mammarella, I. C., Hill, F., Devine, A., Caviola, S., Mammarella, I. C., Hill, F., et al. (2015). Math anxiety and developmental dyscalculia: A study on working memory processes. *Journal of Clinical and Experimental Neuropsychology*. http://dx.doi.org/10.1080/13803395.2015.1066759.

McArdle, J. J. (2009). Latent variable modeling of differences and changes with longitudinal data. *Annual Review of Psychology*, *60*, 577–605. http://dx.doi.org/10.1146/annurev.psych.60.110707.163612.

McGregor, E. B. (1994). Economic development and public education: Strategies and standards. *Educational Policy*, *18*(3), 252–271. http://dx.doi.org/10.1177/0895904894008003001.

Metcalfe, A. W. S., Ashkenazi, S., Rosenberg-Lee, M., & Menon, V. (2013). Fractionating the neural correlates of individual working memory components underlying arithmetic problem solving skills in children. *Developmental Cognitive Neuroscience*, 1–14. http://dx.doi.org/10.1016/j.dcn.2013.10.001.

Miller, H., & Bichsel, J. (2004). Anxiety, working memory, gender, and math performance. *Personality and Individual Differences*, *37*(3), 591–606. http://dx.doi.org/10.1016/j.paid.2003.09.029.

Miyake, A., & Shah, P. (1999). Toward unified theories of working memory: Emerging general consensus, unresolved theoretical issues, and future research directions. In A. Miyake & P. Shah (Eds.), *Models of working memory: Mechanisms of active maintenance and executive control* (pp. 442–481). Cambridge, UK: Cambridge University Press.

Morris, L. W., Davis, M. A., & Hutchings, C. H. (1981). Cognitive and emotional components of anxiety: Literature review and a revised worry-emotionality scale. *Journal of Educational Psychology*, *73*(4), 541–555. http://dx.doi.org/10.1037/0022-0663.73.4.541.

Morris, L. W., Kellaway, D. S., & Smith, D. H. (1978). Mathematics anxiety rating scale: Predicting anxiety experiences and academic performance in two groups of students. *Journal of Educational Psychology*, *70*(4), 589–594. http://dx.doi.org/10.1037/0022-0663.70.4.589.

National Mathematics Advisory Panel. 2008. *Foundations for success: The final report of the national mathematics advisory panel*.

Ng, L. K. (2012). Mathematics anxiety in secondary school students. In *35th annual conference of the mathematics education research group of Australasia incorporated (MERGA 2012) on "Mathematics education: Expanding horizons, Singapore* (pp. 570–577).

Oberauer, K. (2009). Design for a working memory. *Psychology of Learning and Motivation*, *51*, 45–100. http://dx.doi.org/10.1016/S0079-7421(09)51002-X.

Owens, M., Stevenson, J., Hadwin, J.a., & Norgate, R. (2012). Anxiety and depression in academic performance: An exploration of the mediating factors of worry and working memory. *School Psychology International*, *33*(4), 433–449. http://dx.doi.org/10.1177/0143034311427433.

Passolunghi, M. C., & Mammarella, I. C. (2010). Spatial and visual working memory ability in children with difficulties in arithmetic word problem solving. *European Journal of Cognitive Psychology*, *22*(6), 944–963. http://dx.doi.org/10.1080/09541440903091127.

Pastor, D. A., Barron, K. E., Miller, B. J., & Davis, S. L. (2007). A latent profile analysis of college students' achievement goal orientation. *Contemporary Educational Psychology*, *32*(1), 8–47. http://dx.doi.org/10.1016/j.cedpsych.2006.10.003.

Preacher, K. J., Rucker, D. D., MacCallum, R. C., & Nicewander, W. A. (2005). Use of the extreme groups approach: A critical reexamination and new recommendations. *Psychological Methods*, *10*(2), 178–192. http://dx.doi.org/10.1037/1082-989X.10.2.178.

Punaro, L., & Reeve, R. A. (2012). Relationships between 9-year-olds' math and literacy worries and academic abilities. *Child Development Research, 2012*, Article ID 359089, 11 pages, http://dx.doi.org/10.1155/2012/359089.

Raghubar, K. P., Barnes, M. A., & Hecht, S. A. (2010). Working memory and mathematics: A review of developmental, individual difference, and cognitive approaches. *Learning and Individual Differences*, *20*(2), 110–122. http://dx.doi.org/10.1016/j.lindif.2009.10.005.

Reeve, R. A., Reynolds, F., Humberstone, J., & Butterworth, B. (2012). Stability and change in markers of core numerical competencies. *Journal of Experimental Psychology, General*, *141*(4), 649–666. http://dx.doi.org/10.1037/a0027520.

Richardson, F. C., & Suinn, R. M. (1972). The mathematics anxiety rating scale: Psychometric data. *Journal of Counseling Psychology*, *19*(6), 551–554. http://dx.doi.org/10.1037/h0033456.

Rubinsten, O., & Tannock, R. (2010). Mathematics anxiety in children with developmental dyscalculia. *Behavioral and Brain Functions*, *6*, 46. http://dx.doi.org/10.1186/1744-9081-6-46.

Ruscio, J., & Ruscio, A. M. (2008). Categories and dimensions advancing psychological science through the study of latent structure. *Current Directions in Psychological Science*, *17*(3), 203–207. http://dx.doi.org/10.1111/j.1467-8721.2008.00575.x.

Suinn, R. M., Edie, C. A., Nicoletti, J., & Spinelli, P. R. (1972). The MARS, a measure of mathematics anxiety: Psychometric data. *Journal of Clinical Psychology*, *28*(3), 373–375. http://dx.doi.org/10.1002/1097-4679(197207)28:3+<373::aid-jclp2270280320>3.0.co;2-c.

Suinn, R. M., & Richardson, F. (1971). Anxiety management training: A nonspecific behavior therapy program for anxiety control. *Behavior Therapy*, *2*(4), 498–510. http://dx.doi.org/10.1016/s0005-7894(71)80096-5.

Susac, A., Bubic, A., Vrbanc, A., & Planinic, M. (2014). Development of abstract mathematical reasoning: The case of algebra. *Frontiers in Human Neuroscience*, *8*, 679. http://dx.doi.org/10.3389/fnhum.2014.00679.

Swanson, H. L. (2006). Cross-sectional and incremental changes in working memory and mathematical problem solving. *Journal of Educational Psychology*, *98*(2), 265–281. http://dx.doi.org/10.1037/0022-0663.98.2.265.

Tolar, T. T. D., Lederberg, A. A. R., & Fletcher, J. M. (2009). A structural model of algebra achievement: Computational fluency and spatial visualisation as mediators of the effect of working memory on algebra achievement. *Educational Psychology*, *29*(2), 239–266. http://dx.doi.org/10.1080/01443410802708903.

Trezise, K., & Reeve, R. A. (2014a). Cognition-emotion interactions: Patterns of change and implications for math problem solving. *Frontiers in Psychology*, *5*, 1–15. http://dx.doi.org/10.3389/fpsyg.2014.00840.

Trezise, K., & Reeve, R. A. (2014b). Working memory, worry, and algebraic ability. *Journal of Experimental Child Psychology*, *121C*, 120–136. http://dx.doi.org/10.1016/j.jecp.2013.12.001.

Trezise, K., & Reeve, R. A. (2016). Worry and working memory influence each other iteratively over time. *Cognition and Emotion*, *30*(2), 353–368. http://dx.doi.org/10.1080/02699931.2014.1002755.

Trezise, K., & Reeve, R.A. (under review). Characterizing math anxiety experienced solving algebra problem using three-step latent variable analysis.

Uusimaki, L., & Nason, R. (2004). Causes underlying pre-service teachers' negative beliefs and anxieties about mathematics. In *Proceedings of the 28th conference of the international group for the psychology of mathematics education* (pp. 369–376).

Van Dillen, L. F., & Koole, S. L. (2007). Clearing the mind: A working memory model of distraction from negative mood. *Emotion*, *7*(4), 715–723. http://dx.doi.org/10.1037/1528-3542.7.4.715.

Van Dillen, L. F., & Koole, S. L. (2009). How automatic is "automatic vigilance"? The role of working memory in attentional interference of negative information. *Cognition and Emotion*, *23*(6), 1106–1117. http://dx.doi.org/10.1080/02699930802338178.

Vermunt, J. K., & Magidson, J. (2013). *Technical guide for latent GOLD 5.0: Basic, advanced, and syntax*. Belmont, MA: Statistical Innocations Inc.

von Eye, A., & Bergman, L. R. (2003). Research strategies in developmental psychopathology: Dimensional identity and the person-oriented approach. *Development and Psychopathology, 15*(3), 553–580. http://dx.doi.org/10.1017/S0954579403000294.

Xenidou-Dervou, I., van Lieshout, E.C.D.M., & van der Schoot, M. (2014). Working memory in nonsymbolic approximate arithmetic processing: A dual-task study with preschoolers. *Cognitive Science, 38*(1), 101–127. http://dx.doi.org/10.1111/cogs.12053.

SECTION III

EMOTIONS IN THE LEARNING AND TEACHING OF MATHEMATICS

SECTION IIIA

LEARNERS IN DIFFERENT EDUCATIONAL LEVELS

CHAPTER

6

Students' Emotional Experiences Learning Mathematics in Canadian Schools

Jo Towers, Miwa A. Takeuchi*, Jennifer Hall[†,a], Lyndon C. Martin[‡]*

*University of Calgary, Calgary, AB, Canada
[†]Monash University, Melbourne, VIC, Australia
[‡]York University, Toronto, ON, Canada

INTRODUCTION

This chapter offers insight into the complex relationship between emotion and learning in the domain of mathematics education. To better understand this relationship, we designed a project to explore (1) students' lived experiences of learning mathematics in Canadian schools; (2) the images of mathematics that students are developing in schools and how they persist over time; (3) the nature of students' mathematical identities, including students' emotional relationships with mathematics and the emotions experienced while doing mathematics; and (4) the role that teachers and schools play in shaping these identities, relationships, and emotions. The first phase of the research, on which we report here, concerned Kindergarten to Grade 9 students' experiences while later phases addressed secondary and postsecondary students' experiences, as well as those of members of the general public. Preliminary findings of Phase 1 of the study, dealing primarily with students' images of mathematics, appear elsewhere (Hall & Towers, 2014; Hall, Towers, & Martin, 2015; Hall, Towers, Takeuchi, & Martin, 2015; Plosz, Towers, & Takeuchi, 2015; Takeuchi & Towers, 2015; Towers, Hall, & Martin, 2015; Towers, Takeuchi,

[a] Research conducted while at the University of Calgary.

http://dx.doi.org/10.1016/B978-0-12-802218-4.00006-6

Hall, & Martin, 2015). In this chapter, we draw on Kindergarten to Grade 9 students' autobiographical accounts of learning mathematics in schools to explore, in particular, students' relationships with mathematics and the emotions associated with doing mathematics.

REVIEW OF THE LITERATURE

Mathematics education research about affective factors has been conducted since the 1960s, with early studies focusing on mathematics anxiety and attitudes toward mathematics (Zan, Brown, Evans, & Hannula, 2006). Two instruments developed in the 1970s, the Mathematics Anxiety Rating Scale (Richardson & Suinn, 1972) and the Mathematics Attitude Scales (Fennema & Sherman, 1976), are still widely used, particularly as affect becomes an increasing focus in mathematics education research. This focus on affect has been attributed to two underlying arguments: (1) attitudes toward mathematics are related to achievement in mathematics and (2) attitudes toward mathematics are significant in their own right (Di Martino & Zan, 2011; Zan et al., 2006). George (2009) posits that "the nature of participation in mathematics impinges on the relationship/identity a person develops with the subject" (p. 207). In addition to research conducted by mathematics education researchers, cognitive psychologists have also investigated affective factors in mathematics education. However, this research tends to be more narrowly focused, both in terms of methodology (quantitative) and topic focus (negative topics, such as mathematics anxiety) (e.g., Ahmed, Minnaert, Kuyper, & van der Werf, 2012; Ashcraft & Kirk, 2001; Young, Wu, & Menon, 2012). Understanding a wider breadth of students' emotional connections to mathematics is essential for designing mathematics instruction and learning environments that foster students' positive dispositions for learning mathematics (Boaler, 2011).

Feelings About Mathematics

As noted, a great deal of research has investigated students' affective relationships with mathematics. Much of this research involves the use of quantitative questionnaires (i.e., those involving closed question types, such as Likert scales), including the Mathematics Anxiety Rating Scale (Richardson & Suinn, 1972) and the Mathematics Attitude Scales (Fennema & Sherman, 1976), though affect has also attracted the attention of researchers who draw on qualitative approaches. For example, in Hall's (2013) research, most participants expressed that they liked mathematics, felt confident in their mathematics abilities, and recognized the importance of mathematics. While Hall did not find any gender differences in these views, the younger participants (Grade 4 students) in her study expressed more positive views than did the older participants (Grade 8 students). This lack of gender difference stands in contrast to other research, which typically reports that boys feel more confident in their mathematics abilities and like mathematics more than do girls, while the age-related differences in views are consistent with other research (e.g., Hall, 2012; Lubienski, Robinson, Crane, & Ganley, 2013; Lupart, Cannon, & Telfer, 2004). Other aspects of students' identities, including racialized identities, can also impact students' relationships with mathematics. Berry, Thunder, and McClain's (2011) research involved gathering mathematics autobiographies from Black boys in Grades 5–7 in the United States who were top achievers in mathematics. In these reflective pieces, the boys shared that their relationships with mathematics were highly influenced by external recognition (e.g., praise from parents and teachers), and their identities as Black boys, both from a racialized and gendered standpoint, caused them to feel a sense of "otherness" as mathematically proficient students.

Images of Mathematics

An important emerging strand of research concerns students' views about the nature of mathematics and mathematical practices. Perkkilä and Aarnos (2009) addressed this topic and explored the views of Finnish children, aged 6–8, by asking them to draw themselves in the "land of mathematics." The drawings tended to show the students alone, working on arithmetic or number facts, in a nature-based setting. These findings indicate that students view mathematics as a solitary pursuit involving only certain types of mathematics. Other researchers have also investigated students' views about the nature of mathematics, though typically without including drawing as a data collection tool. For example, research conducted in New Zealand (Young-Loveridge, Taylor, Sharma, & Hawera, 2006), involving interviews with elementary students, found that students tended to view mathematics as being related to number and/or operations, as opposed to other mathematical topics, a finding that is consistent with our own research and that we have described elsewhere (e.g., Plosz et al., 2015; Towers, Hall, et al., 2015). Similarly narrow views of mathematics were also found in Hall's (2013) research with elementary school students in Canada. Students in Hall's (2013) research, which involved questionnaires, drawings of mathematicians, and focus group interviews with media prompts, tended to conceptualize mathematics as comprising either number sense and numeration concepts, particularly arithmetic, or financial mathematics. Hall suggested that these narrow views were influenced by the students' interactions with their teachers, parents, and popular media.

Suggestions to Improve Students' Emotional Relationships with Mathematics

Recently, researchers have started to examine the conditions and instructional contexts in which students' emotional reactions to mathematics can be altered (Evans, Morgan, & Tsatsaroni, 2006; Hannula, 2002; Vandecandelaere, Speybroeck, Vanlaar, De Fraine, & Van Damme, 2012). Various suggestions have been made for classroom practices that may help to better engage students in mathematics and consequently improve their emotional relationships with the subject area. In some examples, the suggestions relate to classroom practices of teaching mathematics (with the implication that improved teaching will lead to improved student experience and emotional relationships with mathematics), whereas other examples suggest addressing students' emotional relationships with mathematics directly. As an instance of the former, recent Canadian research in a Grade 10 classroom (Liljedahl, 2014) found that using random groupings during mathematics class led to increased student interest in and more positive views of mathematics because students enjoyed the element of randomness and surprise in each day's mathematics class. Conversely, Stogsdill (2013) recommends directly addressing students' negative relationships with mathematics, particularly mathematics anxiety, by having students write mathematics autobiographies. Stogsdill suggests that students will be able to share their stories with each other, thus validating their experiences and providing support to help each other begin to move on from their anxiety. While these are interesting suggestions, investigations into students' affective relationships with mathematics are still mostly limited to single case studies (e.g., Evans et al., 2006; Hannula, 2002) or survey-based studies (e.g.,Vandecandelaere et al., 2012). Our study extends these earlier investigations through the analysis of a large number of autobiographical narratives from students from Kindergarten to postsecondary levels.

THEORETICAL FRAMEWORK

This research is framed by enactivism, a theory of embodied cognition that emphasizes the interrelationship of cognition and emotion in learning (Kieren & Sookochoff, 1999; Maturana & Varela, 1992; Varela, Thompson, & Rosch, 1991) and troubles the positioning of self and identity as purely individual and static phenomena (Varela et al., 1991). Enactivism recognizes teachers, students, and the learning environment as structurally coupled (Maturana & Varela, 1992) and therefore learning, in this frame, is seen as reciprocal activity. Working in concert with the mathematical environment, the teacher brings forth a world of significance with the learners.

Students' mathematical identities are therefore not determined (solely) by the teacher, but they are dependent on the kind of teaching experienced and the kind of mathematical milieu in which students are immersed. Enactivist thought reorients us to the significance of this mathematical milieu in shaping not only what students learn in school but also their emotional connections with the discipline. This enactivist frame, then, prompts us to seek to understand how students come to have particular relationships with mathematics, how students come to be with mathematics, and what being mathematical means to them.

RESEARCH DESIGN

Our research design is framed by a narrative inquiry of identity and experience. Narrative inquiry is a "profoundly relational form of inquiry" (Clandinin, 2007, p. xv). It is "sensitive to…subtle textures of thought and feeling" (Webster & Mertova, 2007, p. 7) at play in an account of lived experiences and allows for the events that have been of most significance to be brought forth in the stories participants tell. Narrative inquiry focuses "not only on individuals' experiences but also on the social, cultural, and institutional narratives within which individuals' experiences are constituted, shaped, expressed, and enacted" (Clandinin & Rosiek, 2007, pp. 42–43). Narrative inquiry was chosen in order to provide the flexibility for participants to tell stories about their experiences learning mathematics without undue direction from the researchers. Guided by our enactivist framing, we wanted to explore not only participants' emotional connections to mathematics, but also the sources (social, institutional, familial, etc.) of those connections. We also wanted to be able to examine each student's personal identification as a learner of mathematics (i.e., whether an individual presented him/herself, autobiographically, as, e.g., "a math person"). Hence, our research design emphasized opportunities for participants to reflect on their emotional connections to mathematics and on the evolution of their identities in relation to the discipline. We took the rather unusual step of asking participants to draw as part of their autobiographical reflections, which is not a typical component of narrative inquiry research. This component of the research added considerable fidelity to the oral and written data, reinforcing through imagery many of the issues that participants discussed. In some cases, though, drawings in response to the prompt of "How do you feel when doing mathematics" revealed emotions that contrasted with participants' stated relationships with mathematics (e.g., a participant may have declared him/herself to be reasonably confident when doing mathematics but created a drawing that depicted a lack of confidence). While a detailed examination of such methodological issues regarding the use of drawings in exploring students' emotional relationships with mathematics is beyond the scope of this chapter, we are in the process of providing such an examination to be published elsewhere (Hall & Towers, 2017).

The data for the study were gathered primarily in the Canadian provinces of Alberta and Ontario, and all participants were volunteers. Parental consent was given for minors. Research participants were invited to provide written, oral, or multimodal "mathematics autobiographies"—firsthand accounts of the experience of learning mathematics—and to provide drawings depicting (1) how they feel when doing mathematics and (2) their ideas about what mathematics is. For Kindergarten to Grade 9 students, these mathematics autobiographies were collected orally during face-to-face interviews with members of the research team. For secondary and postsecondary students and members of the general public, we drew on a wider range of data collection techniques, including soliciting written mathematics autobiographies as well as multimodal mathematics autobiographies submitted through an online environment. Findings concerning these alternative forms of mathematics autobiographies are in process (see, e.g., Takeuchi, Czuy, & Towers, 2016). Data collection for this project is nearing completion. We have conducted 124 face-to-face interviews with students from Kindergarten to postsecondary (94 Kindergarten to Grade 9 students, 24 Grades 10–12 students, and 6 postsecondary students) and, through an online environment, collected written or multimedia mathematics autobiographies from 94 members of the general public, most of whom were also postsecondary students. In this chapter, we focus on interviews with the 94 students from Kindergarten to Grade 9 classrooms in Alberta (46 Kindergarten to Grade 3 students, 30 Grades 4–6 students, and 18 Grades 7–9 students).

Transcriptions of the interviews and the written and multimedia mathematics autobiographies were analyzed using the qualitative analysis software, NVivo. Qualitative analysis of transcripts was conducted by a group of researchers in the team, which includes the authors of this chapter as well as graduate student research assistants. As an entry point to students' emotional experiences with mathematics, we began by analyzing participants' verbal responses to the interview question, "How do you feel when you are doing mathematics?" For the first round of analysis, we focused on distinguishing students' positive and negative relationships with mathematics. Positive relationships were identified when students reported that they felt good and/or confident about mathematics, liked mathematics, and/or found mathematics to be fun, enjoyable, and/or exciting. In contrast, negative relationships were identified when students reported that they felt confused about mathematics, did not like mathematics, and/or reported themselves to be struggling and/or bored. We used the positive/negative binary to give a fairly coarse-grained initial overview of the emotional relationships students constructed with regard to mathematics. In order to visually and artistically represent these relationships, we constructed Wordle images mainly focusing on verbs and adjectives used in students' responses to the question, "How do you feel when you are doing mathematics?"

From the overview, we moved on to a more nuanced analysis, which is sensitive to subtle textures of students' feelings. This phase of analysis involved examining students' narratives in their entirety, including the drawings that students generated, in order to pay careful attention to details of their experiences. Here, we focused specifically on the complex and occasionally contradictory emotional relationships that students have with mathematics. Our analysis in this phase revealed a mixture of positive and negative feelings toward mathematics and, for some participants, a change in their emotional relationship with mathematics over time.

Drawings were analyzed thematically, with repeated sortings, using guiding themes developed by the research team. These themes included analyzing the drawings by grade level, by the gender identity of the student, and by the category of features included in the drawing (e.g., representations of school mathematics environments or representations of individuals vs groups of students), to name just a few of the analytical themes we addressed. Drawings were also analyzed in relation to the interview narratives, with particular attention paid to differences between the emotions described in speech and those portrayed through drawing.

FINDINGS

Our analysis revealed compelling findings concerning participants' emotional relationships with mathematics. In response to the question, "How do you feel when you are doing mathematics?" 43.6% of the Kindergarten to Grade 9 participants ($n=41$) described their relationships with mathematics as positive, while 17% ($n=16$) reported negative relationships. However, for many students, their broader narratives could not be categorized solely into this positive-negative binary. Even when students reported that they felt positively or negatively about mathematics in response to this particular interview question, some provided more nuanced (and sometimes contradictory) accounts elsewhere in the interview. Also, the ways in which students described their positive or negative feelings toward mathematics were not monolithic—there were many subtleties in students' characterizations of their relationships with mathematics. In the sections that follow, we will first document how students described positive and negative relationships with mathematics, respectively. Subsequently, we will illustrate more complex and subtle relationships that students developed: mixed feelings and changing/changed relationships.

Positive Relationships With Mathematics

The Wordle image (Fig. 1) illustrates participants' descriptions of their positive relationships with mathematics. These students reported that they feel good, confident, and happy when doing mathematics. Interestingly, these students' answers tended to be shorter and less descriptive, compared to students who reported negative relationships with mathematics. Few of the students with reported positive relationships with mathematics elaborated on how they built their current relationship with mathematics.

There were common characteristics observed across students' narratives. One commonality that stood out to us was how students' relationships with mathematics were tied up with their beliefs about the nature of mathematics and/or mathematics learning. As discussed in more detail elsewhere (Plosz et al., 2015; Towers, Hall, et al., 2015; Towers, Takeuchi, et al., 2015), the majority of students held an image of mathematics as number and/or operations. Many students described their positive relationships with mathematics by narrowly focusing on their competence in number sense. For example, students said: "I feel happy because I like numbers" (Grade 6 boy) or "(I feel) proud. Because I do a really large number for division... and multiplication, also sum" (Grade 5 boy).

For many students, feelings about mathematics were grounded in external validation (e.g., teacher and/or parental praise, test scores, validation for speed of completion of work, public acknowledgement of their competence). Representative comments from students include the following: "My mum thinks [mathematics] is pretty good and that ah, I should, when I get better [at mathematics] I need to stay that way" (Grade 5 girl); and "Math, it feels good to do because I'm really good at it" (Grade 5 boy). In describing a favorite aspect of mathematics, another Grade 5 girl said, "I like it [addition] and I do it really fast."

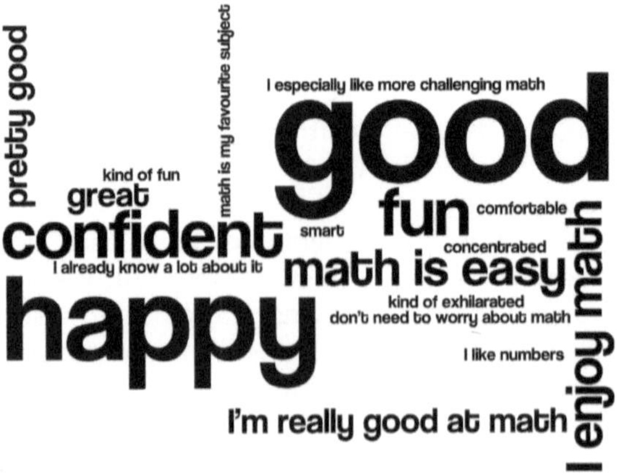

FIG. 1 Wordle image of participants' positive emotional relationships.

There were a few students who embraced incremental challenges in problem solving. These students enjoyed the challenging nature of mathematics. For example, a Grade 7 girl said, "I kind of want to do more and I want harder questions that make me think about it and go into it. I want to be in the problem." Similarly, a Grade 6 boy described his passion for mathematics: "I got it in fifth grade because the equations were getting harder and I love more challenge in math." This student wished for more challenging problems in mathematics and hoped that "there could be more websites that could be used to make math challenging but more fun." It is noteworthy that these kinds of narratives were rather rare in our sample. The majority of students liked mathematics because they found it to be easy or because they "understand everything" (Grade 7 boy).

In describing how they feel when they are doing mathematics, many students, across all grades, drew a picture of a smiling face and explained that it represented happy feelings (as seen in Fig. 2 [left]). This iconic image was the most commonly observed visual representation among students who reported positive feelings. Some students added a commentary to their drawing to emphasize their positive feelings (e.g., "I'm happy," "I feel good," "Math is fun," and "I love math"), as shown in Fig. 2 (center). Some students drew a specific moment when they felt good about mathematics. For example, as seen in Fig. 2 (right), a Grade 5 student illustrated a moment when an idea popped up and she felt "smart and bright."

FIG. 2 Student drawing (*Left*: Grade 8 girl; *Center*: Grade 2 boy; *Right*: Grade 5 girl).

Negative Relationships With Mathematics

The following Wordle image (Fig. 3) provides a snapshot of how students described their negative relationships with mathematics. These students experienced feelings of confusion, struggle, and worthlessness when learning mathematics.

Many students who reported a negative relationship with mathematics noted unfavorable comparisons with others, such as "I'm not very good at

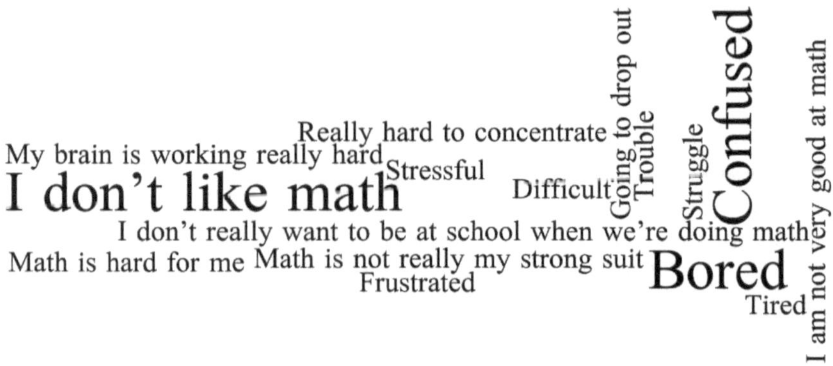

FIG. 3 Wordle image of participants' negative emotional relationships.

math. Because there are people in my class that are much better than me" (Grade 5 girl) and "I must be really dumb if everyone else can do it and I can't" (Grade 8 girl). A Grade 7 boy shared his fear of dropping out from school, because mathematics was repeatedly emphasized as a gatekeeping subject: "Like, how teachers say math is in everything, I'm just having trouble in math. I'm, like, maybe just going to drop out or something or quit."

Attributing their negative relationships with mathematics to their personal characteristics and personal suitability was another common theme. These students believed that they were simply not suited to mathematics. They reported feelings of incompetency and inadequacy that developed over time when learning mathematics. For example, a Grade 8 girl said, "[I feel] dumb… because I can never figure it out." This student reported that mathematics is the only subject that makes her feel that way. She explained that she felt unproductive in mathematics because she did not understand: "[I'm] not getting much work done because I don't get it." Another student (Grade 7 boy) said, "I've always felt that math is difficult for me. It's definitely not one of my best subjects. Um, but the feelings have been the same through the years, …since I'm bad at it." Among the students who perceived mathematics ability as an innate competence, some reported that they were not making efforts to improve. For instance, a Grade 5 student told us: "I don't really like math too much because it's not really my strong suit. I don't practice it much."

The drawings that these students created illuminate some of the memorable moments in their learning of mathematics. As seen in Fig. 4 (left), negative feelings were often attached to a test or a quiz that they were taking. Students described the feeling of frustration, and some conveyed an unexpressed internal voice (e.g., "I'm frustrated") or shouting (e.g., "Arrr! or "Grrr!") in their drawings. Many students' drawings focused on an image of a head. One student (Grade 8 girl) depicted her head shrinking in

FIG. 4 Student drawing (*Left*: Grade 5 girl; *Center*: Grade 5 boy; *Right*: Grade 8 girl).

front of a mathematics quiz or test (Fig. 4 [right]). These head-focused images are consistent with students' verbal narratives:

> I feel like my brain is working really hard. Because I'm not very good at math. (Grade 4 girl)
>
> I think that's one of the reasons why math is kind of hard—'cause it's not really hands-on but I mostly have to use my brain. I like hands-on things' cause I want to be a mechanic when I get older but I know there's going to be a lot of math in there. (Grade 8 boy)

These "negative" drawings were more descriptive and detailed than the drawings students made to represent positive feelings. As discussed earlier, many students simply drew a smiling face in describing their happy feelings. When drawing negative feelings, many students illustrated specific moments when they developed their negative feelings toward mathematics.

Mixed Relationships With Mathematics

Many narratives did not clearly fit into a binary of positive and negative relationships with mathematics, so, in our more nuanced analysis, we developed categories of "mixed" and "changing/changed" relationships with mathematics. Quite a number of participants ($n=24\%$, 25.5%) showed mixed responses by reporting both positive and negative relationships with mathematics, depending on the topics or contexts. We were surprised that more than half of these participants described a mixed reaction to learning mathematics that was highly topic-based, even when those topics were (for us) highly interrelated (e.g., multiplication and division). These mixed dispositions related to similar mathematical topics were particularly prevalent for many of the older participants. The following excerpts from participant interviews illustrate this point:

> With multiplication I feel pretty good but with division I think I need to work on it a bit. (Grade 5 girl)

If it's something that I know first—like the square roots were pretty easy for me. I felt really confident in that but some other things like multiplying fractions, like whole numbers with fractions, they're pretty hard. (Grade 8 boy)

Like right now we're doing algebra and I really don't like doing that [laughter] but I like stuff with like patterns and things like that. (Grade 8 girl)

I like division.

[Interviewer: What would be your least favourite thing about math?]

Fractions. (Grade 5 girl)

The kinds of pairings these students identified (e.g., multiplication and division, algebra and patterns, fractions and division) are pairs of interrelated topics. For example, lessons on patterns are considered a precursor for learning algebra. It is also noteworthy that participants conceptualized "division" and "fractions" as separate disciplines and reported mixed relationships with such interrelated mathematical topics.

This mixed feeling toward mathematics is represented vividly in one of the student drawings (Fig. 5). In describing her drawing, the student said:

The heart is supposed to represent what I love and hate; that's why it's divided. Uh…it's supposed to represent what I feel about it, so over here, I hate positives, negatives, percents, and fractions, like I told you, and my semi-like would be, um, just equations, like many equations like these, divide, multiply, division, uh, sorry, subtraction, addition, those kind of things. (Grade 8 girl)

As a contributor to this divided relationship, participants mentioned the lack of a strong connection across various mathematical concepts that they were learning. When a connection was made between new concepts and previously learned concepts, students could develop a positive emotional relationship, as seen in the following excerpt.

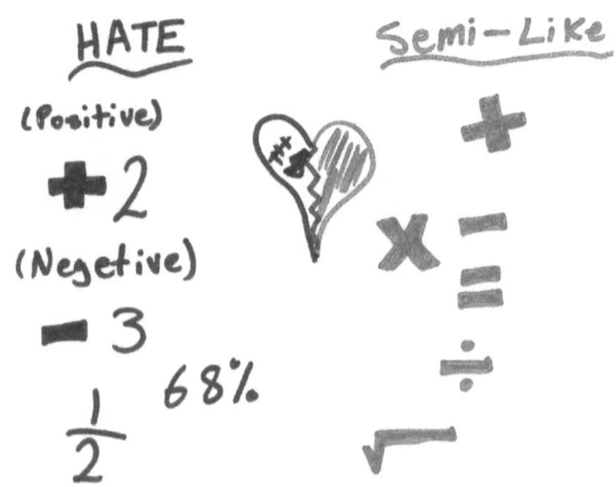

FIG. 5 Student drawing (Grade 8 girl).

That's sort of what I feel when I'm doing math. Confusing. Sort of confusing when you are learning a new one, like something new. If you can't make relations with it, it gets complicated. Like, with square roots, I knew it because, well, I didn't really knew it, but we did, like, um, areas and perimeters before and that really helped me a lot because that I really did good on that so with square roots I just made a relation and I did pretty good. (Grade 8 boy)

Having an understanding of mathematical concepts was related to participants' emotional relationships with mathematics. Many students described that they felt happy, confident, and good for concepts that they understood, whereas they felt irritated, frustrated, and nervous when they did not understand a particular concept. These students described their mixed feelings toward mathematics in ways that cannot be captured merely as static and monolithic, and they used dynamic metaphors in their drawings. For example, as seen in Fig. 6, one of the students used a metaphor of weather to describe this mixed feeling. He noted that the sun represents when "I understand it," the cloud represents when "it's confusing," and the raindrops represent when "it's frustrating that you don't get it."

There were a few students who perceived the state of frustration and uncertainty in a positive light. These students described frustration and uncertainty as a process, and they talked about a sense of accomplishment and passion for the new stage of knowing as follows:

If it's a simpler question, I feel good because I can get it. If it's a tougher question, I will get slightly aggravated, but then when I solve it, you feel really proud and you feel like you've just conquered a massive obstacle. (Grade 9 boy)

How I feel? Ah, frustrated [laughter] a lot of times. Like because when I don't get something, I get frustrated and I tend to want to know. Like I get more passion about it. I feel really good when something comes easy to me. (Grade 9 girl)

FIG. 6 Student drawing (Grade 7 boy).

For these students, the state of frustration was not perceived as simply negative; rather, it was part of the process of understanding and acquiring new mathematical concepts and engaging in problem solving.

Changing/Changed Relationship

Another complexity observed in participants' narratives concerned how their relationships with mathematics have changed over time or are currently changing, through experiencing different mathematics pedagogy, classroom environments, and/or assessment systems at school. Through their narratives, 25.5% of students ($n=24$) described a change in their relationships with mathematics. Almost all the participants who reported a change were older students, above Grade 5.

In describing how negative relationships transitioned to positive relationships, participants named some of the teaching practices that helped them. For example, in the following excerpt, a student explains how mathematics teaching that focuses on sense-making helped him to understand mathematics better.

Student:	Um, I'm really liking the teacher that we have right now for math.
Interviewer:	Why's that?
Student:	He puts it in easier ways to understand. Um, and I also liked my Grade 3 teacher for math. She, ah, helped to make, to help me make sense of it because when you are given just, ah, equations of just numbers and no one really explains it to you it's, like, "What am I supposed to do?"
Interviewer:	And has that happened to you?
Student:	Um, it's happened a little bit, like we'll be given a worksheet and we'll have a little explanation of the first question but it doesn't help enough and then you end up going through and getting everything wrong because they didn't explain it right. (Grade 6 boy)

Another student described how reviewing and making a connection between previously learned content and new content was helpful. When asked what made this year's mathematics class go better than last year's mathematics class, this participant (Grade 5 boy) replied: "Because we are going over it again and we are reviewing everything we did in Grade 4 and a bit more."

Another commonality across many participants' narratives was how their teachers helped them learn mathematics by identifying the challenges that they were facing. These teachers also knew how they could help students to overcome these challenges.

> Right now in my grade, she, I notice, she helps me a lot with the struggles that I have…where other teachers might not get my problem that I had with math. But she and my Grade 6 teacher more understood the problem that I had and how to help me. (Grade 7 girl)

These teaching practices helped to change students' negative relationships with mathematics to positive relationships.

Some students described how their efforts "to concentrate" and getting "help from peers" could make a change. If students receive appropriate support and help, they believe that they can overcome the stage where they feel "I can't do it" or "I can't think." Eventually, they can gain more confidence. For example, as seen in Fig. 7, a student drew a stairway to illustrate her changing feelings toward mathematics. She explained:

> It's kind of like my steps to do math, like I can't do it, and then I can't think, I don't know how, and then I want to try and then I'm getting more confident and then I reach the top and I'm like, "Yes, I can do it!" [laughter] (Grade 7 girl)

In contrast, students also described some of the teaching practices that changed the positive relationships with mathematics that they used to have early in their lives to negative ones. For instance, the following quotation describes how the practice of "just giving the answers" did not help the student to learn and eventually changed her relationship with mathematics.

> Um, ah, my teachers when I was littler obviously helped to teach me the way that would be easiest for me to learn, um, with pictures and stuff like that and, um, then

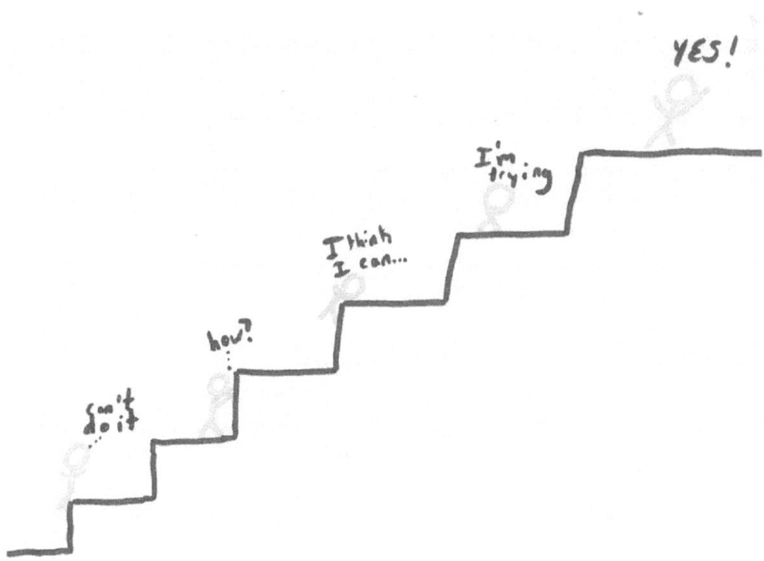

FIG. 7 Student drawing (Grade 7 girl).

some of, some of my teachers really push multiplying and dividing and stuff like that, like, my sixth grade teacher did. But my fifth grade teacher just gave us multiplying tables and that didn't help at all because she was just giving us the answers to everything instead of helping us to learn it. (Grade 8 girl)

Students also reported external pressure as another factor that disturbed their positive relationships with mathematics. The popular perception of mathematics as a gatekeeper can put enormous pressure on students, rather than motivating them to learn more. The following quotation from a Grade 7 student is informative on this point.

Interviewer:	Could you tell me, um, what you think maybe brought you to the current attitude you have about math? Is there someone or something you think has influenced how you think about math?
Student:	Well, my math teacher I have right now…she sometimes gives us lectures about if we don't do well in school, we can just end up working at [a fast food restaurant] [*laughter*] for the rest of our lives and how hard it is to get a job and get into college. (Grade 7 girl)

Many students commented that as they became older, the mathematics concepts they have to learn become complicated and difficult. If students cannot obtain appropriate support, they try their best to overcome the difficulty on their own. In the following exchange, a Grade 7 student describes how she overcame her mathematics problems by using a memorization strategy. However, this strategy did not lead to the construction of a positive relationship with mathematics. She found herself bored learning mathematics because she felt it was simple repetition.

Student:	Um, well usually I get, I think I have the answer but I try it over and over again and I keep getting it wrong.
Interviewer:	Now is that what has happened all along or has it changed a bit?
Student:	It's been changing.
Interviewer:	Okay, and what makes things change?
Student:	Um, I just kind of memorize questions and stuff now.
Interviewer:	Okay, so like times tables and that kind of thing?
Student:	Yeah, I memorized times tables so now when I go home, I practice double-digit times tables.
Interviewer:	And how do you feel when you are doing math?
Student:	Feel kind of bored.
Interviewer:	Okay, and what is boring about it?

Student:	It seems like we are doing the same stuff over and over again. Although we have to memorize it, even if we have memorized it… you have to do it over and over again. It's boring. (Grade 7 girl)

Overall, students' narratives about their changing relationships with mathematics show how certain teaching practices and teachers' messages can impact how relationships with mathematics form and change over time.

DISCUSSION AND IMPLICATIONS

In this section we discuss the findings documented above by drawing attention to a number of themes that we feel summarize our findings and point to significant areas for further research in the study of students' emotional experiences learning mathematics.

A Nuanced Emotional Landscape

While our reading of the existing literature concerning mathematics learning (e.g., Brown, Brown, & Bibby, 2008), and to some extent our own experiences working with elementary preservice and practicing teachers, had led us to expect a predominance of strongly negative emotional responses, our data reveal a more nuanced emotional landscape. While some students certainly did convey through their narratives and drawings strongly negative emotions (as encapsulated in the Wordle image in Fig. 3), for others, learning mathematics had generated a mixture of emotions or strongly positive emotions. Indeed, in our analysis, the largest cluster of students comprised those who expressed relatively positive emotions in relation to learning mathematics. We note, however, that in the phase of our investigations reported here our dataset is skewed toward the early elementary rather than junior high school years, with, for instance, far more Kindergarten students in the participant pool than Grade 9 students. We noticed that, in general, the older students had less favorable relationships with mathematics, a finding that consistent with previous research (as cited in *Feelings About Mathematics* section). In our interviews with high school students, quite a number of students mentioned that they started to dislike mathematics after Grade 4 or 5. Before this turning point, they said they liked mathematics and felt they were good at it. This pattern itself is troubling because it suggests that the more exposure students have to school mathematics, the less they like it. The pattern also suggests that there may be topics that emerge beyond these grades that students find particularly unappealing. However, we also recognize a methodological issue here in that students who are already uncomfortable with mathematics are probably less likely to volunteer for a research study in which they know they will be asked to talk about mathematics, hence further skewing the data toward the positive responses. Nevertheless, given that topics such as mathematics anxiety and students' struggles to learn mathematics dominate the research literature in this field, we feel that it is important to document and explore the experiences of students who have strong relationships with mathematics and we have begun to do so (see, e.g., Hall, Towers, Takeuchi, et al., 2015). Our findings revealed variety in the ways in which students described positive relationships with mathematics. For instance, many students liked mathematics when they found it easy, while fewer students embraced and enjoyed the challenges that mathematics presents. Further research is needed to examine the contexts and trajectories through which students develop particular relationships with mathematics and, in particular, those that generate a positive disposition toward the challenges inherent in mathematics learning.

A further element of the nuanced emotional landscape that we have documented concerns students' mixed responses to learning mathematics. As discussed, we were surprised by the number of students who expressed strongly differentiated emotional responses to mathematics topics that we saw as mathematically interrelated. This topic-based emotional response extends the current field in an important way, suggesting that there is scope for an analysis of the specific topics in mathematics that generate the most concerning negative responses, so that teachers can be alerted to these curriculum "hotspots." This finding also suggests to us, though, that students are not being exposed to classroom experiences that emphasize the connectedness of mathematical concepts. That students can "like" multiplication but "hate" division indicates that their interactions with these topics are not enabling them to see the two concepts as inverse processes. We suspect that teachers' continued emphasis on standardized procedures of arithmetic associated with these topics rather than the underlying conceptual ideas would tend to reinforce such responses. We also note, though, that our analysis indicates that students who had been enabled to see topics and concepts within mathematics as interconnected had developed more consistently positive relationships with mathematics, which suggests that pedagogies focused on emphasizing such interconnections have great potential.

Comparisons With Others/External Validation

Many students who reported a negative relationship with mathematics stated that they compared unfavorably with others in relation to mathematics competence. These students described others as being "much better" at mathematics and described themselves as "dumb" (because others could do the mathematics and they couldn't). Considering that those who built a good relationship with mathematics often drew from external validation and evaluation as a source for positive emotions, the narratives of the students who reported feeling "dumb" highlight the other side of the same evaluation system perpetuating in the school context. Classroom practices and schooling structures (such as streaming, in-class ability grouping, or in-class structures that aim to "seed" small groups with at least one "strong" student) that emphasize (even implicitly) hierarchies of "good" and "not so good" mathematics students are likely to feed such comparisons. The literature, however, contains accounts of mechanisms that, deliberately or as a fortuitous side effect of other changes, eliminate or diminish such hierarchies (e.g., Boaler, 2011; Liljedahl, 2014; Towers, Martin, & Heater, 2013). The widespread presence in our dataset of student narratives that reference comparisons with others and/or external sources of validation of competence suggests that such comparisons are a source of strong emotions, many of which are strongly negative, and that further research into this component of the emotional landscape of doing mathematics would be worthwhile.

The Head as the Location of Mathematizing (and Emotional Focus)

From an enactivist point of view, in which learning is seen as embodied, it is notable that in their drawings many students focused on the image of their own head whereas relatively few of the drawings included students' bodies. When the students' bodies were included, they were almost always shown seated, motionless, at an individual desk. Rarely were groups of learners drawn, and drawings featuring individual bodies were featured doing worksheets or writing on paper (usually at desks) rather than manipulating objects or moving around. The head-centered drawings, of which there are many within our dataset, are indicative of students' learning experiences at school that are centered around their heads (or brains) rather than being active, embodied experiences. These head-focused images are also consistent with students' verbal descriptions of their experiences learning mathematics in school, such as, "I feel like my brain is working really hard. Because I'm not very good at math" (Grade 4 girl) and "I think that's one of the reasons why math is kind of hard—'cause it's not really hands-on but I mostly have to use my brain" (Grade 8 boy). These narratives suggest that despite decades of educational reforms rooted in constructivist discourses, which promote active student engagement in the classroom, and contrary to evidence that, when given the opportunity, children can and do "think and learn through their bodies" (Kim, Roth, & Thom, 2011, p. 207), school mathematics remains a disembodied, mental experience for many. We are interested in understanding why, despite educational reforms and widespread reports in the mathematics education literature of innovative practices occurring in school classrooms, these disembodied experiences persist, and we suggest that further exploration of this phenomenon would be fruitful. We see the methodological vehicle of drawings as particularly valuable in this respect.

Emotions and Schooling

One dominant feature of the data is the place-based aspect of the stories and drawings that we collected. Almost every emotion-laden account relating to learning or doing mathematics, and by far the majority of the drawings, were connected to schools. This is the case despite that fact that during the interviews we explicitly asked students about their experiences with mathematics outside of school and invited them to comment on meaningful encounters with mathematics in their everyday experiences. When students did comment on an out-of-school experience with mathematics, it was consistently related to school-assigned homework. This finding is perhaps not surprising, but it is concerning. Most of our Kindergarten to Grade 9 respondents struggled to offer ideas about how and where mathematics might be used in the world beyond very basic uses of arithmetic (e.g., in shopping scenarios). It seems that their entire experience of mathematics beyond arithmetic is school-bound. We suggest that, despite efforts to popularize mathematics (e.g., in television shows and movies), curricula that increasingly mention uses of mathematics, different mathematical fields of study and mathematical careers, and (as our data showed) parents and teachers' admonitions that mathematics is "needed for life," students still see mathematics as a school subject with limited relevance to their lives and, as we discuss elsewhere (Takeuchi et al., 2016), practically no intrinsic or esthetic value. Based on our findings, we suggest that further exploration of these anomalies and of ways to broaden students' experiences with the intrinsic beauty and structure of mathematics and its esthetic, creative, and practical uses is warranted.

Narratives of Change

Many students reported that their relationship with mathematics had changed over time. These narratives offer some insight into the kinds of structures and practices that help students develop stronger relationships with mathematics, or turn them away from the discipline. Comparisons between the practices of different teachers featured heavily in these narratives, and students were clear that pedagogy focused on sense-making enabled them to learn mathematics and hence allowed them a more positive emotional response to doing mathematics. Students were also clear that pedagogy that focused on rules (e.g., memorizing multiplication tables in the absence of context or conceptual support) or that leveraged threats as an incentive to learn mathematics (e.g., "If you can't do mathematics, you will end up working at [a fast food restaurant]") contributed strongly to changing students' emotional responses to mathematics from positive to negative. Because many of these turning points seemed to be associated with a particular teacher's practices (students frequently named the grade and teacher and spoke passionately about the shift in their emotions associated with these experiences), we are reminded of the critical importance of teachers and the learning contexts that they create in the shaping of students' emotional relationships with mathematics, a finding that is consistent with recent research conducted by Andersson, Valero, and Meaney (2015). Viewed through an enactivist lens, we see the students' comments as acknowledging that learning is inextricably bound up in teaching and dependent on the kind of relationships (student-student, teacher-student, student-mathematics, etc.) that teachers build in the classroom. Hence, the kind of learning environment that teachers create plays a key role in the kind of relationships students are enabled to develop with mathematics, and in turn with the emotions they experience when doing mathematics, with the way they come to be with mathematics, and with how they see themselves as mathematical (or not). Further research that explicitly investigates the teacher's role in the critical turning points in students' emotional relationships with mathematics may help us to better understand, and communicate with teachers about, the practices that enhance positive emotions.

CONCLUSION

Our analysis has shown that the landscape of students' emotional relationships with mathematics is much more complex than a simple binary of "love" or "hate" (or even "like" or "dislike"). It is clear to us that while some students do have these kinds of strong responses, many others like or dislike only some aspects of mathematics (such as particular topics) or like mathematics only when they find it easy or when they receive external validation for their efforts. We were disappointed that so few students seemed to have developed a love of mathematics for its own sake. Through our analysis, we were also alerted to some surprising features of students' emotional relationships with mathematics—the curious preponderance of head-centered drawings and the overwhelming presence of school-related imagery (both drawn and related orally) when students were asked to talk about how they feel about mathematics or about significant mathematical events in their lives. Together, these findings suggest that students may be internalizing a very limited image of what mathematics is—something that one does inside one's head and only in school—and that this may be contributing to some students' negative feelings about mathematics. Finally, when students described how their relationships with mathematics had changed over time, we noted that many of these turning points were dependent on experiencing a particular kind of teaching and hence we see the learning environment as a critical element in students' developing emotional connection with mathematics. We feel each of these dimensions of students' emotional relationships with mathematics is worthy of further attention and we hope that mathematics education researchers will expand their focus on the emotional landscape of mathematics learning.

References

Ahmed, W., Minnaert, A., Kuyper, H., & van der Werf, G. (2012). Reciprocal relationships between math self-concept and math anxiety. *Learning and Individual Differences, 22*(3), 385–389. http://dx.doi.org/10.1016/j.lindif.2011.12.004.

Andersson, A., Valero, P., & Meaney, T. (2015). "I am [not always] a maths hater": Shifting students' identity narratives in context. *Educational Studies in Mathematics, 90*(2), 143–161. http://dx.doi.org/10.1007/s10649-015-9617-z.

Ashcraft, M. H., & Kirk, E. P. (2001). The relationships among working memory, math anxiety, and performance. *Journal of Experimental Psychology: General, 130*(2), 224–237. http://dx.doi.org/10.1037/0096-3445.130.2.224.

Berry, R. Q., III, Thunder, K., & McClain, O. L. (2011). Counter narratives: Examining the mathematics and racial identities of black boys who are successful with school mathematics. *Journal of African American Males in Education*, 2(1), 10–23. Retrieved from http://journalofafricanamericanmales.com/.

Boaler, J. (2011). Changing students' lives through the de-tracking of urban mathematics classrooms. *Journal of Urban Mathematics Education*, 4(1), 7–14. Retrieved from ed-osprey.gsu.edu/ojs/index.php/JUME/.

Brown, M., Brown, P., & Bibby, T. (2008). "I would rather die": Reasons given by 16-year-olds for not continuing their study of mathematics. *Research in Mathematics Education, 10*(1), 3–18. http://dx.doi.org/10.1080/14794800801915814.

Clandinin, D. J. (2007). Preface. In D. J. Clandinin (Ed.), *Handbook of narrative inquiry: Mapping a methodology* (pp. ix–xvii). Thousand Oaks, CA: Sage.

Clandinin, D. J., & Rosiek, J. (2007). Mapping a landscape of narrative inquiry: Borderland spaces and tensions. In D. J. Clandinin (Ed.), *Handbook of narrative inquiry: Mapping a methodology* (pp. 35–75). Thousand Oaks, CA: Sage.

Di Martino, P., & Zan, R. (2011). Attitude towards mathematics: A bridge between beliefs and emotions. *ZDM, 43*(4), 471–482. http://dx.doi.org/10.1007/s11858-011-0309-6.

Evans, J., Morgan, C., & Tsatsaroni, A. (2006). Discursive positioning and emotion in school mathematics practices. *Educational Studies in Mathematics, 63*(2), 209–226. http://dx.doi.org/10.1007/s10649-006-9029-1.

Fennema, E., & Sherman, J. A. (1976). Fennema-Sherman mathematics attitude scales: Instruments designed to measure attitudes toward the learning of mathematics by females and males. *Journal of Research in Mathematics Education, 7*(5), 324–326. http://dx.doi.org/10.2307/748467.

George, P. (2009). Identity in mathematics: Perspectives on identity, relationships, and participation. In L. Black, H. Mendick, & Y. Solomon (Eds.), *Mathematical relationships in education: Identities and participation* (pp. 201–212). New York, NY: Routledge.

Hall, J. (2012). Gender issues in mathematics: An Ontario perspective. *Journal of Teaching and Learning*, *8*(1), 59–72. Retrieved from http://ojs.uwindsor.ca/ojs/leddy/index.php/JTL.

Hall, J. (2013). *Societal views of mathematics and mathematicians and their influence on elementary students (Unpublished doctoral dissertation)*. Ottawa, Canada: University of Ottawa.

Hall, J., & Towers, J. (2014). *Tracing student experience using mathematics autobiographies*. In S. Oesterle, C. Nicol, P. Liljedahl, & D. Allan (Eds.), Proceedings of the joint meeting of the 38th annual meeting of the International Group for the Psychology of Mathematics Education and the 36th annual meeting of the North American chapter of the International Group for the Psychology of Mathematics Education (Vol. 6, p. 91). Vancouver, Canada: PME.

Hall, J., & Towers, J. (2017). *Using drawings in mathematics education research*. Manuscript in preparation.

Hall, J., Towers, J., & Martin, L. C. (2015). *Using I poems to illuminate students' mathematical identities*. In: Paper presented at the annual meeting of the American educational research association. Chicago, IL.

Hall, J., Towers, J., Takeuchi, M., & Martin, L. C. (2015). *Investigating students' positive relationships with mathematics*. In K. Beswick, T. Muir, & J. Wells (Eds.), Proceedings of the 39th annual meeting of the International Group for the Psychology of Mathematics Education (Vol. 1, p. 164). Hobart, Tasmania: PME.

Hannula, M. S. (2002). Attitude towards mathematics: Emotions, expectations and values. *Educational Studies in Mathematics*, *49*(1), 25–46. http://dx.doi.org/10.1023/A:1016048823497.

Kieren, T. E., & Sookochoff, S. (1999). *Braiding of acting and emotioning in mathematical knowing*. In Paper presented at the annual meeting of the american educational research association. Montreal, Canada.

Kim, M., Roth, W., & Thom, J. (2011). Children's gestures and the embodied knowledge of geometry. *International Journal of Science and Mathematics Education*, *9*(1), 207–238. http://dx.doi.org/10.1007/s10763-010-9240-5.

Liljedahl, P. (2014). The affordances of using visually random groups in a mathematics classroom. In Y. Li, E. Silver, & S. Li (Eds.), *Transforming mathematics instruction: Multiple approaches and practices* (pp. 127–144). New York, NY: Springer.

Lubienski, S. T., Robinson, J. P., Crane, C. C., & Ganley, C. M. (2013). Girls' and boys' mathematics achievement, affect, and experiences: Findings from ECLS-K. *Journal for Research in Mathematics Education*, *44*(4), 634–645. Retrieved from http://www.nctm.org/publications/journal-for-research-in-mathematics-education/.

Lupart, J. L., Cannon, E., & Telfer, J.-A. (2004). Gender differences in adolescent academic achievement, interests, values, and life-role expectations. *High Ability Studies*, *15*(1), 25–42. http://dx.doi.org/10.1080/1359813042000225320.

Maturana, H. R., & Varela, F. J. (1992). *The tree of knowledge: The biological roots of human understanding*. Boston, MA: Shambhala.

Perkkilä, P., & Aarnos, E. (2009, June). "Me in the land of mathematics": Children's drawings and gender issues. *IOWME Newsletter*, *23*(1), 6–11.

Plosz, J., Towers, J., & Takeuchi, M. (2015). *Students' images of mathematics explored through drawings*. In T. G. Bartell, K. N. Bieda, R. T. Putnam, K. Bradfield, & H. Dominguez (Eds.), Proceedings of the 37th annual meeting of the North American chapter of the International Group for the Psychology of Mathematics Education (p. 590). East Lansing, MI: Michigan State University.

Richardson, F. C., & Suinn, R. M. (1972). The mathematics anxiety rating scale: Psychometric data. *Journal of Counseling Psychology*, *19*(6), 551–554. http://dx.doi.org/10.1037/h0033456.

Stogsdill, G. (2013). A math therapy exercise. *Journal of Humanistic Mathematics*, *3*(2), 121–126. http://dx.doi.org/10.5642/jhummath.201302.09.

Takeuchi, M., Czuy, K., & Towers, J. (2016). *Pre-service teachers' multimodal mathematics autobiographies: Emotion and learning and teaching mathematics*. In Paper presented at the annual meeting of the Canadian society for the study of education. Calgary, Canada.

Takeuchi, M., & Towers, J. (2015). *Immigrant students' experiences learning mathematics in Canadian schools*. In T. G. Bartell, K. N. Bieda, R. T. Putnam, K. Bradfield, & H. Dominguez (Eds.), Proceedings of the 37th annual meeting of the North American chapter of the International Group for the Psychology of Mathematics Education (p. 592). East Lansing, MI: Michigan State University.

Towers, J., Hall, J., & Martin, L. (2015). Students' images of mathematics. In S. Osterle & D. Allan (Eds.), *Proceedings of the 2014 annual meeting of the canadian mathematics education study group/Groupe Canadien d'Étude en Didactique des Mathématiques* (p. 285). Burnaby, Canada: CMESG/CGEDM.

Towers, J., Martin, L. C., & Heater, B. (2013). Teaching and learning mathematics in the collective. *Journal of Mathematical Behavior, 32*(3), 424–433. http://dx.doi.org/10.1016/j.jmathb.2013.04.005.

Towers, J., Takeuchi, M., Hall, J., & Martin, L. C. (2015). *Exploring the culture of school mathematics through students' images of mathematics*. In T. G. Bartell, K. N. Bieda, R. T. Putnam, K. Bradfield, & H. Dominguez (Eds.), Proceedings of the 37th annual meeting of the North American chapter of the International Group for the Psychology of Mathematics Education (pp. 570–573). East Lansing, MI: Michigan State University.

Vandecandelaere, M., Speybroeck, S., Vanlaar, G., De Fraine, B., & Van Damme, J. (2012). Learning environment and students' mathematics attitude. *Studies in Educational Evaluation, 38*(3-4), 107–120. http://dx.doi.org/10.1016/j.stueduc.2012.09.001.

Varela, F. J., Thompson, E., & Rosch, E. (1991). *The embodied mind: Cognitive science and human experience*. Cambridge, MA: MIT Press.

Webster, L., & Mertova, P. (2007). *Using narrative inquiry as a research method: An introduction to using critical event narrative analysis in research on learning and teaching*. New York, NY: Routledge.

Young, C. B., Wu, S. S., & Menon, V. (2012). The neurodevelopmental basis of math anxiety. *Psychological Science, 23*(5), 492–501. http://dx.doi.org/10.1177/0956797611429134.

Young-Loveridge, J., Taylor, M., Sharma, S., & Hawera, N. (2006). Students' perspectives on the nature of mathematics. In P. Grootenboer, R. Zevenbergen, & M. Chinnappan (Eds.), *Identities, cultures and learning spaces Vol. 2*. (pp. 583–590). Adelaide, Australia: MERGA.

Zan, R., Brown, L., Evans, J., & Hannula, M. S. (2006). Affect in mathematics education: An introduction. *Educational Studies in Mathematics, 63*(2), 113–121. http://dx.doi.org/10.1007/s10649-006-9028-2.

CHAPTER

7

"I Did Use to Like Maths...": Emotional Changes Toward Mathematics During Secondary School Education

Paul Hernandez-Martinez, Maria Pampaka†*

*Loughborough University, Loughborough, United Kingdom
†The University of Manchester, Manchester, United Kingdom

INTRODUCTION

In recent years there has been considerable interest in investigating students' attitudes toward mathematics because of its significance in shaping aspirations and decision-making processes in relation to careers and future life opportunities. Emotions are an important component of these attitudes. Hence, it is of particular relevance to understand how emotional aspects of attitudes toward mathematics change throughout schooling and how their change or stability is related with other aspects of students' school trajectories (such as teaching practices). New understandings along these lines will shed more light into which mathematical pedagogies (and how) can help students form positive emotional links toward the subject and as a result create a more mathematically-disposed society.

In this chapter we present results from a recent large longitudinal study in England, focusing on the emotional aspects of attitudes in secondary school mathematics education. By using a mixed methodology, we aim to provide evidence of differences and changes in students' dispositions toward mathematics and to explore how students make use of affective discourses (in particular those containing emotional reactions) to explain their dis/engagement with mathematics, identifying differences and similarities based on age (among the year groups), as well as associations with

http://dx.doi.org/10.1016/B978-0-12-802218-4.00007-8

various aspects of the pedagogical practices that they have experienced in their (mathematics) educational trajectories and their aspirations for the future.

REVIEW OF RELEVANT LITERATURE AND THEORETICAL APPROACH

Emotions are closely related to the study of students' attitudes and beliefs, which has gained considerable interest in mathematics education over the past 40 years. A very recent edited book on beliefs (Pepin & Roesken-Winter, 2015) and a special issue on attitudes also attest the continuous importance of the topic (e.g., Chinn, 2012; Lazarides & Ittel, 2012; Mata, Monteiro, & Peixoto, 2012). Looking at this literature over the years, one can easily come across the complexity of relevant terms and sometimes inconsistency on the main definitions. We introduce here the most relevant notions for our work within this chapter.

Before the 1980s, psychological studies on attitudes dominated the area but the publication of the book *Affect and Mathematical Problem Solving* by McLeod and Adams (1989) is considered by many a turning point for this type of research. Di Martino and Zan (2015) in their review of the literature on the concept of attitude, note the following:

> For the first time, affective constructs are used not only to prove the existence of a numerical correlation with an outcome (mathematical achievement), but also to interpret a process (the interactions between affective and cognitive aspects in problem-solving activities). (p. 64)

A useful starting point for the conceptualization of attitudes in mathematics education is Ruffell, Mason, and Allen's (1998) decomposition of the concept into three subcomponents, namely cognitive, affective, and conative. In this work they conclude the following:

> We now see attitude as at best a complex notion, and we conjecture that perhaps it is not a quality of an individual but rather a construct of an observer's desire to formulate a story to account for observations. (p. 1)

Hannula (2002) is another significant contributor to the field, with his mainly qualitatively focused thesis where he offered a critical reconceptualization of "attitude." Starting from the everyday notion of attitude as basic liking or disliking of a familiar target, he considers emotions as "always present in human existence" (p. 28). For Hannula, emotions have three mutually independent readouts: adaptive-homeostatic arousal responses (realizing adrenalin in blood), expressive displays (smile), and subjective experiences (e.g., feeling sad), and as he states, "there are only a few basic emotions: happiness, sadness, fear, anger, disgust and interest. The more complex emotions are based on this" (p. 28). This served as a starting point of our analysis as we will detail later.

The concept of emotions has also been recently reported as having undergone a shift in learning research (Niemi, 2009). The "expansions" or changes listed under this conceptualization are different metaphors (referring to the work of Anna Sfard), sociocultural approaches and activity theory, learning as a self-regulated process, mastery and performance orientations in learning, and emotions and motivation as a basis for learning.

In this chapter we take a sociocultural theoretical position on learning and knowing, a framework with which we then try to analyze how students talk about their emotional reactions and relevant changes of dispositions toward mathematics. By adopting this innovative framework, we present emotions as part of a dynamic process in contrast to the dominant views that treat emotions as states or products. Our theoretical perspective is presented next.

Learning Mathematics as a Sociocultural Practice: Attitudes, Dispositions, and Emotions

From a sociocultural theoretical viewpoint, students participating in the practice of school mathematics and learning is seen as a change in that participation; this change involves "becoming" a different person in relation to the (sociocultural) practices of the activity, a transformation of the self (Lave & Wenger, 1991; Wenger, 1998). Within this framework, sociocultural factors are seen as an intrinsic part of the process and not something that can be studied separately. According to Radford (2015), affective factors in general, and emotions in particular, are sociocultural in nature:

> The point that is missed here is that the affective domain in general and motives and motivation in particular are not only subjective but also sociocultural phenomena. They are subjective and sociocultural in the sense that on the one hand, motives are the motives of a concrete and unique person but, on the other hand, they relate to a sociocultural and historical world that transcends the individual. In its transcendence, the sociocultural historical world indirectly – albeit in a decisive manner – shapes and organizes the individual's motives and emotions. (p. 26)

Roth (2007) argues that emotions shape the way in which we understand the world and ourselves; these are *judgments or appraisals* we make "about the world, about other people, and about ourselves and our place in our world" (Solomon, 1978, p. 186). Through emotions we make meaning of our world and this involves concepts of the self, entailing moral and ethical dimensions. Emotions are hence related to motives in a time-projection manner: they relate to the possibility to succeed (or to fail) to reach the object of activity (Radford, 2015).

These perspectives imply that, in education (and in life in general) sociocultural factors, and concomitantly, affective factors are inseparable from the learning that occurs. These factors are not mere influences on the learning process but are constitutive of it. Hence, as Gresalfi (2009) argues,

> Rather than focusing on knowledge as an object that an individual can acquire, this conceptualization of learning shifts attention to the kinds of practices in which people participate, and the ways people come to relate to one another within a particular activity setting. (p. 329)

The importance of the above for our argument in this chapter is that through their participation in the pedagogical practices of school mathematics, learners will develop certain "*dispositions*; that is, ways of being in the world that involve ideas about, perspectives on, and engagement with information that can be seen both in moments of interaction and in more enduring patterns over time" (Gresalfi, 2009, p. 329). We are here, therefore, interested in looking at these "enduring patterns," how students develop them through their participation and engagement in the pedagogical practices of secondary school mathematics and the role that emotions have in creating these patterns.

Note that the term "dispositions" in many occasions is used in the literature as interchangeable with the term "attitudes," but for us there is a difference. Edwards and D'Arcy (2004, p. 148) define dispositions as "a capacity to engage, which is embedded in social practices which enable that engagement." This definition of dispositions emphasizes the interconnection that exists between individuals as *acting* subjects and the affordances or restrictions that their sociocultural contexts pose on them. While attitudes seem to refer to a more static, predefined response to a given situation (and hence closer to "beliefs"), dispositions evoke a dynamic and *relational* link between the subject and the context in which he/she participates. This stresses the possibility that dispositions can change, if that relationship is such that individuals are allowed to participate in an activity in different ways than before—for example, if they participate in a more social pedagogy. So, for example, we have argued elsewhere (Hernandez-Martinez et al., 2011) that transitional moments such as between school and college can allow students the space to become more mature, or to "start afresh" in a new environment and therefore renegotiate their relationship (and dispositions) with mathematics. This is more in line with the theoretical perspective taken in this chapter, and therefore we will use the term dispositions in this sense.

Two important questions that are pertinent to the study of dispositions and emotions are (1) how easy or difficult it is to change someone's disposition/emotional stance toward mathematics, and (2) can pedagogical practices throughout school help shift them? We shall discuss these questions from our theoretical point of view in the next section.

Changing Dispositions and Emotions Within Sociocultural Educational Theories

Evans and Zan (2006) argue that recent research focuses on emotions as *transitory* rather than being restricted to measures of "durable" individual characteristics. This, they claim, allows in turn for a description of an activity as a contextual, dynamic process. However, even though there is a recognition that individuals can and do change according to the contexts in which they participate, there is no agreement as to how easy or difficult it is to change (or how durable dispositions are) or which practices can facilitate this change.

In his sociology of culture, Bourdieu (1990) uses his construct of "habitus" to describe the socialized norms or tendencies that guide behavior or thinking. He defines habitus as

> A system of dispositions to a certain practice, is an objective basis for regular modes of behaviour, and thus for the regularity of modes of practices, and if practices can be predicted... this is because the effect of the habitus is that agents who are equipped with it will behave in a certain way in certain circumstances. (p. 77)

Nolan (2012, p. 204) explains that "habitus is a whole body experience, but it is not one that can be attributed solely to an individual since dispositions are created and recreated through social interaction and tradition." The concept has also been utilized by Kleanthous and Williams (2013) in their investigation of parental influence on students' mathematics dispositions. Swartz (2002) claims that habitus does not determine one's actions but only predisposes us to act in certain ways:

> The term dispositions is key for Bourdieu, for it suggests a way of thinking about habit that is clearly different (more active) from the more popular idea of sheer repetition or routine. (...) Moreover, the idea of disposition suggests that past socialization "predisposes" individuals to act out what they have internalized from past experience but does not "determine" them to do so. The dispositions of habitus shape and orient human action; they do not determine it. (p. 63S)

However, for Bourdieu, the habitus is not easily changed. Although individuals have the capacity to critically reflect on their taken-for-granted views and ways of acting, and therefore try to change them, this is the exception rather that the norm.

If we only consider individuals as rational, logical subjects, we would agree with Bourdieu that dispositions are not easily changed, but we suggest here that emotions, as dynamic and volatile, carry the *potential* for change. Vygotsky (1999, p. 244) explains that emotions develop and "appear in new relations with other elements of mental life," and Leont'ev (cited in Radford, 2015, p. 35) adds that they become related to the motives of the individual; for example, "in the form of interest, desire or passion." Other Soviet Activity Theory scholars such as Gal'perin and Davydov also pointed out the importance of the motives of engagement and how these can be oriented and transformed (Davydov, 2008; Gal'perin, 1978/1992). Therefore, emotional moments can change the motives of engagement of an individual in a certain activity, such as when we hear a highly inspirational, emotional speech, when a teacher introduces an "out of the ordinary" activity that "connects" with our interests, or when someone experiences a "Eureka moment" that is accompanied by an emotional state of happiness. Some of these moments can be highly emotional. These are known as critical moments or turning points and are referred in the literature as spaces where core assumptions of how the world works are challenged and reflexivity can emerge (Chapman-Hoult, 2010; Drake, 2006; Thomson et al., 2002). In fact, Vygotsky used the Russian word *perezhivanie* to describe those cathartic, highly emotional life experiences that "stick in our minds" and shape our development (Blunden, 2016; Roth & Jornet, 2016). Because, as Radford (2015, p. 45) argues, "emotions and thought come to constitute a unity in ontogenetic development." Emotions carry the capacity to change other aspects of our "mental life," such as our beliefs or dispositions. Therefore, it is important to notice this close relationship between emotions and cognition, and the role that emotions have in potentially shifting our thoughts. At the same time, emotions are simultaneously both subjective and cultural phenomena, a way to make sense of the world. They are part of the ever-changing project of the self, by which we position ourselves within the practices in which we participate. Therefore, we see dispositions as the way in which we position ourselves with respect to the cultural practices in which we engage, our way to see things in the world. They are part of our social identity along with our emotions and other parts of our "mental life."

In the final section of our literature review we present an overview of the main instruments related to attitudes and emotions that influenced our own instrument development. Here we focus on attitudes because the vast majority of measurement instruments relate to this construct, while the concept of dispositions has received very little attention.

Measuring Emotions and Attitudes Toward Mathematics

A lot of instruments have been proposed and used in order to measure attitudes toward mathematics, many influenced by the widely cited Fennema-Sherman scales (Fennema & Sherman, 1977). Each of those instruments attempted to capture one of the many dimensions or constructs associated with "attitudes toward mathematics": beliefs, values, identities, engagement, affect, emotions, motivation, confidence, and self-efficacy are only a few on the list.

Grootenboer and Hemmings (2007), for example, in an attempt to measure attitudes to mathematics, developed the "Kids Ideas about

Maths" instrument, which was also employed for moderating effects of attitudes on achievement (Hemmings & Russell, 2010). This instrument was also linked to a model of conceptions of the affective domain from Grootenboer's previous work (Grootenboer, 2003), which includes values, beliefs, attitudes, emotions, or feelings. From a slightly different perspective, the "attitudes toward mathematics inventory" (Tapia & Marsh, 2004) includes four subscales: enjoyment of mathematics, motivation to do mathematics, self-confidence in mathematics, and perceived value of mathematics. A shorter version of this instrument was later revalidated and used in a study in Singapore (Lim & Chapman, 2013).

Our group has also previously designed and validated instruments focused on students' mathematics dispositions (Pampaka et al., 2013) and self-efficacy (Pampaka, Kleanthous, Hutcheson, & Wake, 2011). These instruments, with additions from previously published instruments on attitudes, led to the construction of our questionnaire, which is central to this study. Therefore, our conceptualization of emotions in this work is situated instrumentally as an aspect of mathematics dispositions in particular and attitudes more generally, in our quantitative work; it is also framed theoretically within sociocultural approaches to make sense of students' interviews. Our methodology is detailed next.

METHODOLOGY

This work is part of the TELEPRISM research study (http://www.teleprism.com) that aimed to investigate the teaching and learning of mathematics throughout (compulsory) secondary school education (Years 7–11 in England, 11–16 years old). The study employed a mixed methods approach, providing evidence from quantitative as well as qualitative data and analyses. In our use of a mixed methodology, we take a pragmatic position, attempting "to fit together the insights provided by qualitative and quantitative research into a workable solution" (Johnson & Onwuegbuzie, 2004, p. 16). For the quantitative measures, we considered the role that emotions have in changing dispositions by including items that related to students' emotional reactions within the measures of "math disposition," as we will show later on in this section. Our overall sociocultural theoretical framework was then employed to analyze evidence from students' interviews and results from their responses to the questionnaires.

Instrumentation and Data Collection

The quantitative aspect of the study involved the design and administration of a questionnaire aimed to measure students' mathematical attitudes (affective, cognitive, and conative dimensions) as well as their confidence with various mathematical topics (self-efficacy), future aspirations, and their perceptions of the teaching they encounter in mathematics. The various sections of the questionnaire have been constructed and expanded based on our previous TransMaths framework (http://www.transmaths.org), where we validated and used instruments for students aged 16 and older (Pampaka et al., 2011, 2013; Pampaka, Williams, Hutcheson, Wake, et al., 2012).

Using these questionnaires, data was then collected from students in the 5-year groups from 40 secondary schools across the United Kingdom (with a starting sample of around 13,500 students), at three time points (DP) over one academic year.[1] The sample's description longitudinally, is presented in Table 1.

There is a noticeable attrition in our sample between the DPs as can be seen in Table 1, which is endemic in such longitudinal designs and could be a source of potential bias. Attrition in this case was due to various reasons, such as attrition at school level, class level, or student level, and in some cases it was also related to students leaving the secondary school at the end of Year 11, which made it even harder to follow them afterwards. Even though elsewhere (Pampaka, Hutcheson, & Williams, 2016) we have suggested advanced imputation techniques for dealing with missing data problems, for the purposes of this chapter this is not feasible because we simply employ descriptive analysis. It is important however to revisit the issue along with our analytical approach to clarify to the reader what (part of the) dataset was used for each analytical step.

TABLE 1 Student Sample Size at Each Data Point Collection, by Year Group

	Group		
Year	**Start of 2011–12 academic year (DP1)**	**End of 2011–12 academic year (DP2)**	**Start of 2012–13 academic year (DP3)**
Year 7	3926	2632	2508 (Year 8)
Year 8	3039	1964	1646 (Year 9)
Year 9	2716	1804	1514 (Year 10)
Year 10	2127	1532	1343 (Year 11)
Year 11	1835	767	143 (Year 12)
Total	13,643	8699	7154

[1] It should be noted that self-report perceptions of teaching were also captured directly by the teachers, through a teacher questionnaire (administered twice during the course of the first academic year of the study, 2011–12—in other words, along students' first and second data point collections).

Qualitative data collection was also carried out through case studies in two secondary schools, involving student and teacher interviews and lessons observations. The qualitative analysis presented in this chapter is based on data from the student interviews, about their views and experiences with school mathematics and their aspirations toward the subject. The sample of students was drawn from various year groups with selection criteria aiming to provide sample diversity in terms of gender, year group, ability set, and socioeconomic background. Table 2 shows the distribution of the interview sample in the two schools across year groups and by gender group at DP1.[2] Longitudinal data were established by interviewing most of the students at another instance, with an average interval between the two interviews being two to three academic terms (for school A DP2 took place during the next academic year). A total of 55 students (22 from School A and 33 from School B, with similar distributions of gender as DP1) were interviewed twice and seven new students took part in DP2, making the total number of interviews 103 and a total of 161 interviews.[3]

Analytical Approach

Quantitative Approach

The first step in our quantitative analytical approach is the validation of the constructed measures, which were developed following previous work (Pampaka, Hutcheson, & Williams, 2014; Pampaka, Williams, & Hutchenson,

TABLE 2 Summary of Interview Data by School, Gender, and Year Group (at DP1)

	Group						
	School A			School B			
Year	Female	Male	Total	Female	Male	Total	Total
7	9	5	14	6	4	10	24
8	2	3	5	5	6	11	16
9	6	3	9	3	8	11	20
10	2	2	4	6	5	11	15
11	8	8	16	3	2	5	21
Total	27	21	48	23	25	48	96

[2] For School B this was about the same time as DP1 of the surveys (Dec., 2011) with the DP2 again at the end of the year (as survey DP2). However, for School A, DP1 for interviews tool place over Mar. of the first academic year, with DP2 taking place the next academic year (about a year after).

[3] Students were interviewed individually, with the exception of one pair and one group interview of three at DP2.

TABLE 3 Constructed Measures and Their Working Definition

Measure name	"Construct" description
Math disposition	Measure related to expressions of behavioral intention for future engagement with mathematics (the higher the score, the more disposed the student is toward further study or engagement with mathematics)
Math self-identification	This measure is constructed based on items that express mainly feelings and preferences toward mathematics (the higher the score, the more positively/strongly the student relates or identifies with mathematics)
Math self-efficacy	Confidence in solving mathematical problems
Teaching variation	The higher the score on this measure, the more diverse the math lessons are from students' perspective
Transmissionist teaching scale	The higher the score, the more "traditional" or teacher-centered the practices are as reported by the students

2012; Pampaka, Williams, Hutcheson, Wake, et al., 2012). This measurement step is based on the assumption that there is an underlying construct behind the items in the questionnaires, which were brought together after studying previous research literature and looking at other relevant instruments. Following this, with our questionnaires we intended to measure constructs such as mathematics disposition, self-efficacy, perceptions of transmissionist teaching practice, etc. Those constructs of interest to this chapter are shown in Table 3 along with their broad working definition.

Given the students' responses to the relevant questions we then attempt to validate these aforementioned constructs: in other words, to check whether they exist as "measures" (or scales), and, if not, if there are other dimensions relevant and useful. Our validation process, thus refers to the accumulation of evidence to support validity arguments regarding the students' reported measures. For this purpose, we employed a psychometric analysis, conducted within the Rasch measurement framework, following relevant proposed guidelines (Messick, 1989; Wolfe & Smith, 2007a, 2007b). In most cases we employed the Rasch rating scale model, which is the most appropriate model for Likert type items. Our decisions about the validity of the measures are based on different statistical indices, such as item fit statistics, category statistics, differential item functioning, and person-item maps (Bond & Fox, 2001). More information about the construction of the main variables of interest herein—that is, those related to attitudes (math disposition and self-identification) and the transmissionist teaching scale—is presented elsewhere (Pampaka & Williams, 2016; Pampaka & Wo, 2014). It should be noted, that for the purposes of these analyses and in order for all the relevant statistics to be checked, we use the pooled dataset that includes all available responses in all three data points.

Under our general analytical framework (e.g., as also described in Pampaka, Williams, Hutcheson, Wake, et al., 2012; Pampaka et al., 2014), further statistical modeling usually involves the use of these constructed measures to respond to focal research questions. For the purpose of this chapter, we decided to first revisit the items related with emotionally-related constructs (such as maths self-identification and dispositions), drawing on the survey responses, and observe the trends in their development over the period of the study (thus presenting change). We then detail emerging themes relating to this change from the qualitative analysis (interview data) and link these with further evidence from the patterns found in the whole survey sample. It should also be mentioned that for the descriptive analysis of change over time for the main construct of interest (i.e., math disposition) we draw on all available responses in each DP (as detailed in Table 1), whereas we restrict the presentation of relationships (between emotions and pedagogic practices) to the responses of DP2 because we found them more closely related to the discussions held during interviews.

Qualitative Approach

The idea that learning is intrinsically sociocultural requires for its study methodologies and tools that are fit to understand this complex process, and "to 'make sense' of the world from the perspective of participants" (Eisenhart, 1988, p. 103), which accords to the approach we employed with our in-depth interviews.

The interview transcripts were initially assigned to the software package NVivo for coding and analysis. An initial coding frame was developed based on the interview schedule topics and expanded with additional free nodes corresponding to new ideas emerging from the interviews. For the purposes of the evidence sought for this chapter, further steps were followed. A first search involved finding all instances of the basic emotions as described in Hannula (2002): happiness, sadness, fear, anger, disgust, and interest. Searching was performed not only for these words themselves but also for other synonyms or words that are normally associated with these emotions, for example, enjoyable, glad, annoying, etc. This was done by the second author following the above agreed criteria. The resulting quotes were then subjected to a thematic analysis (Boyatzis, 1998) by the first author, guided by the initial coding frame and resulting in the themes and subthemes that are described later on Table 4. Finally, quotes were recoded for positive and negative emotions depending on the intentions discerned by the context and ways in which they were told in the interview. So, for example, when a student said, "I just felt really happy because I could sit and actually do the work," the quote was classified as positive, while an expression such as "I was just happy that the lesson finished" would have been classified as negative. Such way of classifying

TABLE 4 Positive and Negative Emotional Reactions in Students' Interviews

	Emotional reactions (year group): # quotes			
Themes	**Positive**		**Negative**	
Pedagogy Classroom activities	(7):1, (8):3, (9):1, (10):2	7	(7):2, (9):1, (10):6, (11):3	12
Pedagogy Social pedagogies	(7):1, (8):2, (9):6, (10):3, (11):1	12	(7):1, (10):1	2
Pedagogy Teachers	(7):1, (8):3, (9)(9):2, (10):3, (11):4	13	(7):2, (8):2, (10):2, (11):5	11
Mathematics General views	(7):2, (8):1, (9):2, (10):2, (11):1	8	(7):3, (8):7, (9):3, (10):1, (11):5	19
Mathematics Views on particular topics	(7):1, (11):3	4	(7):1, (8):2, (9):1, (10):2, (11):4	10
Aspirations	(7):3, (8):3, (9):4, (10):1, (11):2	13	(9):1	1
Total		57		55

emotions is in line with our theoretical position in which emotions are intrinsically related to the sociocultural context in which these happen. This coding and its interpretation were confirmed and agreed by both authors. We will now discuss our results.

RESULTS

Declining Dispositions Toward Mathematics in Secondary School Education

We first present the results of the quantitative analysis on attitudes toward mathematics. In Pampaka and Wo (2014) we have reported how we performed the preliminary validation of the measures related to mathematics attitudes, following the measurement approach described earlier. The results from this validation did not support our initial assumption of the existence of an overall measure of "mathematical attitudes"; in addition, dimensionality analysis (Smith, 1996) also supported the existence of two dominant dimensions, which we further analyzed, and resulted in two measures, which we called "mathematics identity or self-concept" and "mathematics disposition" (as described in Table 2). Both constructs included items related to emotions toward mathematics. Fig. 1 shows

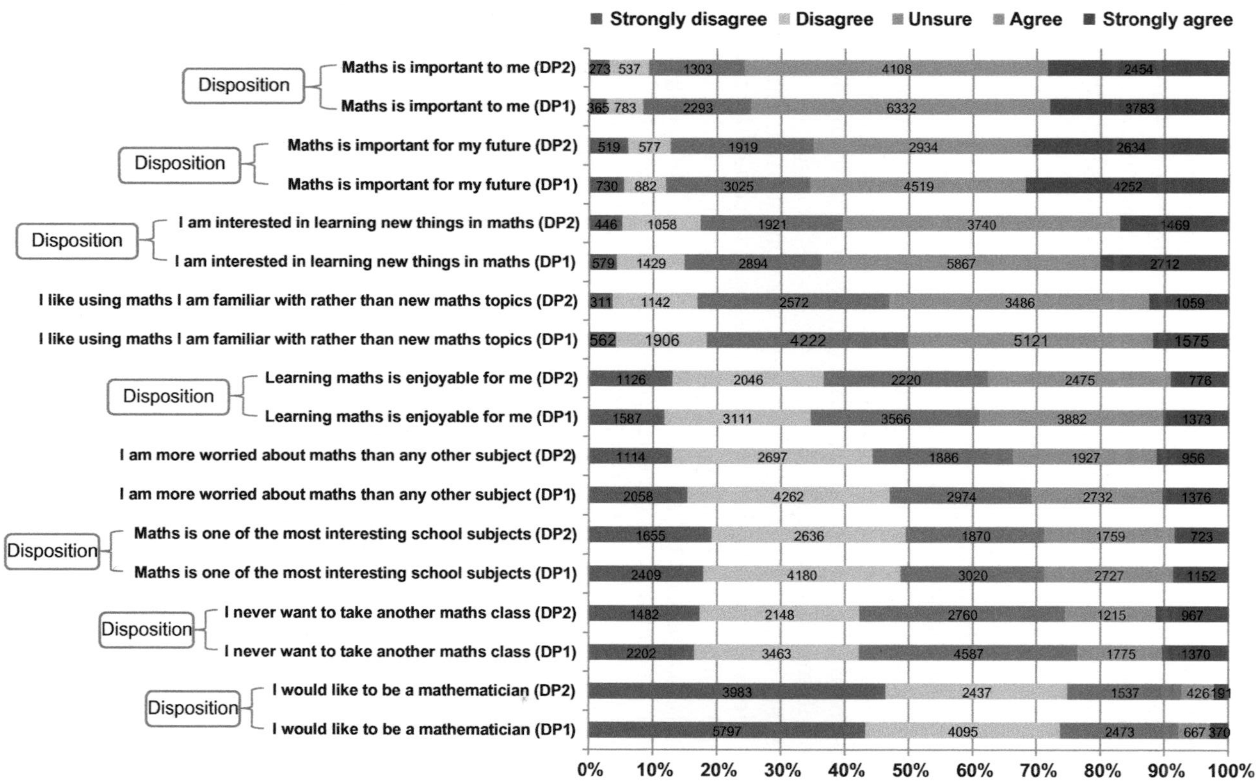

FIG. 1 Distribution of students' responses at DP1 and DP2 on items related to emotional reactions to mathematics.

the distribution of students' responses in these items (plotted in order of agreement for easier interpretation). Some of these items are analyzed in more detail later on in this chapter where we focus on emotions, along with the qualitative analysis.

Most of these items fall under the disposition measure, as indicated with Fig. 1; exception to this is item "I like using maths I am familiar with rather than new maths topics," which comes under the self-identification measure and item "I am more worried about maths than any other subject," which did not fit either of the two measures. Therefore the measure of mathematics disposition is explored in more detail herein. It should be noted that apart from the seven items presented in Fig. 1, the construct was based on two more items ("I would prefer my future studies to include a lot of maths" and "I would look forward to studying more mathematics after school") that were considered more as indicative of preferences rather than emotions. Fig. 2 shows the resulting measurement scale (on the left) with the students' scores and items' difficulties.

At the left end of the figure the logit scale is shown; this is the common measurement scale for both items used to construct the scale and the persons who responded to these items (i.e., students). As a term, "logit" comes from the term "log-odds unit" and ranges from $-\infty$ to $+\infty$. The term odds (of success) is simply defined as the probability of success divided by the probability of failure. The construction of a logit scale then, based on the Rasch model (as a theoretical base), is given through the following equality that brings together respondents' ability and items' difficulty on this common measurement scale:

$$\log\left(\frac{\text{Probability of success}}{\text{Probability of failure}}\right) \equiv \text{Ability} - \text{Difficulty}$$

At the left side of the map the students' distribution in the scale is shown (each # represents 124[4] students, each "." represents 1–123 students). The higher the place of a student in that scale, the more mathematically disposed the student is. On the right-hand side of the students' "histogram" the items that constitute the scale are presented, ranging from the easiest to agree with (bottom) to the most difficult.

As shown in Fig. 2 items related to emotional reactions (such as items 4, 5, and 8) are positioned in the mid-range of this scale, in between easier items to endorse, such as those related to the importance of mathematics,

[4] It should be noted that for validation purposes (i.e., in order to test for and ensure measure invariance over time) the data from all data points are pooled together.

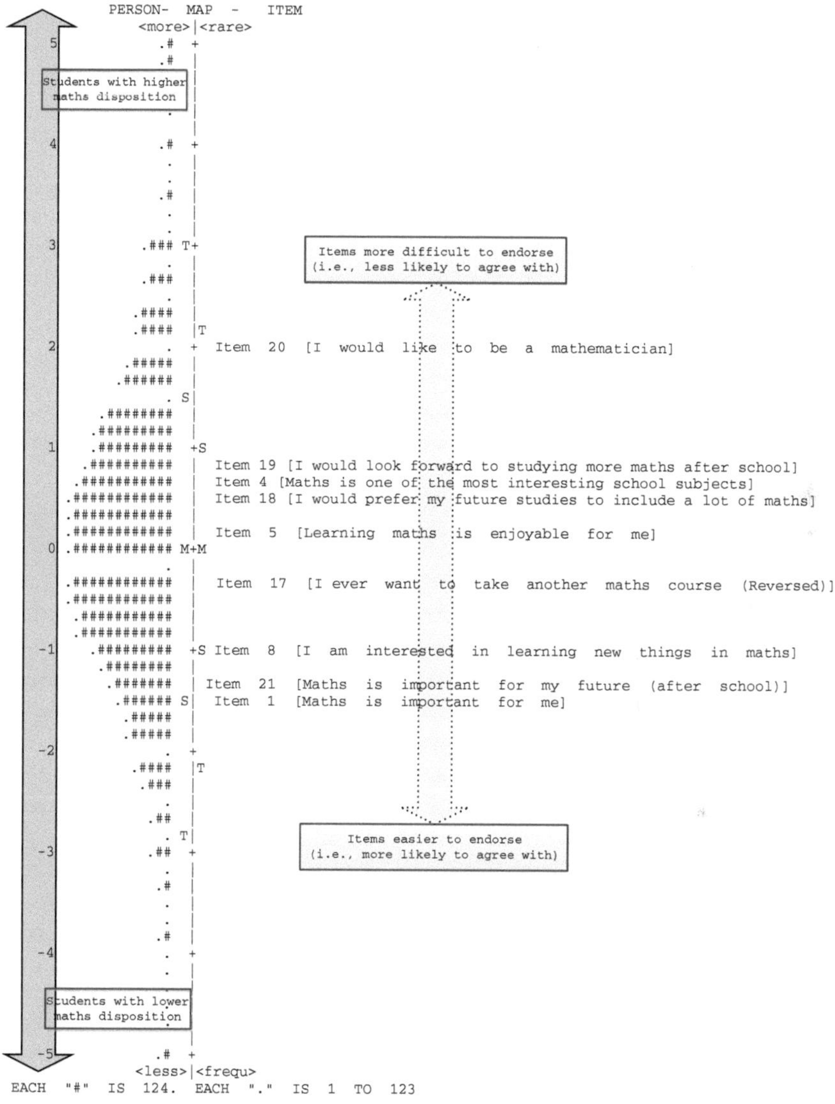

FIG. 2 The hierarchical scale (person-item map) of mathematics disposition.

and the most difficult items to endorse, such as those denoting intention to study more mathematics or even be a mathematician.

Further analysis using the derived measure of math disposition provided objective evidence of a decline in students' dispositions toward mathematics throughout secondary school education, shown in Fig. 3. Within the same academic year (from DP1 and DP2 of the

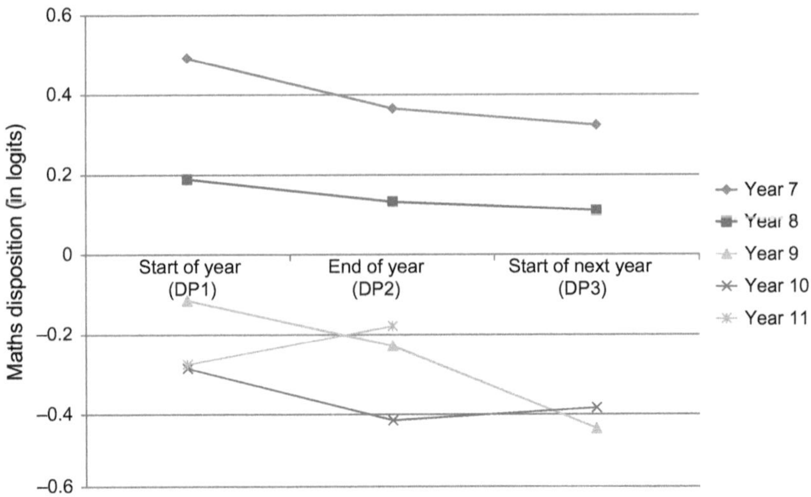

FIG. 3 Students' average math disposition by year group and DP.

study) there is also a considerable drop when looking at year groups, which provides evidence that students in older year groups present lower math dispositions. Exception to that is Year 11,[5] which also presents some interesting patterns along with Year 10. We see, for example, that at the start of the first academic year, students in Year 11 score about the same in these measures as students in Year 10 and in contrast to all other year groups their scores increase during the same academic year. An increase is also observed with the Year 10 students, as they move into Year 11 (i.e., from DP2 to DP3 of the study). These patterns might have to do with decisions already made (e.g., early entrance to the GCSE exam in Year 10), as well as teaching practices related to exams (e.g., revision of past papers, mock examinations, and grades).

The decline in dispositions was also found to be statistically significant within more advanced longitudinal models that are presented elsewhere (Pampaka & Troncoso, 2014), and it is also in agreement with patterns observed with older students in previous work (e.g., Pampaka, Williams, Hutcheson, Wake, et al., 2012): these findings are not explored more in this chapter due to scope and space limitations. Rather, we turn to the student narratives to explain the reasons of this decline.

[5] For Year 11, DP3 sample size was very small, (~100) coming from only one very high attaining school; therefore, it was not used for this analysis since it is not representative. DP2 for Year 11 might also be biased since, as shown in sample description at Table 1, there is more than 55% dropout in the sample.

Students' Emotional Reactions to Mathematics in Secondary School Education

We now present a more focused investigation of emotions through the qualitative analysis, combined with evidence from the questionnaires. Following the analytical approach described in the "Methodology" section, the qualitative analysis resulted in 112 relevant quotes from students' interviews, categorized into the themes shown in Table 4.

Theme 1: (Pedagogy) Classroom Activities

Analysis showed that mathematics classroom activities that supported personal interests, and helped the process of learning, were described in terms of positive emotions. The following is an example (Year 8 student):

Student: I couldn't do like... my times tables were out the window, everything, didn't understand maths at all. The only thing I could do were small numbers add and minus. I couldn't do divide I could do like my five times tables, my tens times tables. I know all my times tables now, which is brilliant, I'm really happy with that 'cos at the end of Primary school what we all done and at the start of High School we all made songs up of like the nine times tables and it just made me... because I like singing, it made me take a bit more interest in it 'cos we made up a song and you want to learn the song because you like singing plus you get a new song into your head and you get the new times table into your head.

Other students mentioned "practical" lessons (e.g., building bridges, Year 10 student) or lessons when something "different" happens; these might or might not have supported their interests but were typically described as "fun," "interesting," or not "boring" and, more importantly, not frequent (e.g., "I have not been in for many"). Some of these were memories of previous experiences from primary school. For example,

Student: I remember in Primary we were told like the nine times table, if you wanted to find out like three nines you put your third finger down and it told you your numbers and I was quite shocked at that.

Other practices described in positive emotional terms were those related to setting (moving up) and grades (getting higher grades than expected). For instance,

Student: In my last test I got a 5B which in my eyes was really good because I've always been getting threes and I'm happy with it. I can't say how happy I was when I got that 5, 5B and in my last test I got 5A. I was absolutely over the moon.

Practices described in negative terms mainly related to "routine" (e.g., "They are all the same"—Year 10 student), "copying" what the teacher does and then doing exercises, which represents the majority of practices in secondary school. The following is a typical description of a mathematics lesson that many students from all years refer to in similar terms:

Student:	We normally copy something off the board and he (the teacher) explains it and gives us a textbook and we have to do a page on it.

Many students, when asked how mathematics can be made more "fun" or less "boring," described practices such as being "relaxed," "interactive (social)," "involved" (i.e., students being more active), and using the computers more. An important feature in these students' discourses was a claim for variety in the practices of school mathematics. For example,

Student:	I just think they should do like... they should make maths more fun because people would want to like come and enjoy it a bit more and be more involved in it, at the moment people are a bit bored.
Int:	Okay, you think maths should become more ...?
Student:	Fun and like interactive and you'd be able to use more. I don't... like more computers and other like techniques and learning, not just writing all the time.

Sometimes, a positive emotion toward mathematics in general can be counterbalanced by practices (e.g., repetition) that evoke negative emotions, like this student from year 8 expressed:

Int:	Overall about maths, do you feel happy doing maths?
Student:	I feel happy, but I feel a bit bored when we are doing things that we went over in earlier years.

The quantitative analysis reflected the fact that students reacted emotionally positive when they perceived variety in their mathematics lessons. Fig. 4 shows the relationship of students' perceived teaching variation and their responses to emotion-related items at DP2.

As can be observed students in classes that they perceived as more varied in regards to teaching practices also tend to report that "learning maths is enjoyable" to them. This pattern is more emphasized for Year 7, 8, and 9 students while this effect is attenuated for students in Years 10 and 11. As already said before, these patterns might have to do with decisions already made (e.g., early entrance to the GCSE exam in Year 10), as well as teaching practices related to exams (e.g., revision of past papers, mock examinations, and grades).

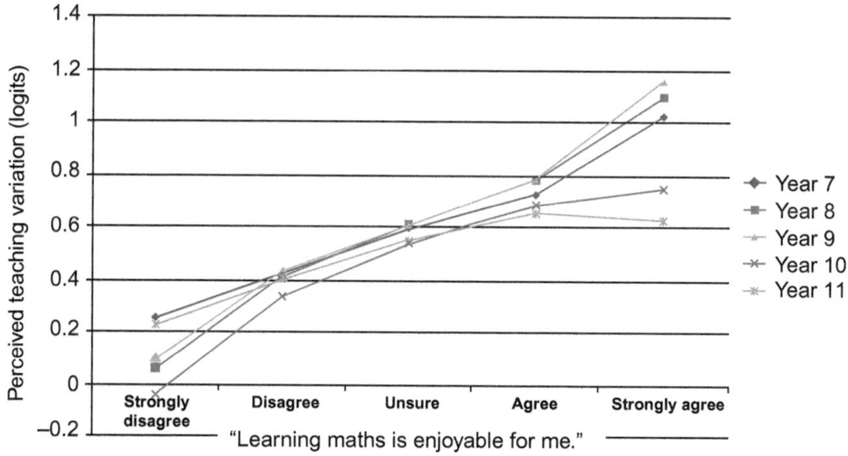

FIG. 4 Students' perception of teaching variation based on responses on positive reactions (item 5: Learning mathematics is enjoyable for me).

Some students relish more challenging work (e.g., "I like looking at problem solving, challenging things like that"—Year 10 student). When this challenge is met, some students have a positive emotional reaction toward mathematics. The following is an example (Year 7 student):

Int:	You found it a lot harder to adapt (to secondary school)?
Student:	But more challenging, so I like it like that.
Int:	So it was a big challenge to adapt?
Student:	From like primary school, it is like a massive leap.
Int:	So what about maths, how does maths fit into your liking?
Student:	Maths seems good, but the work is challenging, which I like, and I think we have moved on more than any other maths group. I would like to use the computers more.

Theme 2: (Pedagogy) Practices That Promote Social Interactions

A theme that was recurrent in students' accounts was that of being able to socialize in the classroom. The majority of students that refer to this practice did so in a positive emotional way. For example (Year 9 student):

Int:	OK, were there any times you felt happy doing maths?
Student:	Yes, when I am sitting with my mates and I can talk. We are allowed to talk and help each other in every lesson.

These students often related the social aspects of a lesson with better learning and having "fun." These characteristics refer back to those rare pedagogical practices that depart from the normal, "boring" lessons. For example (Year 10 student),

Int:	Great; so going back to the lessons this year, what was the most enjoyable lesson you can remember this year? Something that may stay in your memory.
Student:	Probably like at the beginning doing team work and like creating posters, like we were doing trigonometry and we had to create a poster and make it colorful and they're up on the walls in there and partnership and we made like... we can do it ourselves and Miss told us like we could do it in groups and we had some examples and we were still working and stuff doing maths, but it was more fun.
Int:	It was fun, more fun?
Student:	Yes because it was more colorful so it helps you remember because you had to be good at designing it, so helps you remember.
Int:	So if you had to change something, the most recent lessons you had, what will that be?
Student:	I'd say probably do more speaking work and more work on the board and probably maybe like more computer work.

There were a few students that expressed negative emotions toward group work or working in a social environment, usually associated with an unruly or chaotic classroom or with a desire to "get on with the work." The following is an example (Year 10 student):

Int:	Why don't you like group work?
Student:	Things take longer in groups; you have to discuss everything and you have to agree on things. I just prefer to get on with it.

Fig. 5 shows that students that perceived their mathematics lessons as more "sociable" also find their mathematics more enjoyable (and vice versa: in other words, those who reported that they never engage in such "social" practices are much less likely to enjoy their maths lessons).

Theme 3: (Pedagogy) Teachers and Their Practices

Teachers were a recurrent theme in students' emotional descriptions of their mathematical learning. A "good" teacher would evoke positive emotions toward mathematics. For example (Year 9 student),

Student:	I feel like happy but sometimes if I don't get what she's asking me to we just have to ask her and then she explains it to us, so it makes me feel happier and a bit more secure doing maths.

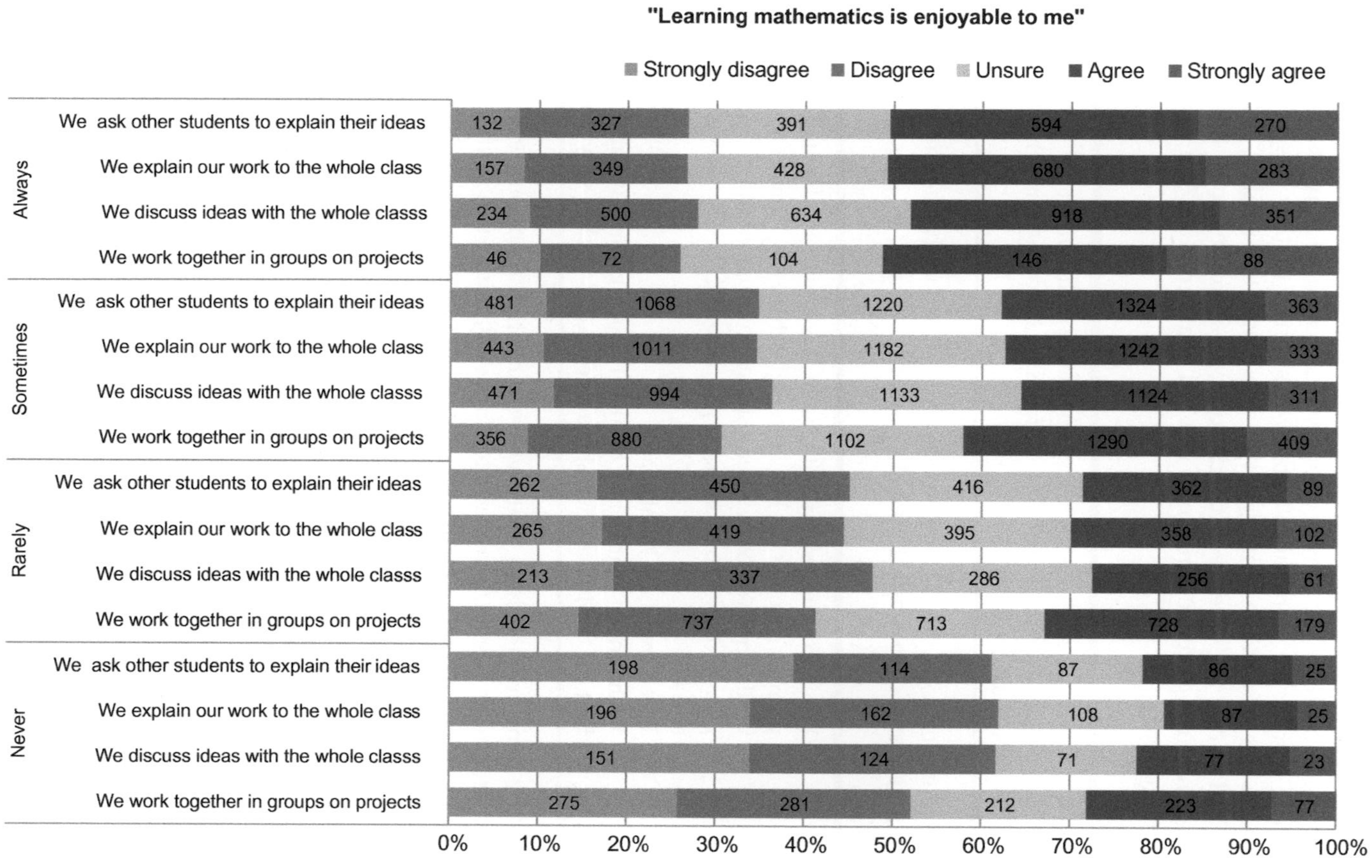

FIG. 5 Distribution of students' answers to item 5 (Learning mathematics is enjoyable to me) vs their perception of some "sociable" aspects within their mathematics lessons.

A good teacher was described with terms like "fun," someone that "explains" well or as many times as needed and usually related to "understanding" (e.g., "She explains it so well so you understand it"—Year 8 student), someone "kind," "patient," or that has a "nice" approach to students, or someone that facilitates the kind of practices described above in "Theme 1: (Pedagogy) Classroom Activities" section as creating positive emotions toward mathematics. For example (Year 8 student),

Student:	All the lessons with Mr. (name) were really enjoyable. He always tries to make it... different; it is not the same thing. It is always different.

In a few occasions, "good" teachers were credited with student achievement in grades. For example (Year 11 student),

Student:	I used to struggle a lot with maths, I was very good at it in primary school, then I went to high school and I was in the top set until about Year 9 and then I started to struggle with it a lot. Then I was moved into Miss (name)'s group and I used to be getting F's and E's.
Int:	When was that? What year?
Student:	Beginning of Year 10 and by Year 11 I was getting A*, mainly because of Miss (name); she is a very good teacher, very good.

On the contrary, teachers that evoked negative emotional responses were described as "mean," "miserable," not explaining sufficiently or efficiently, or having a strict, almost unreasonable, control over the classroom. For example (Year 11 student),

Student:	Yes, I like to sit near people I feel comfortable with, and if I am struggling, I can ask them what to do; but normally they (the teachers) will say, "Sit on your own, sit on this seat," and you end up... if you can't do a question, you end up not being able to try it because you can't ask someone the way to do it. Miss will normally get mad at you and she will shout, "We have gone through this before" and then you are like "I just forgot it." You can't remember everything you do, because I try my best to get everything right in my working out. I will end up forgetting something, because there is only so much that you can keep hold of.

Sometimes, a "bad" teacher can counterbalance a positive disposition toward mathematics. For example (Year 7 student),

Student:	I actually do like maths, but not the way she (the teacher) runs the classroom. It's annoying, you don't learn basically; we just do one thing for a lesson. We don't carry on learning about it, so we do tests about it, assessments.

It is worth noticing that the number of positive quotes and negative quotes in themes "Theme 1: (Pedagogy) Classroom Activities" (classroom activities) and "Theme 3: (Pedagogy) Teachers and Their Practices" (teachers) are almost the same; students will encounter good and bad mathematics practices and teachers throughout their education, but the general declining trend in dispositions (Fig. 3) seems to suggest that there is something more than just practices and teachers that make mathematics unappealing to many. Also, the teaching practices that encourage social interactions were found to be highly related to positive emotions (12 positive quotes against 2 negatives), and therefore they could influence dispositions in a positive way, but these are very rare (see Pampaka, Williams, Hutcheson, Wake, et al., 2012). We now consider students' emotional reactions to mathematics itself.

Theme 4: Mathematics

Discourses of dispositions toward mathematics itself contained emotional aspects related to students' views of mathematics in general or their views of particular mathematical topics, but all were related to their perceived ability to do certain mathematical tasks.

When students expressed a positive emotion toward mathematics itself, they did so in relation to a perceived personal ability or in relation to "being able to do the work." For example (Year 9 student),

Student:	I find it (mathematics) pretty much easy because even if I don't know it, when the teacher first teaches it because the teacher asks you if you understand it from the last lesson and she goes over it. So after one or two lessons you kind of get the hang of it and since it has to do with numbers, even if it is just algebra with the letters, I still find I can still remember numbers easily. So I find I don't forget maths easily; that is why it is one of my favorites.

And this Year 11 student said,

Student: I like algebra. I do it now because I can do it. It sounds really silly but because I can do something I really enjoy doing it because it is like, "That is hard, but I can do it," and it feels like you have achieved something. I know it sounds really silly, but I used to love to sit at home and do it because we had all these revision papers and at first I looked through them and I couldn't do much, but then as it progressed I could do loads and I just felt really happy because I could sit and actually do the work. It was great!

The quantitative analysis reflected the fact that students that perceived themselves as "able to do the maths" enjoyed mathematics more often. Fig. 6 shows the distribution of students' answers to item 5 (learning maths is enjoyable to me) in relation to their perceived ability in mathematics.

In addition to talking about "being able to do the maths," many of these students related this to "understanding," something that provoked a positive emotional reaction. The quantitative analysis shows that students enjoy mathematics most when they feel their lesson is "about right." Fig. 7 shows the distribution of students' responses to item 5 (learning

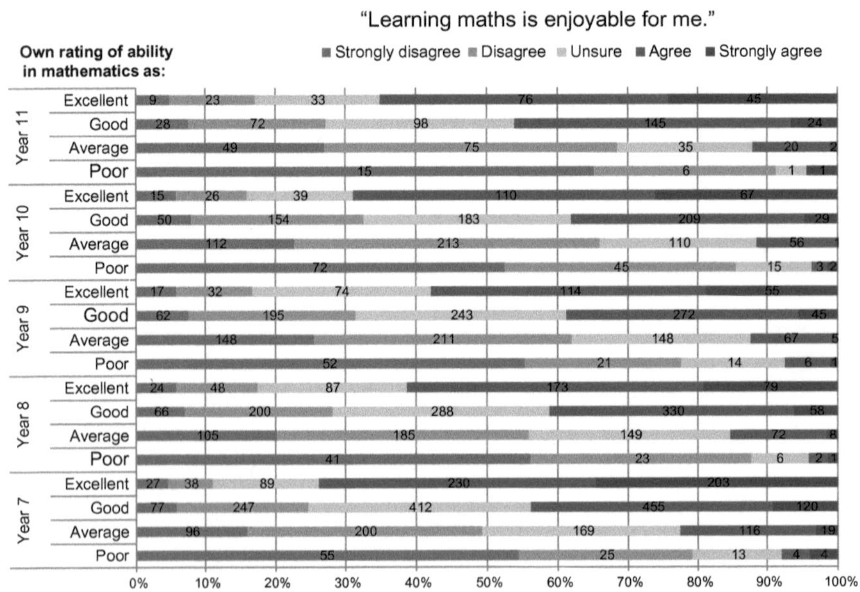

FIG. 6 Distribution of students' answers to item 5 (Learning mathematics is enjoyable to me) vs their perceived ability to do mathematics, per year group (data from DP2).

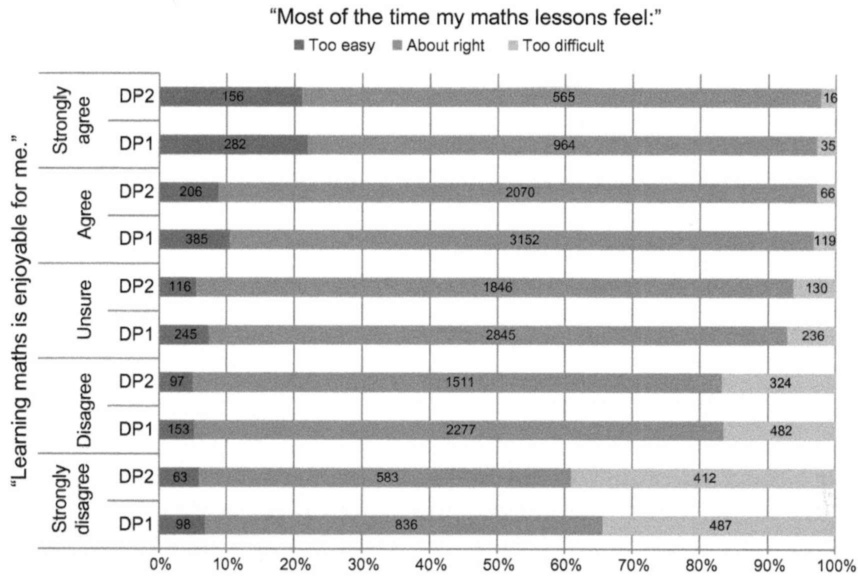

FIG. 7 Distribution of students' responses to item 5 vs their perception of difficultness in their mathematics lessons (data from DP2).

mathematics is enjoyable for me) against their perception of difficultness in their mathematics lessons.

However, we could not find an instance where this understanding was related to something other than grasping a procedure or being able to follow how to solve an exercise or answer a question (related to the test). These students did not talk about understanding a concept (e.g., What is a number?) or how a particular concept related to other concepts within or outside mathematics; even though students had a positive emotional reaction to the fact that they could do the work required from them by their teachers/school, they did not have the same reaction to mathematics itself as a subject. The following quote from a Year 8 student illustrates this point:

Int:	Can you tell me which of the mathematics you found ...?
Student:	I found them all pretty easy.
Int:	What about that (an exercise in her textbook)? What will be the answer to that one, number twenty: Calculate minus six minus plus three?
Student:	Do the ones in the brackets first and then... (thinking) Erm... I don't know what it is but I know I can do some of them.
Int:	Do you like this kind of exercises?
Student:	Erm... no, not really but I can do them.

A further example comes from a Year 10 student:

Int:	So how is this "problem solving"... can you describe an incident when you were doing problem solving?
Student:	Well, with like algebra and stuff like, solving equations.
Int:	Solving equations... you like that?
Student:	Not particularly enjoy it but I don't mind it.
Int:	Do you feel happy in maths?
Student:	I feel happy when I accomplish it.
Int:	When it's over?
Student:	Yeah, like when I see... like understand the task and do well in it and I can say I understand how to do this.

It is not surprising therefore that many students describe mathematics as "ok, but it's not one of my favourite" (Year 11 student): they can do it, and they are happy about that, but mathematics as a subject itself is not something that evokes particular positive emotions for many. The number of negative emotional quotes clearly surpassed the positive ones in our analysis. This point was also evident in the quantitative results when students were asked to report their favorite or least favorite subject in school. Table 5 shows the combined results from all students' responses (all DPs and all year groups) in our secondary school sample, which show that mathematics tops the list of least favorite school subjects with considerable difference from the next entry.

TABLE 5 Frequencies of Top 10 Choices of Favorite/Least Favorite School Subjects

Favorite subject	Count	Least favorite subject	Count
PE	8308	Maths	5067
Art	4661	English	3293
English	1958	RS	3195
Maths	1924	Science	2714
Drama	1703	French	2620
Science	1635	Geography	2249
ICT	1143	History	1940
History	1115	PE	1296
Music	1110	Art	865
Technology	705	ICT	826

Some negative emotional reactions to mathematics relate to the lack of conceptual understanding that we have already talked about here. For instance (Year 11 student),

Student: I don't know why we do simultaneous equations. They are really annoying.

Doing something without a clear rationale for doing it evokes negative emotions. This was particularly the case with statistics, a topic that many students did not care about. For instance, this Year 11 student said the following:

Student: If I could change something it would be like statistics... not, not to do it, because if you are struggling with other subjects in the school, you could have a free lesson to do extra stuff in that, instead of doing statistics, which you don't really care about; you are just being made to do it.

In fact, when some students expressed negative emotions toward mathematics they did so very strongly, as in the quote in the title of this paper (Year 11 student):

Student: I did used to like maths, but then I found it's just really difficult... yes, I liked it in secondary school at one point because when I work in the cafe I am really good at adding numbers up and I don't need a calculator or anything, and because I am good at when people say: "I don't want this," you can easily take it away. It is that and I have got my own system to do it, but when the teachers say: "You use this system, it gets you more marks," it quite confuses you. Because if the person who is marking your test doesn't know what you mean, you will lose marks because he thinks you have done the system working out wrong. And if you can't do it, then you get the answer wrong; that is annoying.

Finally, "not understanding" evoked negative emotions for many students. For example (Year 8 student),

Student: Like if I don't understand something it will like put a bad perspective on it, something like that part of maths I don't like it anymore.

THEME 5: ASPIRATIONS

Some students talked about their aspirations, what they imagined the role of mathematics would be within those aspirations, and the type of emotions that this cause in them. Many talked of mathematics as needed for their future plans and being potentially happy if this involved some of the mathematics they are able to do. For example (Year 9 student),

Int:	But would then... if what you were doing in the future involved maths, would that put you off in any way or would you...
Student:	It depends what kind of maths... like if it was drawing shapes and planning out stuff and like doing loads of area stuff I wouldn't like... be happy, but if it was just the stuff I can do, I'd be alright with it.

It is not surprising that the quantitative analysis showed that students that perceive themselves as "able to do the maths" have positive dispositions toward the subject and would consider studying more mathematics in the future, as shown in Fig. 8.

For a few, however, knowing that mathematics could be an important part of what they envisage themselves doing in the future was a motivator

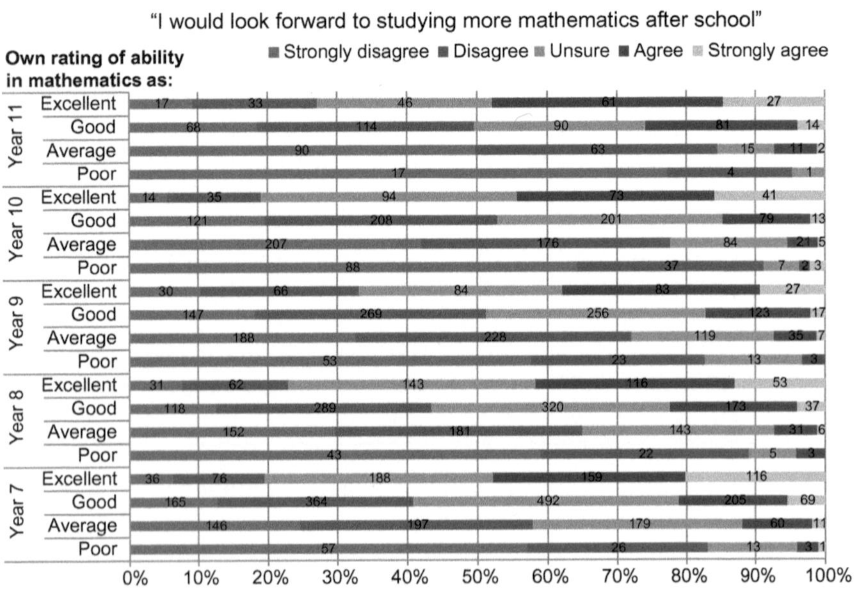

FIG. 8 Distribution of students' answers to item 19 (I would look forward to studying more mathematics after school) vs perceived ability to do mathematics, per year group (data from DP2).

for engagement. For example, a Year 8 student that wants to become a teacher said

Student:	If I do become a maths teacher I want to work hard and make sure that I know everything.

And a Year 11 student who wants to become an architect said

Student:	Well, maths is quite important for me, because I want to be an architect.
Int:	You want to be an architect, so you know from now on…
Student:	Yeah, so I know maths is really important and I'm taking it at college.

Even if a student had not yet decided their future career, the value of mathematics might dispose them positively toward mathematics, like this Year 9 student:

Int:	Is it important for you to be the best at maths?
Student:	I wouldn't say that, it's just easier in later life. And like for jobs you have to be good at maths like hard jobs, like good jobs so…

Fig. 9 shows the distribution of students' answers to item 21 (Maths is important for my future (after school)) in relation to their perceived ability to do math. We conclude that this relation supports the previous interpretation that knowing that mathematics will be important for a future career can be a motivator for engagement with mathematics, creating a positive disposition toward the subject. Therefore, we suggest that the relationship between perceived ability and dispositions toward studying more mathematics (aspirations) is a relational one; in other words, one influences the other and vice versa.

DISCUSSION AND CONCLUSIONS

Our analyses show how emotional aspects were measured as part of students' attitudes to mathematics, and in particular their mathematics dispositions. We then provided evidence of students' declining dispositions toward mathematics throughout their secondary school education. This was apparent in both differences between year groups, with a progressive drop as students get older (from Year 7 to 11) as well as during one academic year, when looking at the averages of students at the start and end of the year within the same year group. In the remaining of the

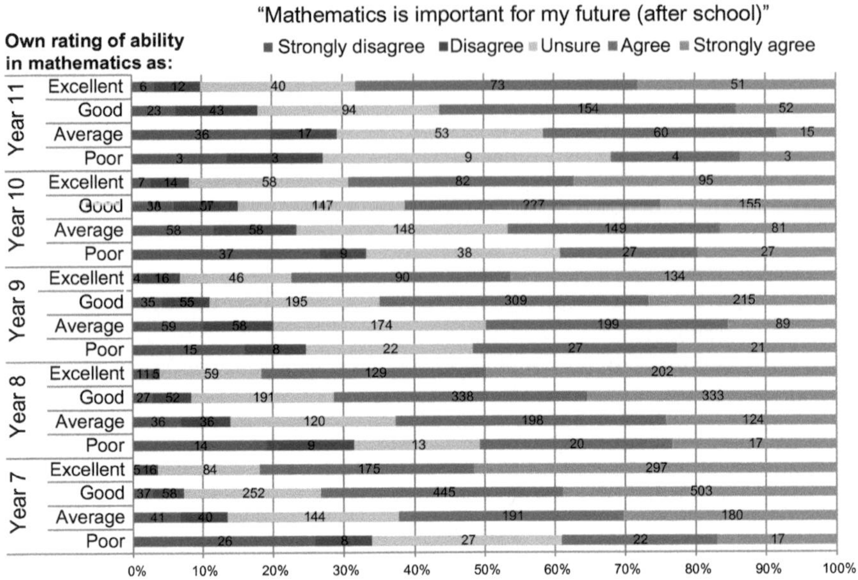

FIG. 9 Distribution of students' answers to item 21 (Mathematics is important for my future (after school)) vs perceived ability to do mathematics, per year group (data from DP2).

chapter we then attempted to provide some explanations for this decline from the students' perspective and focus on their narratives regarding emotional reactions. In line with our sociocultural viewpoint, we see emotions as dynamic and carrying the potential to change dispositions, which are our "way to see the world" in which we participate (Gresalfi, 2009), and that "world" (i.e., the sociocultural context) in turn afford or restrict that participation (Edwards & D'Arcy, 2004).

When students participate in the world of education, they engage in different practices. As presented, most mathematics school practices were described by students as routine and boring, and such practices were situated within negative emotional terms. In contrast, when students talked in positive emotional terms about pedagogy ("varied," "fun," "relaxed," "involved," or "social"), they did so in relation to practices that did not happen very often. This could be responsible for many students ending their compulsory mathematics education with a disposition that "mathematics is ok, but not my favourite subject." Our results provide more evidence in support of Gresalfi's (2009) argument that *"social, affective, and motivational factors... are central to and inseparable from the learning process"* (p. 327).

Indeed, it was noticeable how many students talked in negative emotional terms about mathematics; and even when they had positive emotional reactions to it (e.g., when they "understood" a piece of mathematics) these referred to instances where they could perform a process rather than experience conceptual understanding. The pedagogical practices in which they participate could only afford them this type of procedural understanding, creating as a consequence students that are "good" at mathematics (in terms of grades and attainment) but that are negatively disposed to mathematics. As we saw in our data, many students' dispositions to mathematics are of the form "I don't particularly enjoy it but I don't mind it," even though these students have achieved the grades that would allow them to continue a mathematical education at college or university. Hence, it is worth noticing that many of those students that, at the end of their secondary school education, continue with their mathematical education do so on the basis of a procedural type of understanding. It is not surprising then that when they arrive at sixth form colleges[6] and later at university they encounter that the transitional gap is too big (Pampaka, Williams, Hutcheson, Wake, et al., 2012), and the nature of the mathematics that they thought they were going to study is very different (Jooganah & Williams, 2010). For some students, this produces such a negative emotional reaction that they end up alienated or even worse, they drop out (Hernandez-Martinez, 2016). Elsewhere in our work (Pampaka, Pepin, & Sikko, 2016; Pampaka & Williams, 2016; Pampaka, Williams, Hutcheson, Wake, et al., 2012) we have shown that connectionist (or less transmissionist) pedagogies are related to more positive student dispositions in mathematics at various levels of their education. These pedagogies usually foster a more conceptual understanding of mathematics (Askew, Brown, Rhodes, Johnson, & Wiliam, 1997) and enable students to connect mathematical ideas.

The implications of this study, as we see them, are clear: pedagogical practices in secondary school (and at every stage of education) should afford students a variety of experiences with the subject, and the particularly should include more spaces for socializing, discussing, justifying, and connecting the mathematics to various interests, aspirations, and ways of being of students. This in turn can produce sustained positive emotional reactions that carry the capacity to change, for the good, students' mathematical dispositions. Teachers have a crucial part to play in this change, but they should/could not be alone on this. Institutions need also to enable possibilities and encourage environments in which teachers themselves feel emotionally positive, allowing them to build more positive dispositions toward all aspects of their profession and mathematics itself, which they can then nurture their students.

[6] In England, students aged 16–19 go to sixth form colleges to study for advanced school-level qualifications.

Acknowledgments

This work was supported by the Economic and Social Research Council (ESRC) through grant award RES-061-025-0538.

References

Askew, M., Brown, M., Rhodes, V., Johnson, D., & Wiliam, D. (1997). *Effective teachers of numeracy* (pp. 28–50). London: King's College.

Blunden, A. (2016). Translating perezhivanie into English. *Mind, Culture, and Activity*, http://dx.doi.org/10.1080/10749039.2016.1186193.

Bond, T. G., & Fox, C. M. (2001). *Applying the Rasch model: Fundamental measurement in the human sciences*. Mahwah, NJ: Lawrence Erlbaum Associates Inc.

Bourdieu, P. (1990). *The logic of practice*. Oxford: Polity Press.

Boyatzis, R. E. (1998). *Transforming qualitative information: Thematic analysis and code development*. Thousand Oaks, CA: Sage.

Chapman-Hoult, E. (2010). The search for an authentic methodology in an interdisciplinary study into resilient adult learning (BERA dissertation award 2010). *Research Intelligence, 113*, 22–23.

Chinn, S. (2012). Beliefs, anxiety, and avoiding failure in mathematics. *Child Development Research, 2012*, 1–8.

Davydov, V. V. (2008). *Problems of developmental instruction: A theoretical and experimental psychological study*. New York, NY: Nova Science Publishers.

Di Martino, P., & Zan, R. (2015). The construct of attitude in mathematics education. In B. Pepin & B. Roesken-Winter (Eds.), *From beliefs to dynamic affect systems in mathematics education* (pp. 51–72). Switzerland: Springer International Publishing.

Drake, C. (2006). Turning points: Using teachers' mathematics life stories to understand the implementation of mathematics education reform. *Journal of Mathematics Teacher Education, 9*, 579–608.

Edwards, A., & D'Arcy, C. (2004). Relational agency and disposition in sociocultural accounts of learning to teach. *Educational Review, 56*(2), 147–155.

Eisenhart, M. (1988). The ethnographic research tradition and mathematics education research. *Journal for Research in Mathematics Education, 19*(2), 99–114.

Evans, J., & Zan, R. (2006). Sociocultural approaches to emotions in mathematics education: Initial comparisons. In J. Novotna, H. Moraiva, M. Kratka, & N. Stehlikova (Eds.), *Proceedings of the 30th conference of the International Group for the Psychology of Mathematics Education, Vol. 3* (pp. 41–48). Prague: PME.

Fennema, E., & Sherman, J. (1977). Sex-related differences in mathematics achievement, spatial visualisation and affective factors. *American Educational Research Journal, 14*, 51–71.

Gal'perin, P. (1978/1992). Stage-by-stage formation as a method of psychological investigation. *Journal of Russian and East European Psychology, 30*(4), 60–80.

Gresalfi, M. S. (2009). Taking up opportunities to learn: Constructing dispositions in mathematics classrooms. *Journal of the Learning Sciences, 18*(3), 327–369.

Grootenboer, P. (2003). *Affective development in mathematics*. New Zealand: Auckland Mathematical Association.

Grootenboer, P., & Hemmings, B. (2007). Mathematics performance and the role played by affective and background factors. *Mathematics Education Research Journal, 19*(3), 3–20.

Hannula, M. S. (2002). Attitude towards mathematics: Emotions, expectations and values. *Educational Studies in Mathematics, 49*, 25–46.

Hemmings, B., & Russell, K. (2010). Prior achievement, effort, and mathematics attitude as predictors of current achievement. *The Australian Educational Researcher, 37*(2), 41–58.

Hernandez-Martinez, P. (2016). "Lost in transition": Alienation and drop out during the transition to mathematically-demanding subjects at university. *International Journal of Educational Research, 79*, 231–239.

Hernandez-Martinez, P., Williams, J., Black, L., Davis, P., Pampaka, M., & Wake, G. (2011). Students' views on their transition from school to college mathematics: Rethinking "transition" as an issue of identity. *Research in Mathematics Education, 13*(2), 119–131.

Johnson, R. B., & Onwuegbuzie, A. J. (2004). Mixed methods research: A research paradigm whose time has come. *Educational Researcher, 33*(7), 14–26.

Jooganah, K., & Williams, J. (2010). *The transition to advanced mathematical thinking: Socio-cultural and cognitive perspectives.* In M. Joubert & P. Andrews (Eds.), Proceedings of the 7th British congress for mathematics education. The University of Manchester: BSRLM.

Kleanthous, I., & Williams, J. (2013). Perceived parental influence and students' dispositions to study mathematically-demanding courses in higher education. *Research in Mathematics Education, 15*(1), 50–69.

Lave, J., & Wenger, E. (1991). *Situated learning: Legitimate peripheral participation.* Cambridge: Cambridge University Press.

Lazarides, R., & Ittel, A. (2012). Instructional quality and attitudes toward mathematics: Do self-concept and interest differ across students' patterns of perceived instructional quality in mathematics classrooms? *Child Development Research, 2012*, 11.

Lim, S., & Chapman, E. (2013). Development of a short form of the attitudes toward mathematics inventory. *Educational Studies in Mathematics, 82*(1), 145–164.

Mata, M. D. L., Monteiro, V., & Peixoto, F. (2012). Attitudes towards mathematics: Effects of individual, motivational, and social support factors. *Child Development Research, 2012*, 10.

McLeod, D. B. & Adams, V. M. (Eds.), (1989). *Affect and mathematical problem solving: A new perspective*. New York, NY: Springer Verlag.

Messick, S. (1989). Validity. In R. L. Linn (Ed.), *Educational measurement* (3rd ed.,)(pp. 13–103). Phoenix, AZ: American Council of Education and the Oryx Press.

Niemi, H. (2009). Why from teaching to learning? *European Educational Research Journal, 8*(1), 1–17.

Nolan, K. (2012). Dispositions in the field: Viewing mathematics teacher education through the lens of Bourdieu's social field theory. *Educational Studies in Mathematics, 80*, 201–215.

Pampaka, M., Hutcheson, G., & Williams, J. (2014). Quality of learning support for mathematics in transition to university. *REDU—Revista de Docencia Universitaria, Special Issue Dedicated to Equidad y Calidad en la Docencia Universitaria: Perspectivas internacionales, 12*(2), 97–118.

Pampaka, M., Hutcheson, G., & Williams, J. (2016). Handling missing data: Analysis of a challenging data set using multiple imputation. *International Journal of Research & Method in Education, 39*(1), 19–37.

Pampaka, M., Kleanthous, I., Hutcheson, G. D., & Wake, G. (2011). Measuring mathematics self-efficacy as a learning outcome. *Research in Mathematics Education, 13*(2), 169–190.

Pampaka, M., Pepin, B., & Sikko, S. A. (2016). Supporting or alienating students during their transition to higher education: Mathematically relevant trajectories in the contexts of England and Norway. *International Journal of Educational Research, 79*, 240–265.

Pampaka, M., & Troncoso, P. (2014). *Modelling students' progression in secondary education with the combination of separate cohorts of repeated measurements.* In: Paper presented at the VI European Congress of Methodology, Utrecht, July 2014.

Pampaka, M., & Williams, J. (2016). Mathematics teachers' and students' perceptions of transmissionist teaching and its association with students' dispositions. *Teaching Mathematics and Its Applications, 35*(3), 118–130.

Pampaka, M., Williams, J., & Hutchenson, G. (2012a). Measuring students' transition into university and its association with learning outcomes. *British Educational Research Journal, 38*(6), 1041–1071.

Pampaka, M., Williams, J. S., Hutchenson, G., Black, L., Davis, P., Hernandez-Martinez, P., et al. (2013). Measuring alternative learning outcomes: Dispositions to study in higher education. *Journal of Applied Measurement, 14*(2), 197–218.

Pampaka, M., Williams, J. S., Hutcheson, G., Wake, G., Black, L., Davis, P., et al. (2012b). The association between mathematics pedagogy and learners' dispositions for university study. *British Educational Research Journal, 38*(3), 473–496.

Pampaka, M., & Wo, L. (2014). Revisiting mathematical attitudes of students in secondary education. In P. Liljedahl, S. Oesterle, C. Nicol, & D. Allan (Eds.), *Proceedings of the joint meeting of PME 38 and PME-NA 36, Vancouver, Canada, Vol. 4* (pp. 385–392).

Pepin, B. & Roesken-Winter, B. (Eds.), (2015). *From beliefs to dynamic affect systems in mathematics education: Exploring a mosaic of relationships and interactions*. Switzerland: Springer International Publishing.

Radford, L. (2015). Of love, frustration, and mathematics: A cultural-historical approach to emotions in mathematics teaching and learning. In B. Pepin & B. Roesken-Winter (Eds.), *From beliefs to dynamic affect systems in mathematics education* (pp. 25–49). Switzerland: Springer International Publishing.

Roth, M. W. (2007). Motive, emotion and identity at work: A contribution to third-generation cultural historical activity theory. *Mind, Culture and Activity, 14*(1–2), 40–63.

Roth, M. W., & Jornet, A. (2016). Perezhivanie in the light of the later Vygotsky's spinozist turn. *Mind, Culture, and Activity*, http://dx.doi.org/10.1080/10749039.2016.1186197.

Ruffell, M., Mason, J., & Allen, B. (1998). Studying attitude to mathematics. *Educational Studies in Mathematics, 35*(1), 1–18.

Smith, R. M. (1996). A comparison of methods for determining dimensionality in Rasch measurement. *Structural Equation Modelling, 3*(1), 25–40.

Solomon, R. (1978). Emotions and anthropology: The logic of emotional world views. *Inquiry, 2*, 189–199.

Swartz, D. L. (2002). The sociology of habit: The perspective of Pierre Bourdieu. *OTRJ: Occupation, Participation and Health, 22*, 61S–69S.

Tapia, M., Marsh, G. E., II. (2004). An instrument to measure mathematics attitudes. *Academic Exchange Quarterly, 8*(2), 16–21.

Thomson, R., Bell, R., Holland, J., Henderson, S., McGrellis, S., & Sharpe, S. (2002). Critical moments: Choice, chance and opportunity in young people's narratives of transition. *Sociology, 36*(2), 335–354.

Vygotsky, L. S. (1999). *Collected works: vol. 6*. New York, NY: Plenum.

Wenger, E. (1998). *Communities of practice: Learning, meaning, and identity*. Cambridge: Cambridge University Press.

Wolfe, E. W., Smith, E. V., Jr. (2007a). Instrument development tools and activities for measure validation using Rasch models: Part I—Instrument development tools. *Journal of Applied Measurement, 8*(1), 97–123.

Wolfe, E. W., Smith, E. V., Jr. (2007b). Instrument development tools and activities for measure validation using Rasch models: Part II—Validation activities. *Journal of Applied Measurement, 8*(2), 204–234.

Further Reading

Di Martino, P., & Zan, R. (2011). Attitude towards mathematics: A bridge between beliefs and emotions. *ZDM: International Journal on Mathematics Education, 43*(4), 471–482.

Drodge, E. N., & Reid, D. A. (2000). Embodied cognition and the mathematical emotional orientation. *Mathematical Thinking and Learning, 2*(4), 249–267.

Hascher, T. (2010). Learning and emotion: Perspectives for theory and research. *European Educational Research Journal, 9*(1), 13–28.

Wilkins, J. L. M., & Ma, X. (2003). Modeling change in student attitude toward and beliefs about mathematics. *The Journal of Educational Research, 97*(1), 52–63.

CHAPTER

8

When Being Good at Math Is Not Enough: How Students' Beliefs About the Nature of Mathematics Impact Decisions to Pursue Optional Math Education

Michelle Hurst, Sara Cordes

Boston College, Chestnut Hill, MA, United States

INTRODUCTION

There has been a recent push to increase the science, technology, engineering, and math (STEM) literacy within the United States in order to have more qualified people entering the STEM workforce and becoming active individuals in our innovation- and technology-driven society (e.g., "Change the Equation" initiative, The White House, 2010; "Startup America" initiative, The White House 2011). Unfortunately, however, as soon as young adults are given the opportunity to choose their coursework (generally during the transition from secondary to postsecondary education (i.e., college/university)), many are opting out of additional STEM education, resulting in very low STEM enrollment in college (i.e., only 28% of bachelor's degree students entering in 2003/04 planned to pursue STEM; Chen, 2013). Furthermore, many students who initially anticipate completing a STEM degree end up without it, changing majors during the course of their college career (Stinebrickner & Stinebrickner, 2013). For example, 48% of students who initially declared a STEM major in college between 2003 and 2009 dropped out of STEM before 2009 (Chen, 2013). Given the need for

http://dx.doi.org/10.1016/B978-0-12-802218-4.00008-X

more STEM-qualified applicants in the workforce and the greater push for innovation and technology, there is a real need to identify ways to increase the likelihood that students enroll and continue to pursue STEM courses.

Increasing the number of qualified graduates in STEM fields has proven to be anything but a straightforward problem. Moreover, given the vast diversity of STEM fields, ranging from chemistry to computer programming to neuroscience, it is difficult to identify a single solution to promote participation across all STEM fields. A unifying component of most STEM fields, however, is a strong reliance upon a foundation in mathematics. As a consequence, course requirements for STEM majors often include higher-level math courses, making preparation in math particularly important for advancement in all STEM fields (e.g., Chen, 2013; Hazari, Tai, & Sadler, 2007; Sadler & Tai, 2001). For example, researchers have found that the number of math courses taken in high school and college is an important predictor of success in other areas of STEM (Chen, 2013; Hazari et al., 2007; Sadler & Tai, 2001). Therefore, encouraging more students to choose optional math education will not only help foster a more math literate society at a general level, but it may also promote an increase in qualified graduates across many STEM disciplines.

Math Content

Unsurprisingly, many students find it difficult to acquire novel mathematical concepts, particularly for higher-level math. To understand the reasons for this, researchers have focused on the relationship between specific mathematical content and math achievement, to determine exactly what concepts are deemed most difficult to grasp and what prerequisite mathematical skills may help to promote the acquisition of these concepts. For example, recent work has emphasized the importance of rational number concepts for the acquisition of more advanced algebra concepts (e.g., Siegler et al., 2012). These investigations focus on the learnability of specific content, addressing the effectiveness of curriculum for a particular area and have resulted in the implementation of recent educational reforms (e.g., Common Core State Standards; National Governors Association Center for Best Practices, 2010) that attempt to improve the numeracy and math education of children in grades K-12.

Although mathematical proficiency is no doubt critical for determining whether to pursue optional, more advanced math education (e.g., Aiken & Dreger, 1961; Aiken, 1970; Le, Robbins, & Westrick, 2014; Singer & Stake, 1986), it is not the sole determinant. Evidence suggests that the low frequency with which individuals pursue math education is not solely a function of low ability (e.g., Anderson & Cross, 2014), but rather individuals must be capable *and* must be *motivated to choose* math education. But what factors, beyond that of basic achievement, shape these motivations to pursue mathematics?

Postsecondary Math Education

Although there are many critical transition moments in the math education pathway (e.g., from the natural numbers to the rational numbers, from arithmetic to algebra, etc.), the transition to postsecondary education can be a particularly daunting one. Students often move away from home for the first time, enroll in larger classes, and are provided with less hands-on and more lecture-based instruction, among other things (e.g., Guedet, 2008; Hong et al., 2009; Seymour, 2001). Moreover, particularly in the domain of college mathematics, these basic environmental changes are coupled with substantial changes in academic content as well. Upon entering postsecondary math courses, students are forced to think more abstractly and often change the way they approach math concepts (e.g., Hong et al., 2009; Gruenwald, Klymchuk, & Jovanoski, 2004; Guedet, 2008). Research suggests that the postsecondary math education system shows a distinct difference from the secondary school education system in terms of their use of definitions (e.g., Selden & Selden, 1999), the abstract use and meaning of symbols (e.g., Tall, 1997), and the concept of formal proof and deduction (e.g., Moore, 1994; Tall, 1997). For example, postsecondary math courses require an abstract use of symbols, which involves understanding both the concept associated with the symbol as well as the procedure you are to use (aka "precept"; Tall, 1997, 2008). However, this understanding of symbols is fairly sophisticated and often not emphasized until postsecondary school. This conceptual understanding and use of formal symbols and notation is used heavily when formal proof is introduced, which tends not to occur (or at least, not regularly occur) until postsecondary school. Thus, whereas secondary and earlier mathematics courses may be taught with a greater reliance upon following a set of procedures, postsecondary mathematics requires an understanding of logic and the basics of mathematics in order to justify these procedures. Changes both in superficial features (i.e., presentation) and in core content can make the transition to post-secondary math education particularly overwhelming, calling attention to the disconnect between the beliefs and expectations students have about postsecondary math and the reality of what postsecondary math actually entails.

As such, investigating students' perceptions of math is particularly important during transitional periods in which the student is able to choose their own educational path, such as in the transition to postsecondary schooling. Although the transition from secondary to postsecondary education may be a time when subjective perceptions become particularly salient, many of the students' beliefs and attitudes that influence their decisions are formed long before they are actually able to make choices about their education. In particular, many students "opt out" of math education much earlier by disengaging from the content and emotionally distancing themselves from the topic even though they are still required to take the course (e.g., Ahmed, van der Werf, Kuyper, & Minnaert, 2013; Meece, Wigfield, & Eccles, 1990). These emotional and behavioral reactions have implications both for the individual's ability and their subjective perceptions of the math field.

Current Review

In this chapter, we will review the literature on students' beliefs about mathematics, aligning literature looking at individuals' judgments of the nature of mathematics (e.g., the role of effort) with literature looking at individuals' subjective, emotionally laden judgments of themselves in relation to the field (e.g., "I'm not good at math"). We will particularly focus on how these two kinds of beliefs (about math in general and about math in relation to yourself) influence students' engagement in mathematics, including their choices to pursue optional math education. In addition, throughout the chapter we will highlight the role of emotion in how these beliefs (e.g., "I'm not good at math") may be related to behavioral outcomes (e.g., feeling helpless and engaging in avoidance behavior).

While reviewing literature on these beliefs about mathematics, we will discuss the ways in which having these beliefs may interact with an individuals' feelings of self-efficacy in mathematics, math anxiety and affect, interest in mathematics, and the value they place on mathematics. While substantial research has investigated the factors impacting the transition from secondary to postsecondary education and a separate body of research has suggested various beliefs that people hold toward the math domain, little research has integrated these two literatures. Thus, we will review both of these literatures while suggesting some possible mechanisms through which beliefs about the math domain and emotional reactions to the math domain (e.g., feelings of belonging and/or confidence) may impact decisions to pursue optional math education. In addition, we will focus on women in math as a case study for how these beliefs about the nature of math, individuals' judgments of themselves, and emotional responses to these beliefs often interact to target groups that are particularly underrepresented in STEM fields.

MATH BELIEFS: NEGATIVE MYTHS IMPACTING ACHIEVEMENT

The perceptions or beliefs that an individual holds about mathematics are obvious contributors to whether an individual chooses to pursue the subject. For example, whether an individual views math as a domain in which ability is fixed (i.e., you are either inherently "good at math" or "bad at math") or malleable (i.e., math ability can improve with hard work) can impact the amount of effort that individual is likely to exert in order to pursue mathematics. Similarly, if pursuing math is perceived to be socially isolating, a socially oriented person may be less likely to pursue this field. Thus, investigating commonly held math beliefs and how they may impact an individuals' decision to engage in optional math education is a key component to increasing student participation in math education. In this section, we will discuss two areas in which people tend to hold inaccurate beliefs about the nature of math: (1) the nature of math ability (e.g., as being fixed or malleable) and (2) the nature of the math domain as a field (e.g., as being socially isolating). In addition, we will discuss the ways in which these beliefs interact with and impact affective experiences, judgments of self-efficacy, value, and interest in the math domain, which necessarily impact their educational and career aspirations (e.g., Eccles, 2009).

The Nature of Math Ability

One belief that has been identified as a contributor to lower levels of math participation is the belief that good performance in mathematics is the result of fixed intelligence; that is, individuals are perceived as either inherently "good" or "bad" at math. Recently, Leslie, Cimpian, Meyer, and Freeland (2015) surveyed scholars (graduate students, post-doctoral fellows, and university faculty) in various fields about whether they thought that being a top scholar in their field of study required a "special aptitude that cannot be taught." Using this measure, mathematics was given one of the highest ratings (second only to philosophy), consistent with the idea that people conceive of success in math to be a result of inherent brilliance, not necessarily of hard work and determination (Leslie et al., 2015). That is, it is believed that math ability is a fixed trait linked to a form of innate intelligence and that those who are good at math do not need to work hard. Unfortunately, this belief generally goes hand-in-hand with the converse belief—if an individual needs to work hard to understand math, then they must not be "cut out for it" and perhaps they may not be able to learn it at all. Endorsing this belief likely leads to feelings of helplessness or anxiety when confronted with challenging math content, which, over time, may result in the formation of negative associations with mathematics.

Unfortunately, thinking about math ability as a fixed trait has also been shown to influence how we interact with children about math. Research indicates that teachers who believe that math ability is fixed tend to judge students as having lower ability more readily than teachers who believe math ability is malleable (Rattan, Good, & Dweck, 2012). Moreover, when teachers with these beliefs try to help struggling students, they often approach their classroom and students in ways that demote effort and engagement in mathematics, such as saying things like "Not everyone is good at math" (Rattan et al., 2012). Unsurprisingly, these interactions result in children similarly endorsing ideas about math ability being fixed, which then reduces the likelihood that they will make additional efforts to overcome difficulties they experience (Dweck, Mangels, & Good, 2004). In this way, these beliefs impact the way students engage in their math classes even before they can officially opt out of math education.

The belief that math proficiency is derived from some sort of innate talents, not from hard work, also necessarily leads to a reduction in student feelings of self-efficacy in math (i.e., the belief that they can have an impact on their math ability) and, for many students, to an increase in negative affect toward math (i.e., math anxiety or helplessness). Importantly, these motivational and emotional factors (i.e., math self-efficacy and math anxiety) are directly related to math achievement and course enrollment. In particular, research suggests that one of the primary mechanisms through which math anxiety and self-efficacy impact both achievement and pursuit of math may be the engagement of avoidance behaviors in math contexts. In particular, low self-efficacy (Hendy, Schorschinsky, & Wade, 2014) and math anxiety (Hembree, 1990; Liew, Lench, Kao, Yeh, & Kwok, 2014) are related to avoidance behaviors that result in poor learning. For example, students who report low math self-efficacy and higher math anxiety also tend to engage in behaviors such as skipping math class and not completing math homework (Hembree, 1990; Hendy et al., 2014). Unfortunately, this creates a downward spiral, such that low self-efficacy may lead to engagement of avoidance behaviors, which then leads to even more poor academic outcomes, providing further evidence for the belief that "I am not good at math".

Although this type of behavioral pattern can be seen in other domains as well, the relationship between experiencing difficulty in a domain and changes in self-efficacy and anxiety may be particularly strong in math. For example, the excessive lesson demands experienced in the math domain are more likely to lead to anxiety than excessive lesson demands in language domains (Goetz, Ludtke, Nett, Keller, & Lipnevich, 2013), potentially reflecting differences in beliefs about the nature of ability (fixed vs malleable) and the role of effort in these domains. Similarly, data indicate that the correlation between emotional experiences associated with a domain (e.g., anxiety, enjoyment) and academic outcomes in that domain is stronger for mathematics than it is for various language domains (Goetz, Cronjaeger, Frenzel, Ludtke, & Hall, 2010). This finding hints at the possibility that our experiences in mathematics (either positive or negative) may elicit stronger emotional reactions than our experiences in other domains, precisely because math successes and failures are generally thought to reflect inherent individual traits (e.g., "I got an A in math so I must be smart") more so than other fields.

Moreover, even for students who have successfully navigated and remained engaged throughout their primary and secondary math education, another hurdle arrives at the transition from secondary to postsecondary math education. In particular, students often have inaccurate expectations about postsecondary mathematics (e.g., Stinebrickner & Stinebrickner, 2013), making this a time of major academic adjustment (in addition to all the other challenges students face during this transition). Stinebrickner and Stinebrickner (2013) surveyed hundreds of entering freshman students at an Ontario university in 2000–2001 and continued to survey them throughout their university careers. Many of the surveyed students reported originally wanting to pursue a math or science degree, but they soon realized that the course work was more difficult than they expected, concluding that they did not have the ability to complete the degree (Stinebrickner & Stinebrickner, 2013). However, what is unclear is whether individuals made fair assessments of his or her own ability—what may be more likely is that these students had some belief about math being a fixed ability, so when the postsecondary mathematics courses were more difficult than expected they may begin to believe they are not good at math. For many students, this transition may be the largest jump in their math education history, making it a critical time during which they need to be resilient to potential difficulties and have an accurate expectation about the role of effort and the nature of math ability. With accurate expectations about the role of effort, students may go from having experiences of hopelessness and anxiety about approaching any challenge in math to feelings of pride for working toward a success, even when they occasionally stumble along the way. In turn, these positive emotional reactions may be more likely to foster a positive relationship between the individuals' judgments of themselves and their abilities within the math domain.

The Nature of the Math Domain

It is widely believed that the math field is socially isolating. The stereotype of the awkward and isolated mathematician or scientist is a popular one seen throughout our media, in hit television series and movies. We can think of many examples where the protagonist either is a socially awkward and isolated scientist scribbling down equations and/or is actively trying to avoid becoming a social outcast, making references to the social decay that necessarily comes with engaging in math (e.g., "You can't join Mathletes, it's social suicide"—Michaels and Watersrs (2004). Mean Girls [Motion Picture]). This stereotype can have a strong impact on individuals' pursuit of optional math education. At the most basic level, the belief that it is "uncool" to engage in math may discourage children from engaging in math in order to prevent social ostracizing. However, more indirectly and less obviously, specific beliefs about the kinds of personalities that fit into the math domain based on the nature of working in math fields also impact whether an individual sees themselves fitting into the math domain. For example, many students report that the atmosphere in post-secondary math courses is cold and intimidating (e.g., Daempfle, 2003). This is referred to as the "chilly climate hypothesis," which contends that a cold and intimidating atmosphere makes students feel uncomfortable and emotionally disengaged in their math courses (Daempfle, 2003). In fact, many high achieving students who do not persist in math and science report that the chilly climate was one of the primary reasons for leaving (Seymour & Hewitt, 1997). Importantly, particularly during the transition from secondary to postsecondary education, when students transition from small, hands-on math courses to significantly larger, lecture-based math courses, students may find a chilly climate to be particularly salient, leading students who may have otherwise enjoyed math to emotionally disengage and seek out other fields of study.

Various frameworks have been proposed to describe the impact of personality-based factors on these career decisions, each of which emphasize that individuals choose environments that match their perceived ability, interests, and the social roles they see themselves playing (e.g., Diekman & Eagley, 2008; Eccles, Adler, & Meece, 1984; Le et al., 2014). In the case of mathematics, it has been proposed that the perceived sociality of the domain plays a role in whether students pursue math. In particular, Diekman and colleagues have used role congruity theory to suggest that since math is not perceived to be a communal discipline, individuals who endorse communal goals (e.g., working with others) are less interested in pursuing math and other quantitative fields (Diekman, Brown, Johnson, & Clark, 2010). Thus, when deciding whether or not to pursue optional math and science education, students often think about whether their ultimate socially oriented goals align with their perceptions of the domain of mathematics. In addition to impacting initial interest in pursuing certain fields, the person-environment fit model (Le et al., 2014) suggests some role of feedback from the environment as well. In particular, when the individual chooses a well-suited environment, they have positive affective experiences including enjoyment and satisfaction, reinforcing their identification with and positive affect toward the environment. However, if the choice is not a good fit (due to any number of reasons, including inaccurate assessments of the environment and/or the self), the individual may experience negative affect (e.g., boredom, anxiety, feelings of discontent) leading to an extreme deidentification with the area. For students who are socially oriented and have communal goals, the chilly climate they encounter in postsecondary math classes may be perceived as a strong indicator that the choice they made is not a good one, thus leading to emotional detachment and deidentification with mathematics.

In addition to considering the alignment between their beliefs and their goals, students will assign varying levels of value to math itself, and this perceived value has been shown to have strong implications for engagement in mathematics. In particular, the perceived value and utility (e.g., whether math will help you ultimately achieve your career goals, whether it is something you will need in life, etc.) of specific content strongly influences decision-making in academic contexts (e.g., Eccles & Wigfield, 2002; Eccles et al., 1984; Meece et al., 1990). As early as 8th grade, an individual's perceived value in science is a significant predictor of their STEM degree completion (Maltese & Tai, 2011), suggesting that perceived value might have real lasting and important effects throughout schooling. In the domain of mathematics specifically, studies have shown that individuals who value performing well in math are more likely to enroll in additional math courses (Eccles & Jacobs, 1986; Meece et al., 1990), and perceived value is a stronger predictor of math course enrollment than whether the individual is interested in math (Updegraff, Eccles, Barber, & O'Brien, 1996).

In addition, it is likely that perceived value, self-efficacy, and emotional experiences interact, potentially in many different ways. For example, research has suggested that the relationship between value and test anxiety may depend upon the individual's self-efficacy (Nie, Lau, & Liau, 2011). Meaning, when an individual does not believe they can do well (has low self-efficacy), they experience greater anxiety on high valued tasks (compared to low-valued tasks), but when they believe they can do well, they experience less anxiety overall, across low-and high-valued tasks. Similarly, it may be that emotional reactions to the outcome of events also interact with our perceived value and self-efficacy. When an individual highly values performing well in mathematics, they may have strong feelings of pride or enjoyment when they perform well, promoting feelings of self-efficacy and self-identification with math. On the other hand, attributing high value to mathematics may result in even stronger negative affect when the individual does poorly. In this case, it may be more difficult for an individual to overcome the negative affective experience of performing poorly, while still endorsing the importance of mathematics. In this case, poorer-performing math students may come to value math less in order to protect themselves from feelings of negative affect. In this way, an individual's emotional reaction to their performance may alter their beliefs about how important math is in their life. Relatedly, when math is not valued, then success (or failure) in math classes may be less likely to result in strong affective responses.

MATH BELIEFS: POSITIVELY IMPACTING CHILDREN'S BELIEFS

Despite many prevalent negative beliefs about the math domain (e.g., math is socially isolating, math ability is fixed, etc.), it may be possible to prevent these beliefs in students before they have the opportunity to emotionally check out of math classes. In this section, we will discuss the ways in which the transition to postsecondary mathematics can be addressed both directly and indirectly, by addressing the ways in which content is presented, which can alter students' emerging beliefs about math ability and math as a field. By addressing children's beliefs early in their math education, we will highlight the ways in which math learning at many educational transitions (e.g., arithmetic to algebra, functions to calculus, calculations to proof, etc.) can be drastically improved to promote a positive learning experience.

Teacher and Caregiver Input

Children acquire many of their attitudes and perceptions from their interactions with caregivers and teachers (e.g., Eccles & Jacobs, 1986). For example, the support and value shown by parents for math and science domains is related to how much children value math and science and how much they are motivated to pursue these fields (Bouchey & Harter, 2005; Leaper, Farkas, & Brown, 2012). Teachers regularly (although not necessarily purposefully) transmit information about usefulness, value, and interest about particular domains (e.g., Eccles & Jacobs, 1986). Thus, fostering positive perceptions and attitudes about the math domain in children requires that we focus on promoting positive input from parents, caregivers, and teachers.

One aspect of this social learning might be through the affective cues provided by teachers and caregivers, especially negative affect (Vaish, Grossmann, & Woodward, 2008). Picking up on these (negative) affective cues might be teaching children that there is *something* negative about the subject, before children are even able to articulate or justify the negativity. In other words, the negative affect expressed by teachers and caregivers about math (either in the form of low enthusiasm or explicit negativity like, "I know this isn't very much fun, but…") may provide children with a negative view, which then corresponds with devalued beliefs about math (i.e., low interest, low value, etc.). For example, students' interest, value, and perceived usefulness of math increased between 6th and 7th grade when they reported going from a low support to high support teacher, but the reverse occurred when the students reported going from a high to low support teacher, with the greatest impact found for low-achieving students (where support is a measure of the teachers' fairness, friendliness, and caring, which are all related to positive affective experiences; Midgley, Feldlaufer, & Eccles, 1989). Thus, in addition to fostering greater math ability, the level of support that parents and teachers provide in the domain of math may also implicitly (and/or explicitly) alter a child's perceptions of whether math knowledge and ability should be valued.

In addition, early experiences in math activities may convey the value of math to children. In particular, evidence suggests that the relationship between participation in early math activities (primary/elementary school) and high school math course enrollment is moderated through expectancies and values of the math courses (even when controlling for grades; Simpkins, Davis-Kean, & Eccles, 2006), suggesting that early math activities may be particularly important precisely because they cause students to value math.

Addressing the Transition Indirectly

As noted, the transition from secondary to postsecondary math education seems to be specifically marked in the domain of mathematics for requiring a jump from concrete to abstract thinking for which many students are not prepared (e.g., Moore, 1994; Selden & Selden, 1999; Tall, 1997). However, theories of the development of advanced mathematical thought (e.g., Edward, Dubinsky, & McDonald, 2005; Tall, 1991) contend that this jump in abstract thinking is not necessarily tied to a change in mathematical content, but instead to the way the content is presented in the classroom. It is not that postsecondary institutions are teaching more abstract content and secondary schools are teaching concrete content, but rather that postsecondary math education shifts to a focus on advanced mathematical thought, something which is not emphasized in secondary classrooms. Thus, according to these theories, the transition to postsecondary math may be alleviated by an earlier emphasis on the abstract nature of math content in secondary math courses in order to promote a more continuous transition to math content at later levels of education.

Math Ability Is Not Fixed

Substantial evidence suggests it is critical to foster growth mindsets regarding the domain of math, such that math ability is perceived as malleable, with effort being a necessary component to performing well and becoming better at math (e.g., Mueller & Dweck, 1998; Rattan et al., 2012). This growth mindset may be key to promoting increased persistence in math education by allowing children to work through difficult math transitions, instead of giving up in the face of difficulty. In practice, emphasizing mastery goals (i.e., education for the sake of learning and mastering the content) as opposed to performance goals (i.e., education for the sake of achievement, often comparing achievement to peers) within the classroom has been shown to promote positive behaviors, like increased help seeking (Schenke, Lam, Conley, & Karabenick, 2015). In fact, one study found that a teacher's report about a child's *ability* in math predicted their interest at the beginning of the semester but not their sustained interest, while teachers' beliefs about the child's *effort* (level of trying and persistence) predicted sustained interest across school (Upadyaya & Eccles, 2014). Similar recommendations come from principles based on constructivist learning. In particular, successful math teachers tend to have beliefs about the nature of mathematics and the role of problem solving that are consistent in guiding the students' construction of knowledge, rather than imparting unquestioned knowledge (Beswick, 2007).

Another potential way to encourage a growth mindset is to humanize the "brilliant" mathematician. It has been suggested that one way in which to bring a human face to the mathematics process is by teaching math content alongside a historical perspective of that topic (e.g., Bidwell, 1993; Karaduman, 2010; Fauvel, 1991). By including the historical context, students are made aware of the struggles associated with math developments (i.e., that mathematical ideas and proofs are the product of hard work, not inherent mathematical brilliance) and apply this growth mindset to their own math approach.

In addition, for those students who begin the transition to postsecondary mathematics, a positive attitude toward math (i.e., the absence of math anxiety; positive experiences in secondary education and earlier) may provide students with some protection against negative experiences at later educational time points. In college students, those who reported a more positive high school experience with math were better able to bounce back from poor math preparation in their first year of university (Kajander & Lovric, 2005), suggesting that experiencing positive affect toward mathematics may prevent students from giving up when faced with difficulty in the subject. These findings suggest that the kinds of attitudes, math experiences, and emotional reaction to the experiences that children have before they are able to make academic decisions are critical in setting up how they will eventually make academic decisions and whether they are likely to persist through math education.

Math Is Valuable for Many Reasons

Since many students perceive math and science as being inconsistent with goals of working with others and helping others (Diekman et al., 2010) and perceive the postsecondary math education environment as being particularly cold and intimidating (Daempfle, 2003), promoting the social aspects of math and science fields may lead to lower attrition of strong students. The humanization of mathematics and mathematicians may provide students with a different perspective of a positive role model as opposed to being cold and intimidating (Hekimoglu & Kittrell, 2010). Furthermore, while the image of a mathematician working alone with numbers is a popular one, it is not necessarily an accurate portrayal of the math domain in general. Although it may be true that some math-related avocations are not particularly social or communal, there are myriad of math-related careers in highly social environments. Even extremely social and communal enterprises (e.g., nonprofit organizations, social services, political campaigns, etc.) require individuals who excel in math and science domains in order to keep organized records, model outcome measures, program technological equipment, among many other things. Thus, by broadening the sense of the kinds of careers that require a strong math education, more students may be encouraged to pursue advanced math education—even as a means of promoting social and communal goals.

CASE STUDY: THE UNDERREPRESENTATION OF WOMEN IN MATH

Although enrollment in optional math courses is generally low, it is disproportionately low for women and certain minority groups (in particular, African Americans and Hispanics; National Science Board, 2012; National Science Foundation, 2015). Increasing enrollment from these underrepresented groups could provide a substantial increase in the number and diversity of people engaged in math-oriented fields. In this section, we will focus on how the general beliefs about the nature of math ability and the nature of the math domain may be strong contributors to the underrepresentation of women in math and science (similar arguments can be made for other underrepresented groups, but we will be focusing on women as a case study).

Sex and gender differences in mathematics ability, attitudes, and educational choices have been discussed for many decades (e.g., Fennema & Sherman, 1977; Hyde, Fennema, & Lamon, 1990; Hyde, Fennema, Ryan, Frost, & Hopp, 1990; Meece, Parsons, Kazcala, Goff, & Futterman, 1982), and they remain relevant today. Although the actual gender achievement gap has consistently narrowed (Jerrim & Schoon, 2014), many more women compared to men are still opting out of optional math and science (e.g., Parker, Nagy, Trautwein, & Ludtke, 2014). These gender differences in enrollment of optional math courses may be due, in part, to differences in attitudes and beliefs about mathematics (e.g., Hyde, Fennema, et al., 1990). This may be because *perceived* achievement differences still exist; women still believe they are performing worse than their male counterparts in math and science and endorse the gender stereotype, even when it is not true (Spelke & Ellison, 2009). Additionally, despite similar performance, women often report lower interest in math careers than men (Catsambis, 1994). Thus, given that women (often inaccurately) perceive their math abilities to be lower and report lower interests in math, it is not surprising that women are less likely to enroll in optional math courses and are less likely to pursue STEM fields than their male counterparts.

If gender differences in math ability are marginal at best, then why do large gender differences in interest and pursuit persist? It has been posited that math-gender stereotypes persist because they are transmitted from parents and teachers to children very early in life, even if inadvertently (e.g., Gunderson, Ramirez, Levine, & Beilock, 2012). For example, evidence suggests that mothers provide more math-talk to their 22-month-old sons than to their 22-month-old daughters (Chang, Sandhofer, & Brown, 2011). Relatedly, Beilock, Gunderson, Ramirez, and Levine (2010) showed that girls in first and second grade who were taught by female math-anxious teachers showed higher endorsement of math-gender stereotypes (i.e., "math is for boys") and had lower achievement. This is particularly important to address given that current reports estimate that over 75% of public school teachers are female (National Education Association, 2015) and females in education courses show the highest levels of math anxiety (Hembree, 1990). Thus, it is deeply concerning (but also not surprising) that even in elementary school, young girls report higher levels of math anxiety (Wigfield & Meece, 1988) and are less likely to identify with math than boys (Cvencek, Meltzoff, & Greenwald, 2011). In fact, not only are girls less likely to identify with math, but by as young as 6 years of age, they associate math with boys both explicitly and implicitly (Cvencek et al., 2011). These early developing gender beliefs may later impact whether they choose to pursue optional math education. In particular, evidence suggests that math self-concept (e.g., believing I am good at math) is a much higher predictor of gender differences in math major selection in college than are gender differences in achievement (Parker et al., 2014). Looking at many math classrooms on postsecondary campuses will clearly highlight this disparity, with many more men than women enrolling in optional math education (National Science Board, 2012; National Science Foundation, 2015).

In addition to explicit math-gender beliefs, more indirect gender differences may arise because of gender differences in specific beliefs about the nature of math ability and the nature of the math domain. Unfortunately, the belief that math ability is fixed (i.e., innately determined), likely coupled with the early emerging belief that math is for boys, can have particularly negative impacts on math participation for women. In support of this hypothesis, the extent to which experts rated a given academic discipline as requiring brilliance was strongly negatively correlated with the proportion of women obtaining a Ph.D. within that discipline (Leslie et al., 2015).

Thus, these beliefs about the nature of math ability and math careers may have particularly strong impacts on the enrollment of women, who are typically underrepresented in STEM fields.

In addition, the socially isolating stereotype of math fields may act as a larger deterrent for females, compared to males. That is, although there are substantial individual differences, in general women tend to endorse communal goals (e.g., working with others) more than men (e.g., Costa, Terracciano, & McCrae, 2001; Schwartz & Rubel, 2005). Yet, as already discussed, math is not perceived to be a communal discipline (Diekman et al., 2010). As such, when controlling for the degree to which individuals endorse communal goals (e.g., working with others), Diekman et al. (2010) no longer showed gender differences in interests to pursue math (and other quantitative careers), suggesting that gender differences in the level of pursuit of math and other quantitative fields may be (at least somewhat) rooted in differences in communal goal endorsement.

Given this evidence, it is likely that changing the way people think about math as a domain—by promoting growth mindsets and reinforcing beliefs that strong math education is helpful in a number of social and communal environments—may help to dissipate gender differences in math domains, while also help to promote participation in math (and other STEM-fields) in general.

MOVING FORWARD

Although the current literature provides us with substantial information about beliefs about the nature of math ability and the nature of the math domain as a field, there is much more that we can gain from continued research into understanding the inaccurate and stereotyped beliefs people hold about the math domain.

One potential avenue is to investigate the ways in which the beliefs people hold toward mathematics are unique to mathematics (and possibly some other quantitative domains). For example, although it is commonplace to hear "I'm not a math person," statements like "I'm not a biology person" or "I'm not a history person" are much less common. That is not to say that people do not perform poorly in biology and history courses; many people may say they are "bad at biology" or "bad at history," but these academic preferences are less likely to be perceived as character judgments (I'm not a X *person*). Thus, investigating whether this is true at a more general level, why this may be the case for mathematics in particular, and what impact these beliefs may have for promoting increased math education are important questions for future research.

In addition to investigating people's beliefs about math, there are several important subjective and emotional variables that are related to these beliefs that may impact whether people choose to pursue optional math education. In particular, as discussed, students often rely on their assessments of their own interest, value, anxiety, and self-efficacy within a domain in order to make decisions about whether to pursue optional education in that domain. However, the ways in which these subjective variables are related to inaccurate stereotypes and beliefs about the nature of mathematics more generally is relatively unknown. For example, it is unclear whether learning that math-related fields could be socially oriented leads to an increase in both value and interest or whether value can increase (because it is necessary and useful) without a corresponding increase in interest (useful, but boring). Thus, further research investigating the change in individuals' subjective reports of their attitudes and emotions toward the math domain in addition to their beliefs about the nature of mathematics more generally may shed light on the mechanisms through which these beliefs impact decisions to pursue (or not pursue) optional math education.

Similarly, there are many open questions surrounding how these beliefs may interact with an individual's emotional responses. For example, an individual who believes math ability is fixed may feel helpless when facing a failure but proud when facing an accomplishment. On the other hand, individuals who believe math ability is malleable may experience less negative affect in the case of failure (e.g., they may feel discouraged but be able to consider behaviors that could change the outcome) and greater positive affect in the case of success (e.g., "earning" the success). Although affect is often implicitly involved in discussions of beliefs and attitudes, more research is warranted to investigate specifically how affective experiences may play a role and whether they may interact with beliefs and behavior in critical ways.

Lastly, the research on math education is often isolated to particular parts of the math education pathway (e.g., primary school, secondary school, etc.). Although negative beliefs about math may be particularly exaggerated during the transition from secondary to postsecondary math education—when students are able to choose their own education and opt out of math education—the only way to prevent these negative attitudes is to implement interventions much earlier. Thus, integrating research involving preschool, primary, secondary, and postsecondary math students is imperative for identifying ways to encourage students to remain in the math education pathway throughout. Without a strong understanding of the expectations children gain in secondary school, we cannot begin to predict how they will react in postsecondary school. Therefore, future research that integrates these transitions may provide important insight into the problems that arise and the possible solutions that may be considered (including at what educational point the solutions may be implemented).

CONCLUSION

All too often do we hear statements from children, adolescents, and adults such as, "I'm not good at math," "I'm not a math person," "Ewe, math gives me hives," and other creative antics all to express that math is not something within their wheel-house, but instead something far away from themselves. These attitudes are not conducive to fostering interest in and increasing the perceived value of math education, which can be seen by looking at the low level of STEM graduates and the general state of math literacy in America (e.g., OECD, 2013, 2014). However, changing math curricula can only take us so far—if people are not engaging in the material or are not taking the courses at all then a better curriculum does us no good. The current review has attempted to illuminate other factors that influence whether an individual pursues optional math education—that is, their beliefs about the math domain, how they fit into that domain, and their emotional reactions to experiences with the domain. We have discussed several ways in which inaccurate beliefs about the math domain impact an individual's emotional reactions and attitudes toward mathematics. By discussing several disparate aspects of math education literature, we have suggested some mechanisms through which general beliefs about the math domain impact specific beliefs about how mathematics fits into the individual's life, thus further impacting whether the individual will choose to pursue optional math education. We further suggest various ways to remedy these inaccurate beliefs and suggestions and ways for future research to better understand how these beliefs develop and interact with other attitudes, perceptions, and affective responses throughout the individual's math education. Specifically, we discuss ways to decrease negative stereotypes associated with math, to increase people's understanding that math is a creative and progressive pursuit, and to emphasize a growth mindset, such that effort is imperative for improving an individual's math performance. By implementing these specific remedies and addressing these open research questions, we can hopefully create a dramatic shift in the collective mindset regarding math and math education.

References

Ahmed, W., van der Werf, G., Kuyper, H., & Minnaert, A. (2013). Emotions, self-regulated learning, and achievement in mathematics: A growth curve analysis. *Journal of Educational Psychology*, *105*(1), 150–161.

Aiken, L. R. (1970). Attitudes toward mathematics. *Review of Educational Research*, *40*(4), 551–596.

Aiken, L. R., & Dreger, R. M. (1961). The effect of attitudes on performance in mathematics. *Journal of Educational Psychology*, *52*(1), 19–24.

Anderson, L., & Cross, T. L. (2014). Are students with high ability in math more motivated in math and science than other students? *Roeper Review*, *36*(4), 221–234.

Beilock, S. L., Gunderson, E. A., Ramirez, G., & Levine, S. C. (2010). Female teachers' math anxiety affects girls' math achievement. *Proceedings of the National Academy of Sciences of the United States of America, 107*(5), 1860–1863.

Beswick, K. (2007). Teachers' beliefs that matter in secondary mathematics classrooms. *Educational Studies in Mathematics, 65*, 95–120.

Bidwell, J. K. (1993). Humanize your classroom with the history of mathematics. *The Mathematics Teacher, 86*(6), 461–464.

Bouchey, H. A., & Harter, S. (2005). Reflected appraisals, academic self-perceptions, and math/science performance during early adolescent. *Journal of Educational Psychology, 97*(4), 673–686.

Catsambis, S. (1994). The path to math: Gender and racial-ethnic differences in mathematics participation from middle school to high school. *Sociology of Education, 67*(3), 199–215.

Chang, A., Sandhofer, C. M., & Brown, C. S. (2011). Gender biases in early number exposure to preschool-aged children. *Journal of Language and Social Psychology, 30*(440), 440–450.

Chen, X. (2013). STEM attrition: College students' paths into and out of stem fields: statistical analysis report. U.S. Department of Education: NCES 2014-001.

Costa, P. T., Terracciano, A., & McCrae, R. R. (2001). Gender differences in personality traits across cultures: Robust and surprising findings. *Journal of Personaity and Social Psychology, 81*(2), 322–331.

Cvencek, D., Meltzoff, A. N., & Greenwald, A. G. (2011). Math-gender stereotypes in elementary school children. *Child Development, 82*(3), 766–779.

Daempfle, P. A. (2003). An analysis of the high attrition rates among first year college science, math, and engineering majors. *Journal of College Student Retention, 5*(1), 37–52.

Diekman, A. B., Brown, E. R., Johnson, A. M., & Clark, E. K. (2010). Seeking congruity between goals and roles: A new look at why women opt out of science, technology, engineering, and mathematics careers. *Psychological Science, 21*(8), 1051–1057.

Diekman, A. B., & Eagley, A. H. (2008). Of men, women, and motivation. A role congruity account. In J. Y. Shah & W. L. Gardner (Eds.), *Handbook of motivation science* (pp. 434–447). New York, NK: Guilford Press.

Dweck, C. S., Mangels, J. A., & Good, C. (2004). Motivational effects on attention, cognition, and performance. In D. Y. Dai & R. J. Sternberg (Eds.), *Motivation, emotion, and cognition: Integrative perspectives on intellectual functioning and development*. New Jersey: Lawrence Erlbaum Associates.

Eccles, J. (2009). Who am i and what am i going to do with my life? Personal and collective identities as motivators of actions. *Educational Psychologist, 44*(2), 78–89.

Eccles, J., Adler, T., & Meece, J. L. (1984). Sex differences in achievement: A test of alternate theories. *Journal of Personality and Social Psychology, 46*(1), 26–43.

Eccles, J. S., & Jacobs, J. E. (1986). Social forces shape math attitudes and performance. *Signs: Journal of Women in Culture and Society, 11*(2), 367–380.

Eccles, J. S., & Wigfield, A. (2002). Motivational beliefs, values, and goals. *Annual Review of Psychology, 53*, 109–132.

Edward, B. S., Dubinsky, E., & McDonald, M. (2005). Advanced mathematical thinking. *Mathematical Thinking and Learning, 7*(1), 15–25.

Fauvel, J. (1991). Using history in mathematics education. *For the Learning of Mathematics, 11*(2), 3–6.

Fennema, E., & Sherman, J. (1977). Sex-related differences in mathematics achievement, spatial visualization, and affective factors. *American Educational Research Journal, 14*(1), 51–71.

Goetz, T., Cronjaeger, H., Frenzel, A. C., Ludtke, O., & Hall, N. C. (2010). Academic self-concept and emotion relations: Domain specificity and age effects. *Contemporary Educational Psychology, 35*(1), 44–58.

Goetz, T., Ludtke, O., Nett, U. E., Keller, M. M., & Lipnevich, A. A. (2013). Characteristics of teaching and students' emotions in the classroom: Investigating differences across domains. *Contemporary Educational Psychology, 38*, 383–394.

Gruenwald, N., Klymchuk, S., & Jovanoski, Z. (2004). Reducing the Gap between the school and university mathematics: University Lecturers' perspective. *The New Zealand Mathematics Magazine, 41*(3), 12–24.

Guedet, G. (2008). Investigating the secondary-tertiary transition. *Educational Studies in Mathematics, 67*, 237–254.

Gunderson, E. A., Ramirez, G., Levine, S. C., & Beilock, S. L. (2012). The role of parents and teachers in the development of gender-related math attitudes. *Sex Roles, 66*, 153–166.

Hazari, Z., Tai, R. H., & Sadler, P. M. (2007). Gender differences in introductory university physics performance: The influence of high school physics preparation and affective factors. *Science Education, 91*, 847–876.

Hekimoglu, S., & Kittrell, E. (2010). Challenging students' beliefs about mathematics: The use of documentary to alter perceptions of efficacy. *PRIMUS, 20*(4), 299–331.

Hembree, R. (1990). The nature, effects, and relief of mathematics anxiety. *Journal of Research in Mathematics Education, 21*(1), 33–46.

Hendy, H. M., Schorschinsky, N., & Wade, B. (2014). Measurement of math beliefs and their associations with math behaviors in college students. *Psychological Assessment, 26*(4), 1225–1234.

Hong, Y. Y., Kerr, S., Klymchuk, S., McHardy, J., Murphy, P., Spencer, S., et al. (2009). A comparison of teacher and lecturer perspectives on the transitions from secondary to tertiary mathematics education. *International Journal of Mathematical Education in Science and Technology, 40*(7), 877–889.

Hyde, J. S., Fennema, E., & Lamon, S. J. (1990a). Gender differences in mathematics performance: A meta-analysis. *Psychological Bulletin, 107*(2), 139–155.

Hyde, J. S., Fennema, E., Ryan, M., Frost, L. A., & Hopp, C. (1990b). Gender comparisons of mathematics attitudes and affect. *Psychology of Women Quarterly, 14*, 299–324.

Jerrim, J., & Schoon, I. (2014). Do teenagers want to become scientists? A comparison of gender differences in attitudes toward science, career expectations, and academic skill across 29 countries. In I. Schoon & J. S. Eccles (Eds.), *Gender differences in aspirations and achievement*. United Kingdom: Cambridge University Press.

Kajander, A., & Lovric, M. (2005). Transition from secondary to tertiary mathematics: McMaster university experience. *International Journal of Mathematical Education in Science and Technology, 36*(2), 149–160.

Karaduman, G. B. (2010). A sample study for classroom teachers addressing the importance of utilizing history of math in math education. *Procedia Social and Behavioral Sciences, 2*, 2689–2693.

Le, H., Robbins, S. B., & Westrick, P. (2014). Predicting student enrollment and persistence in college STEM fields using an expanded P-E Fit framework: A large-scale multilevel study. *Journal of Applied Psychology, 99*(5), 915–947.

Leaper, C., Farkas, T., & Brown, C. S. (2012). Adolescent girls' experiences and gender-related beliefs in relation to their motivation in math/science and english. *Journal of Youth and Adolescence, 41*, 268–282.

Leslie, S., Cimpian, A., Meyer, M., & Freeland, E. (2015). Expectations of brilliance underlie gender distributions across academic disciplines. *Science, 347*(6219), 262–265.

Liew, J., Lench, H. C., Kao, G., Yeh, Y., & Kwok, O. (2014). Avoidance temperament and social-evaluative threat in college students' math performance: A mediation model of math and test anxiety. *Anxiety, Stress, & Coping: An International Journal, 27*(6), 650–661.

Maltese, A. V., & Tai, R. H. (2011). Pipeline persistence: Examining the association of educational experiences with earned degrees in STEM among U.S. students. *Science Education, 95*(5), 877–907.

Meece, J. L., Parsons, J. E., Kazcala, C. M., Goff, S. B., & Futterman, R. (1982). Sex differences in math achievement: Toward a model of academic choice. *Psychological Bulletin, 91*(2), 324–348.

Meece, J. L., Wigfield, A., & Eccles, J. S. (1990). Predictors of math anxiety and its influence on young adolescents' course enrollment intentions and performance in mathematics. *Journal of Educational Psychology, 82*(1), 60–70.

Michaels, L., & Waters, M. (2004). *Mean girls [motion picture]*. United States: Paramount Pictures.

Midgley, C., Feldlaufer, H., & Eccles, J. S. (1989). Student/teacher relations and attitudes toward mathematics before and after the transition to junior high school. *Child Development, 60*(4), 981–992.

Moore, R. C. (1994). Making the transition to formal proof. *Educational Studies in Mathematics, 27*(3), 249–266.

Mueller, C. M., & Dweck, C. S. (1998). Praise for intelligence can undermine children's motivation and performance. *Journal of Personality and Social Psychology, 75*(1), 33–52.

National Education Association. (2015). *Rankings and estimates 2014/2015.* Washington, DC: National Education Association. Available at: www.nea.org/home/rankings-and-estimates-2014-2015.html.

National Governors Association Center for Best Practices, Council of Chief State School Officers (2010). Common core state standards for mathematics. Retrieved from: http://www.corestandards.org/assets/CCSSI_Math%20Standards.pdf.

National Science Board (2012). Science and engineering indicators 2012. Arlington VA: National Science Foundation (NSB 12-01). Available at: www.nsf.gov/statistics/seind12/.

National Science Foundation, National Center for Science and Engineering Statistics (2015). Women, minorities, and persons with disabilities in science and engineering: 2015. Special report NSF 15-311. Arlington, VA. Available at: www.nsf.gov/statistics/wmpd/.

Nie, Y., Lau, S., & Liau, A. K. (2011). Role of academic self-efficacy in moderating the relationship between task importance and test anxiety. *Learning and Individual Differences, 21*, 736–741.

OECD. (2013). *OECD skills outlook 2013: First results from the survey of adult skills, OECD publishing*. http://dx.doi.org/10.1787/9789264204256-en.

OECD. (2014). *PISA 2012 results: What students know and can do—student performance in mathematics, reading and science (volume I, revised edition, February 2014)*. PISA: OECD Publishing. http://dx.doi.org/10.1787/9789264201118-en.

Parker, P., Nagy, G., Trautwein, U., & Ludtke, O. (2014). Predicting career aspirations and university majors from academic ability and self-concept: A longitudinal application of the internal-external frame of reference model. In I. Schoon & J. S. Eccles (Eds.), *Gender differences in aspirations and achievement*. United Kingdom: Cambridge University Press.

Rattan, A., Good, C., & Dweck, C. S. (2012). "It's ok—not everyone can be good at math": Instructors with an entity theory comfort (and demotivate) students. *Journal of Experimental Social Psychology, 48*, 731–737.

Sadler, P. M., & Tai, R. H. (2001). Success in introductory college physics: The role of high school preparation. *Science Education, 85*, 111–136.

Schenke, K., Lam, A. C., Conley, A. M., & Karabenick, S. A. (2015). Adolescents' help seeking in mathematics classrooms: Relations between achievement and perceived classroom environmental influences over one school year. *Contemporary Educational Psychology, 41*, 133–146.

Schwartz, S. H., & Rubel, T. (2005). Sex differences in value priorities: Cross-cultural and multimethod studies. *Journal of Personality and Social Psychology, 89*(6), 1010–1028.

Selden, A., & Selden, J. (1999). Tertiary mathematics education research and its future. *Department of Mathematics Technical Report, 1999*(6).

Seymour, E. (2001). Tracking the processes of change in US undergraduate education in science, mathematics, engineering, and technology. *Issues and Trends, 86*(1), 79–105. http://dx.doi.org/10.1002/sce.1044.

Seymour, E., & Hewitt, N. (1997). *Talking about leaving: Why undergraduates leave the sciences.* Boulder, CO: Westview.

Siegler, R. S., Duncan, G. J., Davis-Kean, P. E., Duckworth, K., Claessens, A., Engel, M., et al. (2012). Early predictors of high school mathematics achievement. *Psychological Science, 23*(7), 691–697.

Simpkins, S. D., Davis-Kean, P. E., & Eccles, J. S. (2006). Math and science motivation: A longitudinal examination of the links between choices and beliefs. *Developmental Psychology, 42*(1), 70–83.

Singer, J. M., & Stake, J. E. (1986). Mathematics and self-esteem: Implications for Women's career choice. *Psychology of Women Quarterly, 10*, 339–352.

Spelke, E. S., & Ellison, K. (2009). Gender, math, and science. In C. H. Sommers (Ed.), *The science on women and science*. Washington, DC: The AEI Press.

Stinebrickner, R., & Stinebrickner, T. R. (2013). A major in science? initial beliefs and final outcomes for college major and dropout. *Review of Economic Studies*, 1–47.

Tall, D. (1991). The psychology of advanced mathematical thinking. In D. O. Tall (Ed.), *Advanced mathematical thinking* (pp. 3–21). Kluwer: Holland.

Tall, D. (1997). From school to university: The transition from elementary to advanced mathematical thinking. In *Australasian bridging conference in mathematics. Auckland University, New Zealand*.

Tall, D. (2008). The transition to formal thinking in mathematics. *Mathematics Education Research Journal, 20*(2), 5–24.

The White House. (2010). *"Change the equation" initiative, September 16th 2010*. Retrieved from: www.changetheequation.org.

The White House. (2011). *"Startup America" initiative, January 31st 2011*. Retrieved from: https://www.whitehouse.gov/economy/business/startup-america.

Upadyaya, K., & Eccles, J. S. (2014). How do teachers' beliefs predict children's interest in math and kindergarten to sixth grade? *Merrill-Palmer Quarterly, 60*(4), 403–430.

Updegraff, K. A., Eccles, J. S., Barber, B. L., & O'Brien, K. M. (1996). Course enrollment as self-regulatory behavior: Who takes optional high school math courses? *Learning and Individual Differences, 8*(3), 239–259.

Vaish, A., Grossmann, T., & Woodward, A. (2008). Not all emotions are created equal: The negativity bias in social-emotional development. *Psychological Bulletin, 134*(3), 383–403.

Wigfield, A., & Meece, J. L. (1988). Math anxiety in elementary and secondary school students. *Journal of Educational Psychology, 80*(2), 210–216.

SECTION IIIB

LEARNERS WITH MATHEMATICAL DIFFICULTIES

CHAPTER

9

Special Needs in Mathematics Classrooms: Relationships With Others

Melissa Rodd

UCL Institute of Education, London, United Kingdom

WHOSE NEEDS ARE "SPECIAL"?

> Christopher: All the other children in my school are stupid. Except I'm not meant to call them stupid, even though this is what they are. I'm meant to say that they have learning difficulties or that they have special needs. But this is stupid because everyone has learning difficulties because learning to speak French or understand Relativity is difficult, and also everyone has special needs, like Mrs. Peters who wears a beige-coloured hearing aid, or Siobhan who has glasses so thick they give you a headache if you borrow them, and these people are not Special Needs, even if they have special needs. (Haddon, 2003, Chapter 71 (the 19th chapter as chapters are indexed by prime numbers) page 56)

This introductory quotation is from a novel. It was chosen to provoke consideration of the nature of "special needs." The novel's narrator, Christopher, a 15-year-old with high functioning autism, expresses frustration with societal norms about "stupidity" (and in implied contrast, "ability"). We all have some special needs, so what is the issue? The quotation sets the scene for this chapter on emotion and "special needs" by its emotion-oriented tone and its suggestion that even to talk about special needs is a challenge. Talking about special needs in the context of mathematics learning is particularly tricky because having mathematical difficulties is often socially acceptable. For example, the *Psychology Today* blog asked, "Why is it socially acceptable to be bad at math?" (Wai, 2012) and this question was asked despite the US National Council for Teachers of Mathematics 2010 campaign to "stamp out the phrase *I am bad at math*" (Shaughnessy, 2010). Furthermore, the introductory quotation also points to the issue of "relationships" writ large.

http://dx.doi.org/10.1016/B978-0-12-802218-4.00009-1

This includes Christopher's relating to subjects like "French" or "relativity," his recognition of accessibility-aides that facilitate relating/communicating, and also the people that he is able to relate to.

In his book *Do You Panic About Maths?* Laurie Buxton (Buxton, 1981) exemplified how emotion can debilitate articulate adults when pressed to perform mathematics. Some of his research participants were able to locate the causes of their "math panic" and, with his help, realize their potential. For instance, the person whose father had repeatedly tested her on times tables was initially "blocked" when asked to calculate 15×12 mentally, but she was able to respond when her panic was reduced by Laurie saying "no hurry" (ibid, p. 7). How much more so might emotion impede children with learning difficulties? And how much more do such children need to rely on good relationships with their teachers?

Emotion, Theoretical Orientation, and Special Needs

Emotion is often experienced beneath conscious awareness, yet people can become aware of emotion through encounters with others. Rom Harré's positioning theory is an example of a theoretical perspective that attends to the social dimension of emotion (e.g., Harré & Moghaddam, 2003). However, positioning theory, which is based on Vygotsky's profound appreciation of thought and language, does not particularly capture phenomena like those experienced in basic emotional responses (Ekman, 1992) that are beneath conscious awareness, nor does it capture unconscious emotion related to defenses, repressions, or other responses attended to by psychoanalytic theory (Waddell, 1998). In seeking to theorize emotion in mathematics classrooms where "special needs"—however understood—play out, the field of psychoanalysis is appropriate to consider because the drawing out of emotional responses from beneath conscious awareness is a cornerstone of psychoanalytic practice. In order to frame understandings of emotion within the context of mathematics classrooms, in this chapter, aspects of psychoanalytic theory are employed, particularly those informed by British Object Relations theory and centrally the work of Donald Winnicott (see *Theoretical Frame: Winnicott's "Facilitating Environment": An Environment of Relationships* section). Winnicott's work is particularly helpful when thinking about mathematics and special needs because his "maternal care environment"—which is at essence an environment of relationships—can be adapted to a "math-care environment" (see *Application: To Special Learners of Mathematics* section). And because Winnicott's work centered on the emotional life of the preverbal child, which develops beneath conscious awareness, in a positive way, his ideas may be adapted to think about learners who may not be particularly articulate also in a positive way. Using a "Winnicottian lens" inevitably means that other aspects of psychoanalytic theory are not foregrounded. The discussion here does not go into, for instance, defense mechanisms for and against mathematics (Nimier, 1993), which is being developed elsewhere in learner and teacher contexts (Black, Mendick, & Solomon, 2009; Rodd, 2013).

Furthermore, although issues of disability—such as sense impairment, medical condition, or physical disability—are very relevant to special needs and education, this chapter's orientation to "special needs in mathematics classrooms" is to focus on those who have difficulties with mathematics and have low attainment. This is not to suggest that those with sense impairment, medical condition, or physical disability, like Stephen Hawking at the pinnacle, do not have special needs, but given suitable access relative to their disability, high attainment is achievable. There is long tradition of designing mathematical tasks for inclusivity: the Association of Teachers of Mathematics (based in the United Kingdom) is but one source for providing accessible materials, and research on task design—particularly for hearing and sight impaired—is ongoing (e.g., Healy, Fernandes, & Frant, 2013). So, while suitable tasks are important for accessibility, the central perspective taken in this chapter is to employ psychoanalytic understandings to theorize educational environments for learners who still "don't get it," despite reteaching and extra help.

A psychotherapist will have been educated to hone their awareness of a client's feeling and this takes many years of study and awareness training. However, day-to-day experiences of relating to others include—skillful or less skillful—intuitive responding to another person's emotion. In particular, a wise teacher can have an intuitive sense of what is needed, as Laurie did when saying "no hurry." In the context of understanding emotion in "special needs and mathematics" then, relationships are central (c.f., Black et al., 2009), and the focus on "mathematics classrooms" indicates that this chapter is more to do with schools and teachers and their assistants, than with parents or extracurricular activities.

Outline of the Chapter

After this introduction, the chapter proceeds (*Orientation to Psychoanalysis and Special Needs in Maths Classrooms* section) by offering an orientation to how psychoanalysis can be a useful frame though which to structure approaches to working with emotion and special needs in mathematics teaching and learning. The chapter continues, in *Theoretical Frame: Winnicott's "Facilitating Environment": An Environment of Relationships* section, by presenting some psychoanalytic theory and explaining why Winnicott's concepts of "facilitating environment" and "transitional object" are particularly pertinent to thinking about emotion and mathematics special needs. Following this, in *Application: To Special Learners of Mathematics* section, the presented theory is used to interpret data from teachers of secondary students with special needs in mathematics. *In Teaching Practice* section is oriented toward pedagogical principles and putting them into practice. The chapter concludes with a discussion and critique.

The central theme is that relationships (as well as physical, linguistic, and other aspects) constitute environments. So relationships with mathematics are to be experienced in mathematics classrooms through emotional connection with mathematics itself, people, or other transitional objects. With learners who "don't get it," a direct relationship to mathematics has not been previously successful, so relationships with the people and material transitional objects of the classroom are "way in" for these learners.

ORIENTATION TO PSYCHOANALYSIS AND SPECIAL NEEDS IN MATHS CLASSROOMS

Psychoanalytic understandings start with the existence and the importance of unconscious processes. In his book for nonexperts, Stephen Frosh (Frosh, 2002) explains this existence and importance by illustrating how, in ordinary life, there is unconscious communication: feelings, experienced by someone else, are also felt by oneself. Thus feelings, for instance, of calm or anxiety are projected or introjected between people without those people being aware, usually, that this is happening. In ordinary language, saying that one is "emotionally close" to another person or nonhuman (animal) has the meaning that one understands the other's feelings without need for (verbal) explanation.

Vignette: Hate and Interplay

For an orientation to a psychoanalytic framework for thinking about special needs in mathematics classrooms, consider this fragment of the reflective journal from a teacher, "Mr. Z," which is reproduced below. The extract recorded his interaction with a 12-year-old student with low prior attainment in his small class of five similarly identified students of similar age. The student, Gemma, displays an emotional response to a mathematical game, which she otherwise previously seemed to enjoy:

> I span the spinner, Gemma got the first go and it landed on 9, she said "I don't want to do it, I hate my 9 times table". I told her she might get it right, "no, I won't, I don't like my 9 times table". So I asked her "what is 9 times 2?" she said "that's easy it's 18". ***Teacher log notes from masters student* Mr. Z.**

What is Gemma doing when she resists, saying "I hate"? What is Mr. Z doing when he chooses to ask her "what is 9 times 2?" (rather than another "more difficult" item in the 9 times table)? Such an interplay of negotiation and emotional charge is not the learning objective or part of the lesson plan. Rather, the student unconsciously acts out defenses, such as "avoidance," while the teacher, also unconsciously, works on repairing the relationship between him and his student by engineering a better relationship between his student and the mathematics through her "success." A psychoanalytic framework recognizes the work that both student and teacher were doing in this encounter and the centrality of affect for both of them. This brief vignette illustrates, in Winnicott's terms, a "subtle interplay" (Winnicott, 1986, p. 172) between the teacher and his student: each of them is attuned to the other's moves; they do not have to express themselves in sophisticated language: the attunement is emotional and relational. It can also be inferred, as indicated, that unconscious processes, like defenses and desires, were activated and that the teacher-student relationship was important to them both.

Mathematical Objects Emerging

Richard Skemp relates a story about a baby boy who "adds' his empty baby's bottle to a line of empty bottles (Skemp, 1971, p. 9). Skemp points to the cognitive achievement of this child and to his "relational understanding" (another bottle, my bottle, is of the same kind as the others). In this chapter, the interpretation of this story focuses on that which was not reported in Skemp's story: the emotional achievement of the child and the child's relationship with the observer. Whether the story relates to Skemp's own child or not is unknown, but the baby was observed by someone primed to notice, value, and attribute meaning to the baby's achievements. The reader can imagine the delight of the observer, father, or friend that is felt emotionally by the child, thus strengthening their relationship through the realized action of lining up the bottles. "Lining up" becomes an emotionally charged memory connecting the adult and child in a profound way and contributing to bringing "and one more," or "+1" into being for the child. The cognitive achievement is more easily observed than the emotional one can be. Indeed, one could design a similar task to "measure" this sort of cognitive achievement by how many objects a child is able—or willing—to line up. However, here, I point to another way of thinking about the "bringing into being" of '+1," which is the emotional connection of the adult and preverbal child, the shared environment of their joint engagement in "+1."

In this chapter, the focus is understanding affective dimensions, in particular the learners' relationships in their environment, for learners for whom measuring cognitive achievement produces low, surprising, or erratic results.

THEORETICAL FRAME: WINNICOTT'S "FACILITATING ENVIRONMENT": AN ENVIRONMENT OF RELATIONSHIPS

Winnicott's term "subtle interplay" was referenced above as a suitable phrase to capture aspects of relationship between the teacher and learner in the vignette. Winnicott was a psychoanalyst associated with the British Object-Relations group active in the mid-20th century that included Melanie Klein, Anna Freud, John Bowlby, and Wilfred Bion. The Object-Relations group extended Sigmund Freud's theories of the unconscious by looking at how drives, instincts, and emotional life emerge from studying their development in very young children. An infant, when newborn, is not a differentiated person; it is a bundle of flesh and instincts, yet an infant's instincts, or drives, are material for mental development. Infants are not consciously aware of their mental processes and cannot be expected to express their feelings in language; nevertheless they are emotional. This Object-Relations view contrasts with other psychoanalytic theories. For example, Lacan's theory asserts that the unconscious is essentially a linguistic achievement (Frosh, op. cit., p. 48).

Object-Relations starts with theorizing infant experience and wondering, for example, how objects come into being. How does meaningfulness get initiated? There are several branches of the Object-Relations school (see, e.g., Mitchell & Black, 1995 for an overview). In this chapter, Donald Winnicott's approach is followed, because his contribution to Object-Relations theory centers on his interest in the ordinary and fully lived life as it comes into being; therefore it is applicable to the "ordinary" situation of learning and teaching mathematics.

Winnicott worked with mothers and babies in pediatric as well as psychoanalytic contexts, and his observations in these contexts enabled him to articulate ideas about a baby's excitement and its caregiver's devotion (Winnicott, 1964, p. 22). Winnicott used positive terms, like "excitement" and "devotion," in his descriptions of preverbal children's states of mind. These terms contrasted with the language used by the more well-known Objects-Relation theorist Melanie Klein, who classifies an infant's states of mind as "paranoid schizoid" or "depressive" (e.g., Waddell, 1998, p. 7; Frosh op. cit., p. 41).

Furthermore, Winnicott theorized how objects come into being from a preverbal state by thinking about a baby's developing "capacity to form symbols through transitional objects" (Caldwell & Joyce, 2011, p. xxii); for instance, the baby's bottle, from the previous section, could be interpreted as transitional to the symbolic "+1." Such a theory has scope to be applied to thinking about what transitional objects in mathematics classrooms might be and how meaningful mathematics might develop with them.

A baby develops its capacity to feel, move, and think as it grows and interacts in its environment. Winnicott's insight into how early thinking develops uses his concept of "maternal care" in which the physical and the emotional are merged: "maternal care" is the infant's environment (Caldwell & Joyce, 2011, p. 155) consisting of the physical and emotional environment of empathetic, daily caregiving that ordinary "good enough" (Winnicott, 1964, p. 24) mothers give. "Maternal care" does not exclude the father or another "devoted" person. When Winnicott asserts that *"there is no such thing as 'an infant'!"* (Winnicott, 1960), he is saying that no baby survives outside "maternal care." Thinking emerges from this environment he calls maternal care in various ways. This insight of Winnicott's—that thinking emerges from a maternal care environment—can be applied to educational contexts, as I aim to show, for special needs mathematics learning, in the rest of this chapter. The application of Winnicott's relational environment has been applied in other, more general contexts. For instance, Martha Nussbaum (2001) refers to "facilitating environments" that afford opportunities for communication of meaning through relationships. The gender- and role-neutral term "facilitating environments" has also been used by Caldwell and Joyce (2011). However, in the context of special needs in mathematics classrooms, the facilitating environment of relationships with others could be called a "math-care environment."[1] Aspects of this math-care environment potentially include adaptations of Winnicott's "transitional phenomena" and "holding" as described in the following section.

Containing Relationships

Wilfred Bion and John Bowlby were also associated with the British Object-Relations school and interested in "containing relationships." Bion's focus was on how frustration of not knowing is tolerated, or contained, and he theorized that not knowing is the foundation for subsequent thinking and learning (Bion, 1962, in Mintz, 2014, p. 19). Bowlby worked on attachment within relationships with people, particularly maternal attachment, and he characterized different forms of attachment patterns (High, 2012, p. 52). Winnicott used the idea of holding as indicative of being in a containing relationship.

A baby's recognition of its caregiver (mother) as a person, an "object" distinct from him or herself, emerges, according to Winnicott, from the experience of being "held" in (the baby's) maternal-care (environment). Being related to by holding thus facilitates the baby's becoming more integrated, more a person with his or her own body and mind (Caldwell & Joyce, op. cit., p. 158). Ideas about "infant holding" in a Winnicottian facilitating (maternal care) environment can be applied to a metaphorical "student holding"

[1] Terminology: Winnicott's term "maternal care" is an environment; using the analogous words, "math care," sounds a bit odd to me, so I've used "math-care environment" as the analogous notion in this chapter.

in a "math-care" educationally facilitating environment where relationships and emotions are part of the fabric from which learning grows.

Transitional Objects

> "There is a direct development from transitional phenomena to playing, and from playing to shared playing, and from this to cultural experiences." (Caldwell & Joyce, 2011, p. 246)

A "transitional object" from a Winnicottian perspective is a baby's first possession, an object evoking intense interest that is neither "internal" (imagined) nor wholly "external" (available to others). Typically it starts life as a soft toy or a piece of blanket that the baby "created" (Caldwell & Joyce, 2011, p. 186) as an object for him/herself. Parents and caregivers collude with the baby in the transitional object's importance and its reality; this collusion is an important part of the baby's maternal care environment. Transitional phenomena consist of objects and also the way those objects are engaged with by the baby, like the baby exploring a bit of cloth with its mouth "accompanied by sounds of 'mum-mum', babbling, anal noises, the first musical notes, and so on" (Caldwell & Joyce, 2011, p. 105).

Winnicott's understanding of the "direct development," quoted above—from transitional phenomena, through different stages of play, to shared meaning—has application to the central investigation of this volume into understanding emotions in mathematics thinking and learning. All early learning from Winnicott's perspective is emotionally driven; he discusses this in terms of integration, unintegration, and disintegration (Caldwell & Joyce, 2011, p. 69), and also in terms of instinctual drives and that maternal care is the environment for development of the baby's personality (Caldwell & Joyce, 2011, p. 176).

The term "transitional object" was extended by Christopher Bollas to include people in the baby's environment. (1987, pp. 13—29, in Nussbaum, 2001, p. 184) and in this chapter the term is further interpreted for educational contexts.

APPLICATION: TO SPECIAL LEARNERS OF MATHEMATICS

> Emotions [that] in their origin and many ongoing functions [are] adaptively rational, may frequently also be irrational in the sense that they fail to match their present objects, as they project the images of the past upon them. (Nussbaum, 2001, p. 179).

Learners who have had mathematical difficulties in their pasts are likely to be wary of mathematics because memories of their experiences of being wrong or confused come back to haunt them, consciously or unconsciously.

Math-Care Environment: An Example

> Every single participant, during interview, named division (or indicated the symbol '÷') as something they found difficult, did not understand, or disliked. ... By changing their relationship with [division], I aimed to increase students' confidence and self-efficacy, and encourage more positive identities in relation to the subject, without 'protecting' them from challenge. (Finesilver, 2014, p. 120).

Carla Finesilver's method of working with students with very low prior attainment in lower secondary school (ages 12–14 years), has been, as indicated in the extract heading this section, to mediate her students' relationships with the mathematical content they are to learn (Finesilver, 2014). Her approach involved withdrawing one or a pair of students from their mathematics class and working with them on mathematical "scenarios." The methodological approach taken was to authentically teach students wanting extra instruction, so data came from this interaction.

Finesilver designed these scenarios as opportunities for developing these students' relationships with multiplication/division through their mark marking, personal narratives associated with working on these scenarios, and her interpretation of their marks and stories in the context of the mathematics, and they were designed to facilitate students letting go of false arithmetic methods and intellectual numbness. One such scenario was nicknamed "biscuits," and it involved sharing a number of biscuits, N, between a number of friends, M, with $N \div M$ such as $27 \div 3$, $28 \div 4$, etc. It is possible to get a sense how students' relationships with multiplication and division are developed within this math-care environment by looking at Finesilver's data of Vince's progression with division from tuition sessions 2 and 5 using the biscuits scenario: In Fig. 1, Vince has arranged the straws, donated by the teacher (Finesilver), to represent a fair sharing of biscuits on the "plates" that were drawn by the teacher.

However, by tuition session 5, Vince is creating his own receptacles that convey a shared understanding with his teacher (Fig. 2).

Finesilver's research was not overtly influenced by psychoanalysis, yet there are three aspects of her method to draw attention to in the context of what a math-care environment might be: (1) meticulous preparation, obscure to the student, that afforded her response to material the student produced to be emotionally intelligent and mathematically wise; (2) evidence of transference (considered in the Object-Relations sense): for instance, "I also included occasional snippets of personal information (e.g., "I found it really hard learning times tables too"), openly positioning myself as an individual with an educational history of my own, *sharing something with them in return for their sharing their thoughts with me*" (my italics, ibid, p. 116). Another example is, "I believe it reasonable to state that they enjoyed

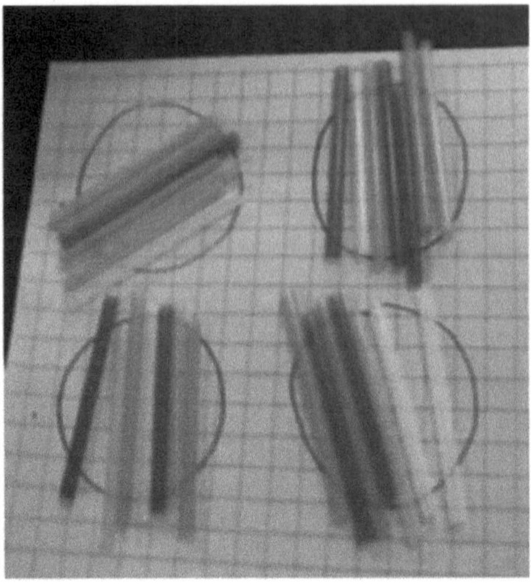

FIG. 1 Vince, 28 ÷ 4, tuition session 2 (Finesilver, 2014, p. 333).

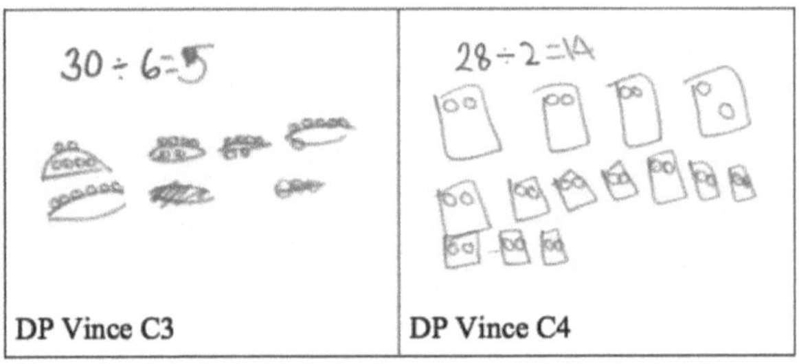

FIG. 2 Vince 28 ÷ 4 tuition session 2 (Finesilver, 2014, p. 339).

having the undivided, patient, nonjudgemental attention of a teacher figure" (p. 115); and the last aspect is (3) an "analytic hour" for the session within which there were no further time constraints.

Subtle Interplay: Example 2

A more visual mathematics education example of subtle interplay, discussed above, can be found in Finesilver's research. The picture (Fig. 3) shows a mathematical artefact "cocreated" (ibid, p. 71) by student and teacher with the student empowered to abstract the mathematics (here, of multiplication) to the extent she can manage in that particular lesson. The teacher-student set-up involved student agency, shared attention, and patience (similar to the parable of the stationary donkey; see *Pedagogical Principles for a "Math-Care Environment"* section). When I discussed this chapter at our informal weekly seminar, some of my

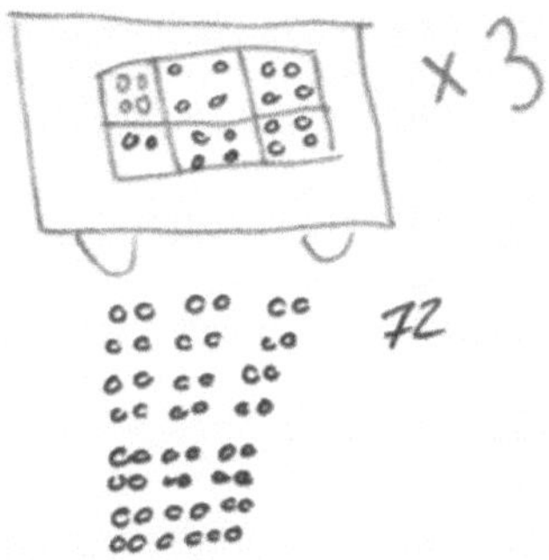

Four bottles per box, six boxes per van, and three vans (CF and Wendy)

FIG. 3 Finesilver's visual subtle interplay (Finesilver, 2014, p. 71).

colleagues were concerned about my using the image of Fig. 3 because "it was not correct" (e.g., bottles are missing from a box in the van). On reflection, this "error" that the teacher need not notice and the child has left behind, is part of the collusion concerning existence, analogous to that in maternal care, and is a good example to illustrate a shared view in a math-care environment.

Containing Relationships and Special Needs Mathematics

> Because I'm working in such a specific environment, my teaching is based on my relationships with my students. So I had to build my relationship first, before they are ready to learn anything at all; before they allow me to teach them at all. Mr Y.

In this section, I present material from Mr. Y., a part-time master's student and full-time teacher in a special school for boys aged 11–16 years who have social and behavioral difficulties. The material is interpreted in terms of the holding, or containing, relational aspects of the "math-educational care" that he indicates he is wanting to cultivate (see the quotation above). The beliefs of individual teachers, like Mr. Y, as well as those of the students, and the ethos of the school and the wider communities, contribute to the emotional climate of math-educational care.

Here is an extract of an interview with Mr. Y (from a research project reported in Crisan & Rodd, 2011). Mr. Y's first response in the interview is the following:

> I work with children who have SEBD (social, emotional and behavioural difficulties). So they usually come from underprivileged families and their home life is very chaotic; so they are looking for structures, even though [pause], unconsciously they

> are looking for structures. So to me, they all enjoy maths; and to me this is because the maths brings structures to their lives. So not only academically, but to their life it just brings structure; because they all enjoy coming to lessons, whether they are good or not that good—they always enjoy maths.

His final statement is the following:

> I keep it as structured as possible, because my boys, they need structure otherwise they fall into pieces; so it would be definitely structured.

Mr. Y has been a part-time student for about 5 years. His commitment to "his boys" has been evident and he takes pride in relating positive career trajectories of some of his students. He is motivated to improve the lives of his students who "due to their social circumstances, difficulties in dealing with their emotions or knowing how to control their behaviour, are condemned to academic low achievement." Mr. Y believes that he provides an attachment figure, in the sense of Bowlby (*Transitional Objects* section), that his students had not previously enjoyed. He also believes that mathematics as a discipline can be a container—he refers to its "structure" —that affords this attachment. This is in the same spirit as the calming function of mathematics discussed in Rodd (2006).

There are layers of "containment" in this material; in tutorials with Mr. Y I listen to accounts of the classroom on two levels: his account of his students' mathematical work and their classroom behavior and to hear about the challenges of his job including his emotional response to its challenges. This indicates that relationships in the math-educational care, the facilitating environment for mathematical learning, extend into supporting communities.

Transitional Objects

The general outline of Winnicott's notion of a "transitional object" was outlined in *Transitional Objects* section. Here, attention is drawn to three instances in mathematics classrooms of phenomena that can be interpreted as "transitional objects." These are (1) students' body parts, (2) number symbols, and (3) the teacher, respectively.

Transitional Objects for Mathematical Learning

From any learner's perspective, mathematical objects—numbers, functions, shapes, and so on—have a period as transitional phenomena. This is because, as a learner, these mathematical objects are "neither me nor not-me"; they can be thought of as created subjectively—mentally—yet shared objectively— culturally. For typical learners in numerate societies, by the age of twelve numbers can be said to be properties of their society's environment, though "special needs" students, like Mr. Z's (quoted above in *Orientation to Psychoanalysis and Special Needs in Maths Classrooms* section), are still on their journey towards numbers as objective. Consider their transitional phenomena for multiplication:

> The first task was to complete the 5 timetables using your fingers and toes.
> R and S [each] got stuck when they reached 20 because they had run out of fingers and toes. So I suggested that they should use each others' toes and fingers. They stopped when they got 40. So I asked them what are they going to do about it. S said: "we'll borrow B's fingers and toes".

This extract from Mr. Z's teaching log (when teaching the same class as mentioned in *Orientation to Psychoanalysis and Special Needs in Maths Classrooms* section) illustrates the playful nature of the task posed and also the explicit prompt the children needed to venture into shared playing. Both task and response fit with thinking about number learning within a Winnicottian framework: the teacher provides an environment for playing together and he is available for them to encourage, or "hold," this playing.

This extract also provides an example of "transitional objects" for mathematics learning, thus interpreting Winnicott's notion for the mathematics classroom. In the extract above, the students begin by playing with their fingers and toes and then imagine other peoples' fingers and toes. In the English language, fingers are also called "digits," suggesting the transitional nature of fingers in counting more widely than in these children's play. The students' emotional achievement is seen in their investment in the relationships their classroom supports (with each other and their teacher). Thus their subjective digit-touching becomes objective number knowledge through such experience and they become part of the culture that understands 40 as 8×5 without reference to their own and their friends' fingers and toes.

Mathematical Objects as Transitional Objects

Learners who do not have much spoken language may be able to make use of numbers and other mathematics as transitional objects. Many years ago, I taught 14-year-old, Len, diagnosed with autism and used very little spoken language indeed, but he would engage with number patterns, writing out list after list of numbers in sequences, including sequences like the triangular numbers that start 1, 3, 6, 10, … . Like a baby stroking its ribbon in satisfaction, the objects of his creating, the lists of numbers in patterns, seemed to soothe him. And while the sequences retained a pattern I could judge as "accurate," there was not an observable shared playing, and I always wondered whether or not Len saw his number patterns as part of "cultural experience." Because I can now interpret Len's behavior in terms of Winnicott's theories, I would now aim to provide, in the classroom environment, more shared playing.

Teacher as Transitional Object for Mathematics

"Chalk and talk" is a phrase used for a teaching style where a teacher stands in front of a class and talks through a procedure as he/she writes that procedure on a board. It is usually used with negative connotation unlike the phrase "modeling" (Ofsted, n.d.), which has similar meaning. While contemporary pedagogical approaches tend to decenter attention from the teacher, a teacher-centered approach (that may include giving information, repeating, patterning, etc.), can position the teacher to be a transitional object for mathematics learning. For example, a high school student said of Mr. W's help, "he explained it (trigonometry) using exactly the same words, three times in a row. Then I got it!" In this case, analogous to the ribbon the baby strokes, Mr. W made himself available for the student to use—by his repeating the same words—in order to objectivize her previously incoherent experiences with trigonometry. It is not possible to go back in time and ask Mr. W whether he deliberately or intuitively repeated his explanation verbatim, and indeed, the very asking of such a question might well illicit a verbal response as a defense that represses the unconscious, intuitive action. Nevertheless it is possible to imagine a teacher-student dyad, like a care-giver-baby dyad, where explaining, like soothing, is repetitive and focused on the relationships (with the student and her discomfort or the baby and its discomfort). When learning specific mathematics, like numbers or trig ratios, people who hold the knowledge (typically teachers but sometimes parents or others) can be thought of as a "transitional object: that the learner uses (metaphorically: "strokes") in order to transition from the muddle (subjectively) experienced to the culture in which this mathematics he/she is learning is cultural knowledge (i.e., objectivized).

IN TEACHING PRACTICE

Pedagogy and Relationship

This section is on realizing "math-care environments," aspects of which were discussed in *Application: To Special Learners of Mathematics* section, and are in two sections. First, two contemporary agendas are briefly discussed: the inclusion agenda and the knowledge agenda. Although these two agendas impact differently on teacher practice in different educational jurisdictions; nevertheless, worldwide, teachers and policy makers need to make decisions about inclusion and determine what knowledge might be useful to them in the domain of special needs. Second, pedagogical aspects of math-care environments are discussed. In particular, how does a teacher provide and sustain such an environment?

Contemporary Agendas

Policy and Culture: The "Inclusion Agenda"

The student, who said that she "got" trigonometry because of Mr. W, was a high attaining student who had become "mentally paralyzed' by being confronted with this new mathematics that was outside her imagined scope. During the period of numbness and rising panic, like Laurie Buxton's participants, her needs were to do with mathematics. She was "special" at that moment; a "good enough" math-care environment was sufficient to get her going again.

As the introduction suggested, what special needs means is both contested and also varies across time and culture. In many countries today, children with disabilities or with special educational needs are educated, when possible, with their peers; suitable adjustments—like tactile materials for blind students—are made so that those with a special need can access the learning and be included in the mainstream. (For a UK-based introduction to teaching mathematics inclusively, see, e.g., Finesilver & Rodd, 2016). An emotional dimension of this inclusion is felt by the individual students as they experience mainstream education—for some it is liberating, but others do not cope well and need some withdrawn special provision, perhaps just for a few years. The agenda for a policy for inclusion, within the United Kingdom at least, was stimulated by the influential Warnock report (Warnock Committee, 1978) that reported, much to the surprise of the general public and educationalists alike at the time, that about 20% of the school population within the United Kingdom had a special need for at least some period over their 11 years mandatory schooling. This societal awareness that to have special needs was not that uncommon can be thought of as another emotional dimension of inclusion, where those with special needs are accepted as part of the spectrum of human variation. How this acceptance plays out in the classroom with individual learners and their teachers depends, inevitably, on the emotional states of the child and his/her teacher, the teacher's knowledge about the child's condition and the mathematics to be taught and, crucially, relationships between the teacher and the child and others in the learning environment. Cultures within different educational establishments will vary; for instance, some will give opportunities for students to have a one-to-one with their teacher, while others will provide other resources: for example, peer-to-peer help or a link to a website. How math-care environments flourish or otherwise, as a function of their institutional establishments addressing inclusion, is a serious question, but it is outside the scope of this chapter.

Research and Scholarship: The Knowledge Agenda

In this era, information appears wirelessly with a tap on our screens as a product of the knowledge economy. Such an environmental affordance positions us, as 21st century teachers, to believe that if we, find that right piece of knowledge about a special need and mathematics—say dyslexia and number facts—then with that information, the child can be taught those number facts either with specific pedagogical approaches or with suitable assistive technology or, typically, a combination. On one hand it is of course a fantasy that such "problems" can be cured through mere "application of knowledge" and yet, on the other, having access to detailed research, practitioner accounts, and technological innovations gives a teacher a much wider repertoire as well as opportunities to communicate with other teachers across the globe who are teaching students with similar special needs. A significant resource for information about mathematics and special needs up to the new millennium has been produced by Olof Magne (Magne, 2003). This bibliography, of about 4300 documents, includes three sections particularly relevant to emotion and special needs mathematics: "Affect and motivation" (291 documents), "Aggressivity, asocial behaviour, criminality" (19 documents), and "Social conditions in mathematics, ecology" (100 documents) (ibid, pp. 14–15). Doubtless, since this bibliography was compiled, internet users will be able to access hundreds, if not thousands more publications on these and related topics. Another significant resource, which reviews the field of affect and mathematics education generally, not just special needs, has been more recently written by Gerald Goldin (Goldin, 2014), who is himself a significant contributor to the field of affect and mathematics education. The bibliography in Goldin's article contains 115 items, each of which gives a lead to original research, pedagogical principles, or further reviews.

The issue of how teachers can engage with such an avalanche of research and scholarship, as well as policy initiatives and emerging practices, can have no general solution. However, a teacher may be able to cultivate a Winnicottian, math-care environment involving working in relationship with the students. In order to do this, as Mr. Y well knows, the teacher needs to attune to his/her own emotional responses in relation to those of his/her students, and this requires the teacher's looking after him/herself too.

Some ideas for classroom practice that include attending to the inclusion agenda, the knowledge agenda, as well as teacher flourishing are proposed in the next section.

Pedagogical Principles for a Math-Care Environment

The parable of the stationary donkey

A donkey was standing, with packs full of goods to be sold at market, but no matter how much the lad pulled at the donkey's rein, it would not move! The lad's grandfather, appreciating the situation, said "stand by his shoulder with a loose hold on the rein, look in the direction you want him to go and wait". Sure enough, after a while, the donkey and lad set off together in the direction of the market. (Adapted from Williams & Penman, 2011, pp. 111–112).

The learners we are considering here are, more or less, stationary. By pausing alongside them, knowing the direction (the mathematics of the curriculum) and traveling with them on their mathematical journey, perhaps they will "get to market." The key idea here, in the discussion of understanding emotion in the special needs mathematics classroom, is that thinking arises from emotional connection suitably held by the environment where that environment includes key people, like the teacher. Actually, although this observation is relevant to students more generally, considerable commitment is required of teachers who teach students with mathematical difficulties. Macnab and Cummine remark the following:

> At the heart ... is a sensitivity on the part of the teacher to be able to feel as the learner feels, to put himself in the learner's shoes, not only in the cognitive sense, but also in the emotional sense. (Macnab & Cummine, 1986, p. 39)

This notion of feeling the emotion of the learner, (as Mr. Z did, when wanting to avoid stoking Gemma's fear of failure), is at the heart of understanding emotions and special needs in mathematics. But what might prepare a teacher to have this capacity? This question is addressed below in the context of Macnab and Cummine's plea for emotional connection with the learner together with a Winnicottian conception of a learning environment being relational thus emotional.

Productive Uncertainty, Subtle Interplay, and Pausing

A proposal, when thinking about principles with which we, as teachers, approach the mathematics education of children with mathematical difficulties, is to work with Joseph Mintz's notion of "productive uncertainty" (Mintz, 2014). Mintz draws on his practice, as a teacher, theories of teacher knowledge like those of Schön (1983) and Brown and McIntyre (1993) and from psychoanalysis, particularly Bion's notions of containment to characterize a teacher's productive uncertainty. He positions productive uncertainty as a positive feature of professional practice that develops theoretically Schön's "reflection" and recognition of "in the moment" decision-making and is crucially based in relationship:

> There is a particular intensity about the relationship between teacher and child, or nurse and patient, or social worker and client. The problems with which these professionals are faced are problems about people, and the uncertainties that they experience involve, to a significant extent, the human mind; certainly an embodied mind, but nevertheless a complex mind that they can only come to know through relationship. (Mintz, op. cit., p. 4)

The notion of productive uncertainty is allied to Winnicott's subtle interplay discussed above because it is a part of the relationship environment that yields something new. An example from Mintz's case studies illustrates how aligned productive uncertainty is with "subtle interplay": One of the teachers, Kathy, was reported as having engaged in "theatrical flourishes" with a student on the autism spectrum for whom this expression of joint interest was exciting for him. Mintz puts the case that "this exertion of energy … served to create some sort of live connection with Mark" (Mintz, 2014, p. 104). This "live connection" is a relational environmental source of learning for Mark.

One of the big challenges in teaching at any level is to pause and "pick up" information from the students. Sometimes in a teaching situation this will be explicit: students are asked questions, and the teacher notes who offers an answer and the quality of that answer. But usually, what people mean by the informal phrase picking up in a classroom context is that there are unconsciously intuited projections and introjections of that educational environment. This happens in a lecture of several hundred and in a one-to-one tutoring session.

Mintz uses the psychoanalytic concept of countertransference, which is a deep theoretical and experiential notion, to theorize how a teacher, like a therapist, unconsciously perceives a student's emotional response (ibid, p. 21) within their teacher-student encounter. However, it is an unusual teacher who has a working understanding of countertransference on a theoretical level together with having the capacity to reflectively engage with countertransference experientially; learning to recognize and work with countertransference is part of a psychotherapist's training. In the context of teacher preparation and continuing professional development, including psychotherapist training is not an appropriate expectation. Although Mintz does report incidents of teacher productive uncertainty, there is a sense in his book that these incidents are happy accidents rather than developing aspects of practice.

How might it be possible then for a teacher to cultivate the pause that productive uncertainty requires without needing to learn about psychoanalysis either as a theory or a practice?

Mindful Preparation for Productive Uncertainty/Subtle Interplay/Pausing

One way to cultivate the capacity for productive uncertainty and the ability to pause is through the contemporary secular practice of "mindfulness." The "Modern Mindfulness Movement" (Stanley, 2012) encompasses a set of meditation practices derived from Buddhism that are designed to release the practitioner from mental affliction, like classroom-related stress. So, in difficult situations where defenses might kick in, "mindfulness practice" can develop a capacity for pausing and developing a distance from the stress or other affliction. The adaptation of Buddhist meditation for a non-Buddhist clientele was successfully shown to be a treatment for pain relief and depressive illness (Baer, 2003; Kabat-Zinn & Hanh, 2009) and has been rolled out to many different populations, including students and their teachers worldwide as any internet search on mindfulness will show.

However, mindfulness training is not a "one off" but has to be sustained as an ongoing daily meditation practice. A central aspect of the training is to be aware *but not judgmental* about what is happening at the present as experienced in the body (Williams & Penman, 2011, p. 34). This sense of training moment-by-moment awareness suggests alignment with Schön's recognition of "in the moment" decision-making.

Mindfulness training aims to develop a practice of a friendly curiosity, joy, equanimity, and, essentially, compassion. This is quite different from the probing and tricky questioning associated with, for instance, the practice of formative assessment. Formative assessment checks the progress of learners against a standard held by the teacher on behalf of an authority (school, exam board, or other community of practice) and communicates that progress "and how you can do better" to the learner in writing or in conversation as part of the teaching process. In the cultivation of mindfulness, as understood as a secular training of calm responding to events with kindness and compassion, an interaction with a learner with special needs has at its heart a quality of connectedness and does not position them as merely a less able "other"; this is the insight from Macnab and Cummine (1986) quoted above.

Mind the Gap: Back to Relationship

What does a teacher have to do to interact "mindfully" with learners? Work on him/herself daily by training his/her mind to be able to pause and reflect in the present moment. The words of the phrase used on the London underground, "mind the gap" (between the train and the platform), can be played in your mind; let there be a gap between feeling negative (e.g., because of stress) and reacting to that feeling. Indeed, stress is a feature of most teachers' professional lives that tends to decrease possibilities for productive uncertainty or subtle interplay because stress contracts the very openness that is required for the productive pause. And teachers working with students who have, in the current euphemism, challenging behavior have additional stresses that include possible physical exchanges. Mindfulness practice can help to provide a gap between a feeling and reacting to that feeling. Thus practicing teachers are positioned to allow the possibility for productive uncertainty to arise and avoid defensive or reactive pulling on the "donkey's rein." And in that gap, there is the possibility to empathize with the suffering of the learner. Compare this with observing the lack (of knowledge, of self-control, of skill) of the learner. That moment of connecting empathy is a moment without self-and-other. And it is that connected feeling to others associated with love and commitment. It is a different feeling from assessment of features of the student's learning (often itemized minutely) that shows up what knowledge has to be "filled in"; this approach exacerbates "othering" of the learner, particularly in the case when the student is low attaining, special needs, or otherwise culturally different from the teacher.

The pedagogical principle then is that minding the gap in productive uncertainty, which is related to subtle interplay, rather than "filling in gaps in knowledge," should be part of the facilitating math-care environment for special needs students. Teacher stress militates against a teacher being able to do this day in day out to sustain a math-care environment. However, mindfulness, or other similar practice, might empower a teacher to pause and feel the student's emotional connection with mathematics, which is at the heart of a relationally-centered math-care environment.

DISCUSSION AND CRITIQUE

In this chapter, the focus has been on conceptualizing the educational environment in Winnicottinan terms that has relationship as central. The quotation from "Christopher" at the beginning drew attention to the variation of what constitutes a "special need." In some sense each learner has his/her difficulties, although attention here has been on learners who have difficulties with basic mathematics. This chapter has taken a psychoanalytical, rather than cognitive approach, with emotion being central in psychoanalytic theories of learning. There are many guides to cognitive approaches to special needs in mathematics; for example, Daniels and Anghileri (1995) give an organizing framework, Dowker (2004) reports on studies of cognitive interventions, and Haylock (1991) provides pedagogical gambits for low prior attaining learners in middle school years. Even when attention is being paid to cognitive attainment and the barriers to achievement that a "special need' presents," issues of emotion still hover beneath the surface. In particular, emotion can be positioned as an unfortunate consequence of an educational environment that is insufficiently facilitating. For instance, Dowker, writing of a student with dyscalculia, says, "It is important that people like Charles be identified as having difficulties early on, so as to reduce the risks of intellectual confusion and emotional frustration and possibly humiliation" (2004, pp. 14–15). In this chapter, I started with the premise that emotion exists in every encounter and used a small portion of a theory that privileges emotion. There are other theories that privilege emotion; for example, in neurophysiology, thinking (appraising, remembering, taking action) is linked to the way emotion arises preconsciously (Damasio, 2003). There are now cross-disciplinary studies, affective neuroscience for example, that link these relevant fields of enquiry into learning, emotion, and special needs. However, from an education perspective, the "grain size" for enquiry is the classroom with its students and teachers; this educational environment has at its heart the relationship between teacher and student and this has been the focus here.

Educational Environment

The notion of "math-care environment" has been used as an application of Winnicott's term "maternal care" (or "facilitating environment"), where people and the relationships between people are environmental. When a teacher says, "I felt guilty that they did not get good grades," his sense of guilt is part of the environment. And what is behind the guilt and the carrying of responsibility is a fear of being considered worthless in the profession in which is imbued so much of that person's identity.

Teacher Defenses

I have not addressed teacher defenses (conceptualized psychoanalytically) much in this chapter; the focus has been on the relationships in the classroom understood as environmental. The way that a teacher's defenses play out in his/her classroom, and the consequences of such for student learning is a difficult investigation as data collection requires skills associated with being psychotherapist. It can be done. Mintz (2014, op. cit.), using countertransference as a research tool, presented case studies (2014, pp. 74–153) that show how teacher defenses and student approach to learning were "held" in different situations. There are also the issues with defenses against mathematics, mentioned above, that are relevant to teacher defenses.

Teacher Education

Other approaches to working with teachers on psychotherapeutic relationships with their special students include those discussed by Helen High and colleagues in the context of educational psychotherapy (High, 2012). High and colleagues build on traditions established at the Tavistock clinic in London in using therapeutic techniques to understand some children's emotional barriers to learning. Because educational psychotherapy requires the same person to teach the school work and be the therapist, they have extended their work to providing in-service courses for teachers so that they might be in a position to use educational psychotherapy in their daily teaching of their students.

Conclusion

Students with low prior attainment will each have had a bad relationship history with mathematics inasmuch as they have not succeeded. Some will say they "don't care," and others will be bored or tired; others will act out antisocial behaviors, and each of these reactions, (and other possible reactions), has psychoanalytic interpretation; some examples are denial, suppressed anger, and fear of annihilation, respectively.

The aim of this chapter was to find ways of understanding emotions in mathematical thinking and learning by looking at those in mathematics classrooms with "special needs" with low prior attainment through a psychoanalytic lens. Not only should this help us, as educators, with the students with mathematical difficulties, but also prime us to see more clearly cues from students who are more successful in the school system because we would be better attuned to the affective dimension of learning.

Acknowledgments

Thanks to my students for permission to share their material—particularly to my former doctoral student, Carla Finesilver PhD. Thanks also to colleagues in mathematics education at UCL Institute of Education for their feedback.

References

Baer, R. A. (2003). Mindfulness training as a clinical intervention: A conceptual and empirical review. *Clinical Psychology: Science and Practice*, *10*(2), 125–143.

Bion, W. 1962. *Learning from experience*. London: William Heinemann.

Black, L., Mendick, H., & Solomon, Y. (Eds.), (2009). *Mathematical relationships: Identities and participation*. London: Routledge.

Bollas, C. (1987). *The shadow of the object: Psychoanalysis of the unknown thought*. London: Free Association Books.

Brown, S., & McIntyre, D. (1993). *Making sense of teaching*. London: Open University Press.

Buxton, L. (1981). *Do you panic about maths?*. London: Heineman Educational.

Caldwell, L., & Joyce, A. 2011. *Reading Winnicott*. London: Routledge.

Committee, Warnock. (1978). *Special educational needs: The warnock report*. London: Department for Education and Skills.

Crisan, C., & Rodd, M. (2011). Teachers of mathematics to mathematics teachers. In C. Smith (Ed.), *Proceedings of the British society for research into learning mathematics*, *31*(3), 29–34.

Damasio, A. (2003). *Looking for spinoza: Joy, sorrow, and the feeling brain*. San Diego, CA: Harcourt.

Daniels, H., & Anghileri, J. (1995). *Secondary mathematics and special educational needs*. London: Cassell.

Dowker, A. (2004). *What works for students with mathematical difficulties. Research report RR554*. London: Department for Education and Skills.

Ekman, P. (1992). An argument for basic emotions. *Cognition & Emotion*, *6*(3-4), 169–200.

Finesilver, C. (2014). *Drawing division: Emerging and developing multiplicative structure in low-attaining students' representational strategies*. PhD thesis University of London.

Finesilver, C., & Rodd, M. (2016). Inclusion in practice: special needs pupils in mainstream mathematics. In S. Johnston-Wilder, D. Pimm, & C. Lee (Eds.), *Learning to teach mathematics in the secondary school: A companion to school experience* (4th ed.). London: Routledge.

Frosh, S. (2002). *Key concepts in psychoanalysis*. London: The British Library.

Goldin, G. A. (2014). Perspectives on emotion in mathematical engagement, learning, and problem solving. In R. Pekrun & L. Linnenbrink-Garcia (Eds.), *International handbook of emotions in education* (pp. 391–414). New York: Routledge.

Haddon, M. (2003). *The curious incident of the dog in the night time*. London: Jonathan Cape.

Harré, R., & Moghaddam, F. M. (Eds.), (2003). *The self and others: Positioning individuals and groups in personal, political, and cultural contexts*. Westport, Connecticut: Greenwood Publishing Group.

Haylock, D. (1991). *Teaching mathematics to low attainers*. London: Sage. pp. 8–12.

Healy, L., Fernandes, S.H.A.A., & Frant, J. B. (2013). Designing tasks for a more inclusive school mathematics. In *Task design in mathematics education. Proceedings of ICMI study, Vol. 22* (pp. 61–69).

High, H. (2012). *Why can't I help this child to learn?*. London: UKCP Karnac Series.

Kabat-Zinn, J., & Hanh, T. N. (2009). *Full catastrophe living: Using the wisdom of your body and mind to face stress, pain, and illness*. New York: Delta.

Macnab, D. S., & Cummine, J. A. (1986). *Teaching mathematics 11-16: A difficulty-centred approach*. Oxford: Blackwell.

Magne, O. (2003). Literature on Special Needs in Mathematics: a bibliography with some comments. *Educational and psychological interactions*: *124* (4th ed.). Malmö Sweden: School of Education.

Mintz, J. (2014). *Professional uncertainty, knowledge and relationship in the classroom*. London: Routledge.

Mitchell, S. A., & Black, M. J. (1995). *Freud and beyond: A history of modern psychoanalytic thought*. New York: Basic Books.

Nimier, J. (1993). Defence mechanisms against mathematics. *For the Learning of Mathematics, 13*(1), 30–34.

Nussbaum, M. (2001). *Upheavals of thought*. Cambridge: Cambridge University Press.

Ofsted (n.d.) https://www.ncetm.org.uk/public/files/725865/Ofsted+key+indicators.pdf. (Accessed 27 February 2017).

Rodd, M. (2006). Commentary: Mathematics, emotion and special needs. *Educational Studies in Mathematics, 63*(2), 227–234.

Rodd, M. (2013). Teacher visualisation loss mid explanation: an issue when teaching geometry. *Proceedings of the British Society for Research into Learning Mathematics, 33*(3), 49–54.

Schön, D. (1983). *The reflective practitioner: How professionals think in action*. New York: Basic Books.

Shaughnessy, J. M. (2010). http://www.nctm.org/News-and-Calendar/Messages-from-the-President/Archive/J-Michael-Shaughnessy/It_s-Time-to-Stamp-Out-the-Phrase-"I_m-Bad-at-Math"/. (Accessed 1 May 2016).

Skemp, R. R. (1971). *The psychology of learning mathematics*. London: Penguin.

Stanley, S. (2012). Mindfulness: Towards a critical relational perspective. *Social and Personality Psychology Compass, 6*(9), 631–641.

Waddell, M. (1998). *Inside lives: Psychoanalysis and growth of personality*. London: Routledge.

Wai, J. (2012). https://www.psychologytoday.com/blog/finding-the-next-einstein/201203/why-is-it-socially-acceptable-be-bad-math. (Accessed 1 May 2016).

Williams, M., & Penman, D. (2011). *Mindfulness: A practical guide to finding peace in a frantic world*. London: Hachette Digital.

Winnicott, D. W. (1960). The theory of the parent-infant relationship. *International Journal of Psychoanalysis, 41*(6), 585–595.

Winnicott, D. W. (1964). *The child, the family and the outside world*. Harmondsworth, Middlesex: Penguin.

Winnicott, D. W. (1986). Holding and interpretation: Fragment of an analysis. *The international psycho-analytical library: Vol. 115* (pp.1–194). London: The Hogarth Press and the Institute of Psycho-Analysis.

Further Reading

Rodd, M. (2010). Organising the teaching of mathematics for children and young people who have additional to typical needs. In S. Johnston-Wilder & C. Lee (Eds.), *Leading practice and managing change in the mathematics department - A resource book for subject leaders in mathematics* (pp. 139–158). St Alban's: Tarquin Press.

CHAPTER

10

The Construct of Mathematical Resilience

Clare Lee, Sue Johnston-Wilder†*

*The Open University, Milton Keynes, United Kingdom

†University of Warwick, Coventry, United kingdom

INTRODUCTION

Current educational research has developed a number of constructs, such as mathematics avoidance, mathematics anxiety, and learned helplessness, each of which focuses on negative aspects and consequences associated with learning mathematics. The literature surrounding these constructs offers ideas and suggestions for treatment once a problem has developed, but it typically does not concentrate explicitly on ways in which mathematics can be learned that do not result in the growth of such negative constructs. We have found that a pragmatic argument can be made for delineation of a positive construct that would enable learners to develop a positive stance toward mathematics. We call this construct "mathematical resilience" (Johnston-Wilder & Lee, 2010a).

Resilience exists in wider psychological literature as a concept to describe the phenomenon of how some young people avoid negative consequences and succeed despite significant adversity. As we developed and worked with the construct of mathematical resilience, we found that four aspects were important and pragmatically worthwhile to focus on. We will explain the four aspects in detail below. Alongside this discussion, we will give some examples of resilient approaches and teaching strategies, together with some evidence for regarding the four aspects as important. Mathematical resilience shares many characteristics with such constructs as self-efficacy, optimism, motivation, and confidence, and we will go on to show how each of these relates to mathematical resilience.

http://dx.doi.org/10.1016/B978-0-12-802218-4.00010-8

The construct of mathematical resilience allows learners to manage and protect themselves from unhelpful emotions that may arise when mathematics becomes difficult to learn. Resilient learners know that while learning mathematics requires struggle, appropriate support can be found and positive emotions that come from success can be experienced. Teaching for mathematical resilience enables learners to use mathematics effectively and to acquire new mathematical skills when needed, to empower their day to day lives and careers. Learners of any age can develop the resilience they need to approach mathematics safely. All learners can learn mathematics in ways that do not cause them to develop negative traits.

THE NEED FOR MATHEMATICAL RESILIENCE

A Worldwide Need

The mathematical resilience construct originated with work carried out in the United Kingdom, where there is an established long tail of underachievement in mathematics and a prevalence of negativity, including anxiety, toward the study of mathematics (Callan, 2015; Roberts, 2002). Negativity toward mathematics seems to be a worldwide, but not global, problem; most countries have negativity toward mathematics within their populations (OECD, 2013). The report noted that "mathematics anxiety increased slightly since 2003" (p. 98).The OECD also measured a related construct, mathematical helplessness, which has an average prevalence of about 30% in participating countries (OECD, 2013, p. 80). It further noted that in Jordan, Thailand, Tunisia, Brazil, Qatar, and Argentina at least 45% of students are reported as feeling helpless when doing a mathematics problem. The OECD report stated that the inverse relationship between mathematics anxiety and mathematics performance was strong among most participating countries and economies. "On average across OECD countries, greater mathematics anxiety is associated with a 34-point lower score in mathematics—the equivalent of almost one year of school." (ibid, p. 18)

OECD (2013) showed that disadvantaged students can succeed despite adverse socioeconomic status, that is, some students were found to be resilient as the term is traditionally used in education. Across OECD countries, 31% of students from disadvantaged backgrounds were found to be resilient in this way, achieving beyond expectations. Such students succeeded by being engaged, motivated, and holding strong beliefs in themselves and their abilities. In many countries, students' motivation, self-belief and dispositions toward learning mathematics were found to be positively associated both with mathematical performance, and relative performance compared to other students in the same school. Furthermore, most mathematics teachers in participating countries believed that the social and emotional development of their students is as important as the acquisition of mathematics skills. Overall, the findings reported by OECD (2013) are consistent with our claim about the importance of developing "mathematical resilience" in learners of all ages.

Negative Influences on the Learning of Mathematics

The existence of negative attributes such as mathematics anxiety, avoidance, learned helplessness, and exclusion is well-established (see, e.g., Ashcraft, 2002; Hembree, 1990; Hernandez-Martinez & Williams, 2013). Negative emotional stress is also said to be frequently engendered by many traditional approaches to teaching mathematics (see, e.g., Nardi & Steward, 2003). Experience and research (e.g., Johnston-Wilder, Lee, Garton, Goodlad, & Brindley, 2013) has shown us that many learners are resilient in diverse aspects of their lives but seem reluctant, or unable, to bring that resilience to the learning of mathematics. Learners have told us (Lee & Johnston-Wilder, 2013), for example, that while they readily talked to friends about any problems they had with their work in English or geography, they felt discouraged from doing so when engaged in mathematics work. Thus a specific notion of *mathematical* resilience is necessary.

Mathematics Anxiety and Avoidance

Mathematics anxiety and mathematics avoidance were important in our thinking about mathematical resilience because, as has been shown, these phenomena are prevalent in many countries (Ashcraft, 2002; Ma, 1999; OECD, 2013). In those countries, there is also a widespread concern about the decline in students prepared to study mathematics or mathematically demanding subjects after compulsory age (in the United Kingdom see, e.g., Roberts, 2002; Johnston-Wilder, Brindley, & Dent, 2014; and elsewhere see the United States: National Academies, 2007; and Australia: Slattery & Perpitch, 2010). Evidence (e.g., Johnston-Wilder, Lee, Garton, & Brindley, 2014) suggests that these two phenomena are linked. The widespread nature of mathematics anxiety and avoidance is also implicated when students drop out of studying subjects that become unexpectedly mathematical (e.g., engineering courses or courses that involve statistics, such as psychology) (Johnston-Wilder, Lee, et al., 2014).

Anxiety and avoidance are acquired phenomena. They may be either directly acquired and rooted in previous firsthand experiences or gained secondhand through talking to others about experiences in mathematics. In a society where mathematics is routinely talked about negatively, such as the United Kingdom, it is common to take a stance of antipathy toward mathematics (Callan, 2015).

Mindset

One pervasive influence that is implicated in many people's negative stance toward mathematics is the prevalent cultural assumption of a fixed mindset (Dweck, 2000). Those with a fixed mindset consider a person's intellectual abilities to have a fixed limit or ceiling beyond which they cannot grow. The terminology associated with a "fixed mindset" (e.g., low-ability pupil, reaching your potential) is part of the language used widely in the field of education in, for example, the United States and the United Kingdom, and this discourse is particularly prevalent in relation to mathematics (Callan, 2015). Many students undertake tests to measure their "ability" in mathematics, which seems to indicate an assumption that students possess a mathematics capability that remains fixed over time and is not susceptible to good teaching. Confusingly at the same time, many students are told that if they try harder, and make the effort, they will increase their attainment. In most secondary schools in England, learners are placed into ability groups, according to the results of specific tests. Such sorting, or stratification, seems to frequently be a self-fulfilling prophesy (Wiliam & Bartholomew, 2004), offering some justification to those who advocate the use of such ideas. The pupils in lower "ability" groups often suffer from lower teacher expectation and tend to be given less challenging work; as a result, the attainment gap between students in different groups widens (Boaler, 1997). Hence the school system may reinforce the idea that some people can "do mathematics," and others cannot, which is the basis of the fixed mindset. Such ideas, once ingrained, can make learning mathematics as an adult very difficult. However, these ideas can be challenged (Yeager & Dweck, 2012); we argue that this does not often happen explicitly enough.

The discourse of a fixed mindset also reinforces a form of elitism (Nardi & Steward, 2003), or exclusiveness, that surrounds mathematics. The assumption of elitism, the idea that only a minority can "do mathematics," means that the helplessness often seen in students of mathematics is accepted by the majority.

Learned Helplessness

Another prevalent phenomenon mentioned by the OECD (2013) is learned helplessness. The helpless person has learned that in a particular situation, certain responses, and outcomes are independent of their own effort (Abramson, Seligman, & Teasdale, 1978). For example, a student may try to remember all the steps in a mathematical procedure inadvertently miss one out, and arrive at the wrong answer. If this happens frequently over time, the student learns that their effort does not achieve the outcome of the correct answer. Thus they conclude that their effort is not enough to achieve success in mathematics and next time the learner expends less effort.

The constituent ideas of fixed mindset, elitism, anxiety, exclusion, and helplessness all result in a large proportion of society that wants nothing to do with mathematics and creates a society in which mathematics avoidance is implicitly condoned, and in which being willing and able to engage with mathematics gives significant additional cultural capital to the elite few.

MATHEMATICAL RESILIENCE

Instead of working in ways that unwittingly develop anxiety and avoidance, learned helplessness, and exclusion, teachers and others involved with the promotion of mathematical learning, such as coaches and parents, can work explicitly to develop mathematical resilience. Rather than focus on the negative, in the spirit of positive psychology and medicine (e.g., Tennant et al., 2007), it seems to be more productive to focus on the positive: attitudes and beliefs that enable learners to approach mathematics positively and that future employers say they want to see. The construct mathematical resilience is designed to work explicitly against negative influences. Mathematical resilience is above all characterized by a "can do" approach to any new mathematics encountered, a willingness to put in effort to develop fluency, and an ability to assemble whatever support is needed to overcome any barriers to mathematical growth. Learned helplessness and mathematics anxiety can be overcome.

Why Mathematical Resilience?

Resilience is particularly needed in learning mathematics, due in part to the characteristics of that subject, the prevalent fixed mindset, and "the overwhelming impression [...] that mathematics is a very difficult subject for most children" (Hart, 1980, p. 209). Any survey of people in the United Kingdom in general will quickly reveal that many not only "hate maths" but also have a story to tell about just why they feel such antipathy. A further and very important reason that mathematics is often shunned is the performativity agenda as identified by, among others, Hernandez-Martinez and Williams (2013).

> Our research does pose this important question, then: given the predominant educational culture of performativity to which schools and, increasingly, universities, are subjected, how can mathematics teaching allow for reflexivity and for the construction of relevant capital for the majority and, ultimately, for the creation of such resilience among all students? ***Hernandez-Martinez and Williams (2013, p. 57)***

For many people, engaging in mathematics is a form of adversity that they are required to confront until the age in which they leave school, but they choose to avoid it as soon as possible thereafter. Much, but certainly not all, teaching of mathematics leads to many learners viewing mathematics as being the rapid performance of technically correct steps with no obvious reason, steps that rely heavily on memory (Ward-Penny, Johnston-Wilder, & Lee, 2011). The creativity and puzzlement, the experimentation and missteps that characterize a working mathematician's life (Burton, 2004) are rarely reflected in the learner's experience. Fluency with procedure is often valued over the understanding of that procedure or knowing how that procedure connects with other aspects of mathematics (see, e.g., Ofsted, 2012). Working memory is reliant upon being calm (Ashcraft & Krause, 2007) and anxiety interferes with the recall that students consider to be required (Nardi & Steward, 2003). The way that mathematics is often presented causes many, but not all, learners to experience adversity, sometimes to the point of trauma. Stories of such adversity that we know of include a student reduced to tears because she could not remember a given procedure; the more she was instructed to "just think" the more impossible it became to think clearly. Thirty years later, this person still recalls how she felt during and after the incident, and she has avoided mathematics ever since.

It would, however, be wrong to say that students should always learn and practice mathematics in a completely smooth and untroubled way. Indeed, one might argue that if a student is always experiencing mathematics without challenge, they will be ill-prepared for struggle later in their studies (Ward-Penny et al., 2011). There are barriers to overcome when studying mathematics and struggle is part of learning. Connections must be actively sought; fluency is necessary and only comes with familiarity. Until a learner can confidently conjure one procedure, often another aspect will remain out of reach (McGowen & Tall, 2010). Stigler and Hiebert (2009) have shown how pervasive path smoothing (Wigley, 1992) is and how some teachers of mathematics in the United States have come to understand that their job is to remove barriers rather than to help learners in their struggle. Wigley (1992) emphasizes that a teacher's role is to help learners build the resilience to engage in the struggle to overcome the barriers that mathematics presents.

Growing Mathematical Resilience

The fours aspects of mathematical resilience (growth mindset, personal value of mathematics, knowing that mathematics requires struggle, and knowing how to recruit support in pursuing mathematical learning) chime with aspects postulated to be important by others, including Bandura (1995), Dweck (2000), Williams (2014), Hernandez-Martinez and Williams (2013), and Ryan and Deci (2000), but are explicitly focused on the learning of mathematics. Kooken et al. (2015) explored the measurement of the construct in people who could be considered to have mathematical resilience. They concluded that there were at least three affective dimensions to becoming sufficiently mathematically resilient to engage in university-level study of subjects with a high mathematical content. The dimensions were the following:

- *Value:* the experience that mathematics is a valuable subject and is worth studying
- *Struggle:* the recognition that struggle with mathematics is universal, even for people who have a high level of mathematical skill
- *Growth:* the belief that all people can develop mathematical skill and that everyone can learn more mathematics with effort and support

Further work in this area (Goodall & Johnston-Wilder, 2015; Johnston-Wilder & Lee, 2010b; Lee & Johnston-Wilder, 2014) has led us to believe that, when considering how to enable people to develop mathematical resilience, there are four aspects that need to be considered, which we will go on to discuss in depth:

1. *Growth mindset:* developing an incremental or growth mindset
2. *Value:* understanding and experiencing mathematics as important in society and also personally. This aspect is also about the value of the individual as part of the mathematical community
3. *An understanding of how to work at mathematics:* that progress in mathematics requires struggle, curiosity, and perseverance as well as learning to manage the emotions that come with learning something new
4. *Knowing how to recruit support:* an awareness of the value of collaboration—the use of conjoined agency to aid in the struggle to grow mathematical knowledge, skills, and understanding

A Growth Mindset

Pupils who have developed mathematical resilience are characterized by a growth mindset (Yeager & Dweck, 2012). Students who believe that intellectual abilities can be developed, as opposed to being fixed at a certain level, tend to show higher achievement (Yeager & Dweck, 2012) and to develop resilience. Encouraging students to develop a growth mindset is ultimately about reminding students continually that when they practice new or different ways of doing mathematics, their brains develop (just like a muscle) and they become more adept at dealing with the challenges of learning mathematics. A growth mindset emphasizes the malleability of the brain, including adult brains. Focusing on the importance of adapting strategies, and learning skills that let learners use their brain in smarter ways, helps develop resilience when faced with novel and challenging problems. Developing a growth mindset needs an emphasis on getting better at something through effort, acknowledging that "you can't do it *yet*" but given time, persistence and appropriate help (not too much) from others, "you will be able to." The brain grows every time a learner makes connections and learns something new (Willis, 2007).

Value

People who have developed mathematical resilience can be characterized by an understanding of the value of mathematics. Mathematics is seen as a way of thinking about topics and issues of interest and a way of modeling the world, as well as a way of accruing personal cultural capital. Resilient learning demands an understanding of the purpose of the skills being learned in terms of personal, employment, and cultural objectives (sometimes historical, sometimes current). Furthermore, resilient learners feel included in the community of those learning mathematics. They understand that their own position as part of the community of mathematicians or people who use mathematics is of value. Those who are studying or are going to study mathematics or mathematically demanding subjects will quickly find that their mathematical qualifications are valued in the jobs market.

It may seem self-evident that mathematical learning is of value but this is not always the case and learners often feel excluded. Brown, Brown, and Bibby (2008) state that among several factors that have an impact on the uptake of mathematics at A-level are that mathematics is perceived to be "hard," "boring," and "useless." For most financial transactions, the use of a calculator is sufficient, and cash registers now work out any change needed, so even basic arithmetic may be perceived as not in demand in employment. The mathematics that will empower winning the best deal, building a stable structure, navigating a small boat, or understanding whether a chart is a fair representation of the facts frequently hides against its background and requires a certain amount of exploration to uncover. Mathematics is hard for some people to value when it cannot be seen in the valued aspects of their life. Sadly many mathematics teachers we have met are content to study mathematics for its own sake, and they are not well-prepared for the question, "when will we use this in real life?" Many teachers of mathematics have not recognized the connection between algebraic skills and choosing a mobile phone tariff or preparing a large meal.

Algebraic thinking can be made more inclusive; for example, Lugalia (2015) had girls running to algebra lessons. The software she used was designed with Bruner's (1966) enactive-iconic-symbolic model as its foundation. This model is about accessibility; if learners are struggling with a symbolic algorithm for factorizing a quadratic expression, they could be introduced to an iconic representation using diagrams, which may make the task more accessible. If learners are struggling with expanding brackets, such as $(x+2)(x+1)$, they may be included by using some pieces of card, squares to represent the x^2 terms, and strips 1 by x to represent the x terms and unit squares. If the learners are then invited to make the shapes into a rectangle and describe its dimensions, they will find themselves enacting the expansion. One teacher, who had taught for many years in two countries, introduced this approach and was astounded at how capable all her pupils at age 11–12 years became with quadratic expansions, when most of her previous students had struggled with a purely symbolic approach throughout five years of secondary schooling. She recognized that she had unwittingly excluded many students from algebraic thinking.

By paying attention to inclusion, all learners are invited to become valued members of a community, engaging in mathematical thinking and thus beginning to experience personal value in mathematics.

Knowing How to Work at Mathematics: The Growth Zone

Learning and using mathematics is not easy, everyone has barriers to overcome, mistakes are made and people get stuck (Mason, Burton, & Stacey, 2010); mathematics needs perseverance (Burton, 2004; Williams, 2014). However, resilient learners know that they are not struggling alone. There is a community ready to offer support and appropriate help—from working alongside, to encouraging, to suggesting resources, to reminding of known but forgotten processes. Mathematics does not need enormous memory power—this is a common misconception (Ward-Penny et al., 2011); connections can be explored, patterns can be seen, and collaboration is very productive (Askew, Brown, Rhodes, Johnson, & Wiliam, 1997; Swan, 2006).

Resilient learners exercise agency in their learning (Lee & Johnston-Wilder, 2013; Ryan & Deci, 2000). Learners develop agency through being allowed choice, independence, and the opportunity to experiment. A facilitator of learning mathematics in a resilient manner will emphasize that there is no one fixed way of doing things, but rather that each learner must start from where they are, using what they know already (McGowen & Tall, 2010), and work to fit the new learning into their existing schemata. Mistakes will be made, which is important because mistakes indicate an opportunity to learn (Mason et al., 2010).

Developing mathematical resilience can be especially hard for those who have previously developed anxiety and learned helplessness. In helping adults and teenage learners to sufficiently overcome such anxieties and to begin to approach mathematical ideas with resilience, we developed a model we call the Growth Zone Model. In developing the model, our thinking was influenced by our study of Vygotsky (1978), but this model is purposely simple in its construction. It is designed to enable any learner to articulate and begin to understand the emotions they may be feeling as they approach mathematical learning (Fig. 1).

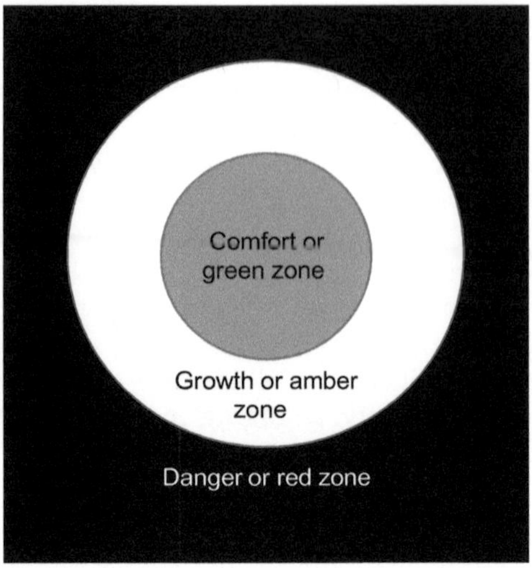

FIG. 1 The growth zone model.

The model consists of three zones: comfort, growth, and danger.

Comfort (green): In the comfort (green) zone a learner will feel comfortable; when working at mathematics they will be doing something that they know about and can do on their own. Thus the green zone is useful for consolidating ideas, developing fluency and the knowledge that you understand and can use a particular skill in mathematics. However if learners stay too long in the green zone they will not learn as much as they could, and it is quite likely that they will suffer boredom and a lack of stimulation.

Growth (amber): In the growth (amber) zone the learner is being challenged to some degree. The more time the learner can stay in this zone the more they will learn and progress. In this zone, the challenge is high but not too high; there is excitement as novel ideas are being explored and there is the satisfaction of overcoming barriers and grasping at understanding. There is sufficient adrenalin to persevere when things seem tough. It is a positive, but somewhat exhausting, place to be. Therefore, when fatigue sets in, learners may choose to resort to the green zone to consolidate the progress made. However there is always the risk that the challenge might get too high, the learners may not feel able to pause and draw breath when they need to, and thus they may begin to feel unsafe and unable to safeguard themselves.

Danger (red): The danger (red) zone is marked not only by anxiety but also panic, and it is sometimes the kind of panic that triggers the "fight, freeze, or flight" mechanism, which means that thinking and reasoning are almost impossible; the prefrontal cortex goes "offline." Many learners we have encountered have so many memories of being in the red zone when faced with mathematics that they find themselves in the danger zone when asked just to talk about doing mathematics.

We have found that this growth zone model enables even the most anxious learner to articulate what they are feeling and therefore to consider how they might overcome their emotional barriers sufficiently to spend just a short time in the growth zone. One insightful learner suggested to her teacher: "the trouble is that in maths my growth zone is too narrow." Collaborative working with conjoined agency, where each member of the group is charged with supporting, as well as being open to receiving support, can enable more people to access and stay longer in their growth zone.

Knowing How to Gain Support

Resilient learners of mathematics know that mathematics requires struggle but they also know how to access help and support in that struggle. Some learners may need to be shown successful models of mathematical thinking and reasoning. All need to be listened to because in formulating thinking to communicate ideas to others, that thinking becomes clarified (Lee, 2006). Furthermore when the "sticking point" becomes obvious, a more capable peer will be able to question the reasoning or offer targeted and appropriate support. Collaborative discourse may enable students to learn to ask themselves the right questions (Lee, 2006).

In order to work collaboratively, learners need to build up sufficient language or vocabulary so that they can use enough of the mathematics register to be able to express and explore mathematical ideas (Lee, 2006). This is not to say that they must use the "right" words but rather that they see a need to express mathematical ideas themselves and thus seek a way to do so effectively. Effective mathematics communicators can explore their own understandings and connections and support others. Mathematical resilience is based within a social constructivist domain (Vygotsky, 1978); expressing mathematical ideas and talking about mathematical learning within a mathematical community are both vital aspects of developing the resilience that allows for learning mathematics.

Digital technologies have a role to play here. Search engines furnish a process or ready-made procedure that may yield solutions to a problem, and software such as Grid Algebra or dynamic geometry programs provide a safe place to express ideas, experiment, think, and learn. The potential to make mistakes and to learn from those mistakes is high when using IT because feedback tends to be immediate and having "another go" is straightforward. The transitory nature of attempts on screen appeals to many people who suffer acute anxiety, while they build their resilience.

RELATING MATHEMATICAL RESILIENCE TO OTHER CONSTRUCTS

Resilience

Resilience seemed an important term to use because it describes the attributes needed to overcome adversity, which many learners were experiencing when learning mathematics. Along with several other authors (e.g., Hernandez-Martinez & Williams, 2013; Knight, 2007) we think it is important to see resilience as a sociocultural construct that communities can address, not an individual psychological construct.

Resilience has many subtly different definitions in literature. The term resilience grew out of the work of developmental psychologists, such as Rutter (1993) and Garmezy (1991), who recognized that, among groups of children believed to be at high risk of developing particular difficulties, many individuals remained undamaged by significant adversity. Others offer clarification; for example, "resilience refers to a dynamic process encompassing positive adaptation within the context of significant adversity" (Luthar, Cicchetti, & Becker, 2000, p. 543). Or resilience is "both the capacity to be bent without breaking and the capacity, once bent, to spring back" (Vaillant, 1993, p. 248). Resilience is further characterized as "...mechanisms and processes that lead some individuals to thrive despite adverse life circumstances" (Galambos & Leadbeater, 2000, p. 291). Commonality in the definitions resides in the capability of some individuals to recover or to continue on a positive path when faced with significant adversity.

Literature surrounding those children who are resilient indicates that there are three sets of factors implicated in the development of resilience: (1) attributes of the children themselves, (2) aspects of their families, and (3) characteristics of their wider social environments (e.g., Masten & Garmezy, 1985; Werner & Smith, 1992).

1. The resilient attributes of the learner can be influenced, either adversely or positively, by the persons who are supporting their learning. Ensuring that this influence is as positive as possible requires the learner and supporter to become aware and mindful of affective traits and emotions as part of developing resilience. This is a crucial part of mathematical resilience.
2. That the family has a significant effect on building resilience is also recognized within our work; parents can learn to enable their child's learning of mathematics more effectively without learning any more mathematics (Goodall & Johnston-Wilder, 2015; also Chapter 11).
3. Resilience is affected by the psychosocial aspects of the environment in which the student is learning mathematics. Where such an

environment is transformative (Mezirow, 2000), focusing on the development of interaction between learners, then resilience is likely to be an outcome for the learner. In a transformative environment, critical and constructive thinking methods enable learners to look deeply into practices, to develop creative ways of thinking, to improve problem-solving skills, and to work collaboratively. "Thus, the learning of resilience fits well with the transformative educational framework" (McAllister & McKinnon, 2009, p. 375).

Growing Resilience

We consider that "resilience is not a quality that some possess and others do not" (Knight, 2007, p. 545), but rather learners can grow their resilience when learning mathematics. All learners need some resilience to overcome the significant challenges that learning can at times present and "there is convincing evidence that individuals can learn or acquire resilient qualities" (McAllister & McKinnon, 2009, p. 374). Schools, teachers, and supporters of learning beyond schools can provide an environment that fosters resilience. For example, Borman and Overman (2004) suggest that a safe and orderly environment and positive teacher-student relationship are likely to promote resilience in poor ethnic minority students. Knight concludes that teachers need to have "a 'resiliency attitude' so that they see children and young people as competent, and focus on their strengths rather than deficits" (Knight, 2007, p. 544) Exactly how the focus on "strengths rather than deficits" is done is important, because if anyone trusted offers comfort to a student that reinforces the idea that a weakness is inherited or to be expected, then that may further contribute to the notion of a fixed mindset and mathematical exclusion (Rattan, Good, & Dweck, 2012).

We see resilience as important for all learners, not only for those who face significant adversity in their lives. Therefore the next question for us was the following: what are the important factors in a learning environment that enables learners of mathematics to also learn to be resilient?

The Importance of Mindset

The growth or incremental mindset is one of the four key aspects of mathematical resilience:

> incremental theorists were more likely than entity theorists to endorse working harder and spending more time studying next time in response to an academic setback, whereas entity theorists were more likely than incremental theorists to report that they would spend less time on the subject and would attempt to avoid it altogether in the future. ***Nussbaum and Dweck (2008, p. 600).***

The fixed or entity mindset seems to have a detrimental effect on many learners, including those who are placed in "high ability" groups. Some "high ability pupils" with a fixed mindset may consider that they are better than everyone else, grow complacent, and fail to make the progress they could, given the expertise they already have. Others become steadily more anxious that they are "imposters" and will eventually be found out (Boaler, 1997). Neither stance is helpful in encouraging young people to continue to study mathematics.

Having a fixed mindset is also a likely cause of anxiety. If every engagement with mathematics has the potential to expose the actual level of your intellectual abilities that you may worry are low, then every step will cause anxiety and every off-the-cuff remark by the teacher is at risk of being taken as confirmation of low abilities. Where every encounter with mathematics is a threat to the student's self-esteem and social acceptance (Nussbaum & Dweck, 2008), defense mechanisms for their self-esteem will conflict with mathematical progress. Nussbaum and Dweck (2008) further posit that those students who hold an entity theory are very unlikely to learn from their mistakes because they react to a loss of self-esteem by comparing themselves with others rather than by actively seeking ways which might allow more learning.

Yeager and Dweck (2012) have demonstrated that it is possible to convince learners to abandon a fixed way of thinking about mathematics and adopt a growth theory:

> Our research and that of our colleagues show that if students can be redirected to see intellectual ability as something that can be developed over time with effort, good strategies, and help from others, then they are more resilient when they encounter the rigorous learning opportunities presented to them. ***Yeager and Dweck (2012, p. 306)***

Thus a learning environment that enables learners to develop mathematical resilience must help learners to see their intellectual ability as something that can be grown.

Optimism and Resilience

The resilience that is required to learn mathematics relates strongly to the idea of optimism (Seligman, 1995), which inclines students to problem solve (Williams, 2014); indeed Seligman explained that optimism is a form of resilience. Optimism is seen as a combination of confidence, persistence, and perseverance. "Optimism (resilience) is a psychological characteristic possessed by students who are inclined to problem-solve" (Williams, 2014, p. 407). Mathematical problem solving is often part of the learning environment that allows for the growth of mathematical resilience.

Appropriately challenging learning opportunities have to be offered so that learners understand that resilience is needed and that mathematics is not just about finding the right answer quickly. However problem solving can be experienced as a situation of adversity because of the many failures that can occur before success is achieved.

Confidence can be explained as a combination of knowing that you have sufficient resources and technique to succeed in completing a challenge and that you have successfully completed similar work before. Consequently, it is a characteristic that can be grown through successful engagement with mathematical learning. Confidence and persistence, which are part of optimism, are characteristics of successful mathematics learners (Martin, 2003), but some confident students are not inclined to problem-solve (Dweck, 2013; Pajares, 1996; Williams, 2013). Thus there can be said to be two types of confidence:

- an optimistic confidence, which allows for perseverance and successful problem solving and the demonstration of resilience and
- a nonoptimistic confidence that is based on an unconnected knowledge of mathematical ideas and does not incline the possessor of this type of confidence to step beyond certain limits

Learners with an optimistic confidence (Williams, 2014) know both how to work at a challenge and how to access the support, techniques, and knowledge they need to meet that challenge. They are open to alternative ideas; they see that others' ideas or suggestions have the potential to extend their own understanding of the situation, affording them an opportunity to see a more efficient way forward and allow their intellectual capabilities to grow even further.

Optimism is used by Seligman (1995) in a nontraditional way, to describe a positive explanatory style where successes are perceived as permanent; the meaning thus differs from the everyday use of the term. Optimism is seen as pervasive and personal; the optimistic child sees failures as temporary, specific to that given situation, and external (Seligman, 1995). The ideas behind optimism help in understanding and explaining resilience (Williams, 2014). However in everyday language the term resilience seems to us to convey more meaning to those for whom it was designed to speak, those who are learning or are helping others learn mathematics. Resilience includes confidence, persistence, and perseverance. Thus, while acknowledging the two terms can be thought of as interchangeable "mathematical resilience" is preferred. Mathematical resilience enables a simple message to be conveyed that public and policy makers will understand, while remaining fully grounded in research and maintaining an inner complexity.

The Role of Persistence and Perseverance in Resilience

Persistence and perseverance, although often regarded as similar or even the same, can usefully be differentiated when thinking about mathematical learning and resilience. Conroy (1999) demonstrated the difference between persistence and perseverance by drawing attention to "adjusting": "Part of the skill of the power of perseverance is to make those adjustments as you persist" (p. 30). Thom and Pirie (2002) distinguished between the two by describing perseverance as …sense (i.e., intuitive and experiential) in knowing when to continue with, and not to give up too soon on a chosen strategy or action, and … knowing when to abandon a particular strategy or action in the search of a more effective or useful one" (Thom & Pirie, 2002, p. 2).

For a mathematically resilient learner it is not sufficient to persist; perseverance is more important. Resilient students report much higher levels of perseverance, intrinsic, and instrumental motivation to learn mathematics, mathematics self-efficacy, and mathematics self-concept and lower levels of mathematics anxiety than nonresilient students (Lee & Johnston-Wilder, 2013).

Self-Efficacy

Self-efficacy is part of developing mathematical resilience. Bandura defined self-efficacy as "the belief in one's capabilities to organize and execute the courses of action required to manage prospective situations" (Bandura, 1995, p. 2). Hence the term "self-efficacy" concerns the belief in an individual's ability to succeed in specific situations. Many of the reasons that adults put forward for not wishing to engage in mathematics are rooted in a belief that they cannot do it; they experience low self-efficacy with regard to mathematics. A strong sense of self-efficacy enables learners to view challenging problems as tasks to be mastered; giving a deep interest in and commitment to the activity and an ability to recover quickly from setbacks (Hoffman, 2010). Sources for self-efficacy are important in our thinking. Bandura (1995) suggests the following four sources:

- *A mastery experience:* achieving success in situations where there are high levels of demand. Generally over time, with effort and sufficient challenge successfully met, self-efficacy is developed
- *Vicarious experience:* observing people demonstrating mastery; if a learner believes they are similar have similar knowledge and skills, then a vicarious experience can prompt the learner to feel more self-efficacious
- *Verbal persuasion:* encouragement and positive feedback can contribute to self-efficacy. The encouragement that is important in self-efficacy

is recognition of making appropriate attempts and persevering rather than getting a correct answer.

- *Physiological and emotional states influence self-efficacy:* someone who regards mathematics with apprehension will not feel self-efficacious and conversely, helping someone overcome their anxieties will help them to become more self-efficacious.

The fourth source of self-efficacy in particular adds to the previous discussions. Anxiety is common in learners when approaching mathematics (Ashcraft, 2002) hence we developed the "growth zone model" discussed earlier as a way to discuss the physiological and emotional state of a learner and support learners to position themselves to feel able to maximize learning.

Motivation

Motivation is also important in learning and mathematical resilience. Motivation, like self-efficacy, is related to belief that the amount of motivation a learner has to work toward a particular goal is dependent on how much they believe that they will be able to achieve a particular outcome. Ryan and Deci (2000) argue that when a student has intrinsic motivation, learning can take place effectively because the individual will seek out ways to fulfill their goals as well as the necessary support. Learners will develop more intrinsic or self-determined motivation to learn, if the learning environment allows for the satisfaction of three innate needs: to feel connected, effective, and agentic when learning. Thus intrinsic motivation can be seen as the result of the learner feeling part of a supportive community and experiencing feelings of agency and control.

Mathematical resilience therefore requires an inclusive environment where all are valued, enabled to execute such actions that will allow all to succeed, and where all are able to exercise appropriate control. Thus when learners within an environment are challenged but given choices in ways to complete that challenge and are supported so that their emotional state remains positive and they succeed with meeting the challenges, that environment can then be said to be developing mathematical resilience.

Thus mathematical resilience involves a growth mindset, self-efficacy, motivation, and an optimistic confidence. All these factors can be grown in a nurturing and transformative learning environment that is inclusive and helps learners understand the personal value of learning mathematics. Furthermore the literature on motivation and confidence shows that resilient learners must learn how to approach their learning and exercise appropriate choice and control.

TEACHING FOR MATHEMATICAL RESILIENCE

There is evidence that teaching for mathematical resilience requires system-wide change in Western education. Teachers are known to make the biggest difference to the outcomes of any learner (Alton-Lee, 2003); however too many teachers are "teaching to the test" in this era of performativity (Rasmussen, Gustafsson, & Jeffrey, 2014). In every learning environment we have worked in, we have shown that a focus on developing mathematical resilience means including all learners and freeing them from a fixed mindset. Teaching to the test may be a major cause of avoidance, anxiety, and learned helplessness which may also be occasioned by path smoothing (Wigley, 1992). A focus on developing mathematical resilience can be achieved without adversely affecting results. For example, external test results improved slightly despite troubled times at a school we worked in. The difficulties were occasioned by a high turnover in mathematics teachers and sometimes no mathematics teachers at all (Johnston-Wilder & Lee, 2010b), but a resilient attitude developed in the pupils and some other staff avoided any deterioration in examination results.

Existing research, experiments, and experience point the way to how to facilitate the development of mathematical resilience. Learning environments that build mathematical resilience allow learners to have agency, independence, and choice so that they can focus on tasks that challenge them. Learners are actively encouraged to work collaboratively, not just work in a group (Swan, 2006), so that they are active and harness their energies constructively. In such environments, learners can articulate their ideas clearly and safely (Lee, 2006) and see mistakes as part and parcel of optimal learning (Mason et al., 2010). They are encouraged to be curious, to try out ideas, make mistakes, and make their own connections (Askew et al., 1997). Learners are likely to experience their mathematical attainment growing. Learners will be encouraged to engage in repetition and consolidation of ideas so that they build familiarity and fluency (Nunes, Bryant, & Watson, 2008). However given appropriate tasks, consolidation and repetition can also be part of exploring ideas.

COACHING FOR MATHEMATICAL RESILIENCE

While good mathematics teachers are needed if society is to continue to grow its technological base, learning mathematics does not always need the presence of a mathematics teacher, as students need time to work on ideas. Successful outcomes for adult and school-age learners have been gained by developing coaches for mathematical resilience who do not purport to "know the answer" but do know how to work resiliently on mathematics (Johnston-Wilder et al., 2013). In this "coaching for mathematical resilience" project, the adult coaches first needed to build their own mathematical resilience before they could begin to support apprentices or adult learners of mathematics; otherwise mathematics anxiety was likely to be "transmitted" (Beilock, Gunderson, Ramirez, & Levine, 2010). The idea of journeying together was used as part of the project, along with the metaphor of safely proceeding up a mountain so that the participants could see that it was important that no one was left behind. It seemed this was a novel idea to the adults; they were used to being allowed to "fall by the wayside" in mathematics. That the others in the group would wait, offer different ways of looking at the ideas, and consider each other's progress to be a matter of conjoined agency was ultimately important in the development of a community of coaches. The coaches' mathematical thinking improved with their willingness to engage in reasoning about and doing mathematics. In order to be in their "growth" zone, many had to wrestle against the panic and acute anxiety represented by the "danger/red" zone. However the conjoined agency of the community meant that "coming out of the red" was a joined endeavor; they took care of each other's emotions.

There is further evidence that parents are well placed to help their children, again not by knowing the answer or what to do mathematically, but by asking questions and helping the learner find the help that they need (Goodall & Johnston-Wilder, 2015); this is discussed in depth in the next chapter.

CONCLUSIONS

The construct "mathematical resilience" describes the positive attributes that learners require in order to be prepared to engage with, learn, and use mathematics both at school and, perhaps more importantly, beyond. It is needed because of the negative emotions and exclusion that have been seen to be engendered by many traditional approaches to teaching mathematics (Ashcraft, 2002, Nardi & Steward, 2003). Mathematical resilience can be developed in learners by the use of specific approaches to teaching and learning but also by offering explicit and targeted support by coaches and parents, who do not know the answer but can encourage resilient learning behavior in seeking to overcome barriers and find solutions.

Mathematical resilience is based in the emotional attributes that are demonstrated by those who have a positive stance to mathematics. There are four aspects to mathematical resilience: growth, value, knowing how to work on mathematics, and knowing how to gain support in pursuing mathematical learning. Those with mathematical resilience have a growth theory of learning; they know that they can grow their mathematical abilities. Those who are happy to engage with mathematics understand its value and know themselves to be a valued, included member of a mathematical learning community. Importantly those learners with mathematical resilience understand that success in learning and using mathematics will involve a certain amount of struggle. They know that mathematics does not always offer a smooth path to an answer and often needs perseverance. However resilient learners also know that they are not struggling with mathematics alone; there is a community from which support can be gained.

The construct of mathematical resilience grew out of the wider psychological construct of general resilience. While acknowledging that all learning requires a certain amount of resilience, we have encountered many learners who show resilience in diverse aspects of their lives but seem unable or unwilling to bring that resilience to the learning of mathematics. Hence there is a need to delineate the specific construct of mathematical resilience.

Mathematical resilience enlists ideas from constructs such as mindset (Dweck, 2000), optimism (Seligman, 1995), self-efficacy (Bandura, 1995), and motivation (Ryan & Deci, 2000) in order to present a meaningful but complex construct in itself. We have found that the ideas encapsulated by mathematical resilience speak to learners, teachers, parents, and others entrusted with overseeing mathematical learning in society. Importantly the simple message of developing mathematical resilience also seems to be gaining the attention of policy makers. More research will be needed before we know for sure how the complexities of mathematical resilience are played out within the whole mathematics education system.

The ideas captured by mathematical resilience are spreading, as more people concerned about mathematics education come to know about and use the construct. As mathematical resilience is built on established ideas and research, this gives reason to hope that the negativity that surrounds mathematics can be dissipated and replaced with a positive image of agency, empowerment, and control.

References

Abramson, L. Y., Seligman, M. E., & Teasdale, J. D. (1978). Learned helplessness in humans: Critique and reformulation. *Journal of abnormal psychology*, *87*(1), 49.

Alton-Lee, A. (2003). *Quality teaching for diverse students in schooling: Best evidence synthesis.* Wellington: Ministry of Education.

Ashcraft, M. (2002). Math anxiety: Personal, educational, and cognitive consequences. *Current Directions in Psychological Science*, *11*(5), 181–185.

Ashcraft, M., & Krause, J. (2007). Working memory, maths performance and math anxiety. *Psychonomic Bulletin and Review*, *14*(2), 243–248.

Askew, M., Brown, M., Rhodes, V., Johnson, D., & Wiliam, D. (1997). *Effective teachers of numeracy*. London: King's College.

Bandura, A. (1995). *Self efficacy in changing societies*. NY: Cambridge University Press.

Beilock, S., Gunderson, E., Ramirez, G., & Levine, S. (2010). Female teachers' math anxiety affects girls' math achievement. *Proceedings of the National Academy of Sciences of the United States of America, 107*(5), 1860–1863.

Boaler, J. (1997). *Experiencing school mathematics: Teaching styles, sex, and setting*. Buckingham, UK: Open University Press.

Borman, G. D., & Overman, L. (2004). Academic resilience in mathematics among poor and minority students. *Elementary School Journal, 104*(3), 177–195.

Brown, M., Brown, P., & Bibby, T. (2008). "I would rather die": Reasons given by 16-year-olds for not continuing their study of mathematics. *Research in Mathematics Education, 10*(1), 3–18.

Bruner, J. (1966). *Towards a theory of instruction*. Cambridge, Mass: Belkapp Press.

Burton, L. (2004). *Mathematicians as enquirers, learning about learning mathematics*. Dordrecht: Springer.

Callan, S. (2015). The fear factor, mathsaction. Retrieved from: http://mathsaction.org.uk/. (Accessed 13 April 2016).

Conroy, W. (1999). Never give up. The power of perseverance. *Vital Speeches of the Day*, 29–31.

Dweck, C. (2000). *Self-theories: Their role in motivation, personality, and development*. Philadelphia, PA: Psychology Press.

Dweck, C. (2013). *Mindset: How You Can fulfil your potential*. New York: Ballantine Books.

Galambos, N. J., & Leadbeater, B. J. (2000). Trends in adolescent research for the new millennium. *International Journal of Behavioral Development, 24*, 289–294.

Garmezy, N. (1991). Resilience in children's adaptation to negative life events and stressed environments. *Paediatrics, 20*, 459–466.

Goodall, J., & Johnston-Wilder, S. (2015). Overcoming mathematical helplessness and developing mathematical resilience in parents: An illustrative case study. *Creative Education, 6*(5).

Hart, K. M. (1980). *Secondary school children's understanding of mathematics: A report of the mathematics component of the concepts in secondary mathematics and science programme*. London: Chelsea College of Science and Technology.

Hembree, R. (1990). The nature, effects, and relief of mathematics anxiety. *Journal for Research in Mathematics Education, 21*(1), 33–46.

Hernandez-Martinez, P., & Williams, J. (2013). Resilience in mathematics students in transition. *British Educational Research Journal, 39*(1), 45–59.

Hoffman, B. (2010). "I think I can, but I'm afraid to try": The role of self-efficacy beliefs and mathematics anxiety in mathematics problem-solving efficiency. *Learning and Individual Differences, 20*(3), 276–283.

Johnston-Wilder, S., Brindley, J., & Dent, P. (2014). *A survey of mathematics anxiety and mathematical resilience among existing apprentices*. London: The Gatsby Foundation.

Johnston-Wilder, S., & Lee, C. (2010a). Mathematical resilience. *Mathematics Teaching, 218*, 38–41.

Johnston-Wilder, S., & Lee, C. (2010b). *Developing mathematical resilience*. Retrieved from www.beraconference.co.uk/downloads/abstracts/symposium1984.htm.

Johnston-Wilder, S., Lee, C., Garton, E., & Brindley, J. (2014). Developing coaches for mathematical resilience: Level 2. In *7th international conference of education, research and innovation, Seville, Spain. Published in CERI2014 proceedings* (pp. 4457–4465).

Johnston-Wilder, S., Lee, C., Garton, L., Goodlad, S., & Brindley, J. (2013). Developing coaches for mathematical resilience. In *ICERI 2013: 6th International Conference on Education, Research and Innovation, Seville, Spain*. Retrieved from www.mathematicalresilience.org/papers.htm.

Knight, C. (2007). A resilience framework. *Health Education., 107*(6), 543–555.

Kooken, J., Welsh, M., Megan, E., McCoach, D., Johnston-Wilder, S., & Lee, C. (2015). *Development and validation of the mathematical resilience scale*. Measurement and Evaluation in Counselling and Development.

Lee, C. (2006). *Language for learning mathematics: Assessment for learning in practice*. Buckingham, UK: Open University Press.

Lee, C., & Johnston-Wilder, S. (2013). Learning mathematics-letting the pupils have their say. *Educational Studies in Mathematics*, *83*(2), 163–180.

Lee, C., & Johnston-Wilder, S. (2014). Mathematical resilience: What is it and why is it important? In S. Chinn (Ed.), *The routledge international handbook of dyscalculia and mathematical learning difficulties* (pp. 337–345). Abingdon: Routledge.

Lugalia, M. (2015). *Pupils learning algebra with ICT in Key stage 3 mathematics classrooms (thesis submitted for the degree of PHD, Coventry: University of Warwick)*. Retrieved from University of Warwick library.

Luthar, S., Cicchetti, D., & Becker, B. (2000). The construct of resilience: A critical evaluation and guidelines for future work. *Child Development*, *71*(3), 543–562.

Ma, X. (1999). A meta-analysis of the relationship between anxiety toward mathematics and achievement in mathematics. *Journal for Research in Mathematics Education*, *30*(5), 520–540.

Martin, A. (2003). Boys and motivation. *The Australian Educational Researcher*, *30*(3), 43–65.

Mason, J., Burton, L., & Stacey, K. (2010). *Thinking mathematically*. Upper Saddle River NJ: Prentice Hall.

Masten, A., & Garmezy, N. (1985). Risk, vulnerability, and protective factors in developmental psychopathology. In B. Lahey & A. Kazdin (Eds.), *Advances in clinical child psychology: Vol. 8* (pp. 1–52). New York: Plenum Press.

McAllister, M., & McKinnon, J. (2009). The importance of teaching and learning resilience in the health disciplines: A critical review of the literature. *Nurse Education Today*, *29*, 371–379.

McGowen, M., & Tall, D. (2010). Metaphor or met-before? The effects of previous experience on the practice and theory of learning mathematics. *Journal of Mathematical Behavior*, *29*, 169–179.

Mezirow, J. (2000). *Learning as transformation*. San Francisco, CA: Jossey-Bass.

Nardi, E., & Steward, S. (2003). Is mathematics T.I.R.E.D.? A profile of quiet disaffection in the secondary mathematics classroom. *British Educational Research Journal*, *29*(3), 345–367.

National Academies. (2007). *Rising above the gathering storm: Energizing and employing America for a brighter future*. Washington: The National Academies Press.

Nunes, T., Bryant, P., & Watson, A. (2008). *Key understandings in mathematics learning*. Retrieved from http://www.nuffieldfoundation.org/key-understandings-mathematics-learning.

Nussbaum, D., & Dweck, C. (2008). Defensiveness versus remediation: Self-theories and modes of self-esteem maintenance. *Personality and Social Psychology Bulletin*, *34*, 599–612.

OECD. (2013). *Mathematics self-belief and participation in mathematics-related activities in PISA 2012 results: Ready to learn - Students' engagement, drive and self-beliefs (volume III)*. Retrieved from http://www.oecd.org/pisa/keyfindings/PISA2012-Vol3-Chap4.pdf.

Ofsted. (2012). *Mathematics—Made to measure*. Retrieved from https://www.gov.uk/government/publications/mathematics-made-to-measure.

Pajares, F. (1996). Self-efficacy beliefs in academic settings. *Review of Educational Research*, *66*, 543–578.

Rasmussen, A., Gustafsson, J., & Jeffrey, B. (Eds.), (2014). *Performativity in education: An international collection of ethnographic research on learners' experiences*. London: E&E Publishing.

Rattan, A., Good, C., & Dweck, C. S. (2012). It's OK-Not everyone can be good at math: Instructors with an entity theory comfort (and demotivate) students. *Journal of Experimental Social Psychology*, *48*, 731–737.

Roberts, G. (2002). *SET for success: The supply of people with science, technology, engineering and mathematical skills*. London: The Stationery Office.

Rutter, M. (1993). Resilience: Some conceptual considerations. *Journal of Adolescent Health*, *14*, 626–631.

Ryan, R. M., & Deci, E. L. (2000). Self-determination theory and the facilitation of intrinsic motivation, social development, and well-being. *American Psychologist*, *55*, 68–78.

Seligman, M. (1995). *The optimistic child*. Adelaide: The Griffin Press.
Slattery, L., & Perpitch, N. (2010). *Mathematics students in serious decline*. Retrieved from http://www.theaustralian.com.au/news/nation/mathematics-students-in-serious-decline/story-e6frg6nf-1225838901032.
Stigler, J., & Hiebert, J. (2009). *The teaching gap*. New York: Free Press.
Swan, M. (2006). *Collaborative learning in mathematics: A challenge to Our beliefs and practices*. Leicester: National Institute of Adult Continuing Education.
Tennant, R., Hiller, L., Fishwick, R., Platt, S., Joseph, S., Weich, S., et al. (2007). The Warwick-Edinburgh mental well-being scale. *Health and Quality of Life Outcomes, 5*(1), 63.
Thom, J. S., & Pirie, S. E. B. (2002). Problems, perseverance and mathematical residue. *Educational Studies in Mathematics, 50*, 1–28.
Vaillant, G. E. (1993). *The wisdom of the ego*. Cambridge, MA: Harvard University Press.
Vygotsky, L. (1978). *Mind in society*. Cambridge, MA: Harvard University Press.
Ward-Penny, R., Johnston-Wilder, S., & Lee, C. (2011). Exit interviews: Undergraduates who leave mathematics behind. *For the Learning of Mathematics, 31*(2), 21–26.
Werner, E. E., & Smith, R. S. (Eds.), (1992). *Overcoming the odds: High risk children from birth to adulthood*. Ithaca, NY: Cornell University Press.
Wigley, A. (1992). Models for teaching mathematics. *Mathematics Teaching, 141*, 4.
Wiliam, D., & Bartholomew, H. (2004). It's not which school but which set you're in that matters: The influence of ability-grouping practices on student progress in mathematics. *British Educational Research Journal, 30*(2), 239–279.
Williams, G. (2013). High performance, confidence, and disinclination to explore: A case study. In *MERGA36 2013 proceedings of the 36th mathematics education research group of Australasia 2013 conference* (pp. 658–665). Melbourne, Vic: AAMT & MERGA.
Williams, G. (2014). Optimistic problem-solving activity: Enacting confidence, persistence, and perseverance. *ZDM, 46*(3), 407–422.
Willis, J. (2007). *Research-based strategies to ignite student learning*. Alexandria, VA: Association for Supervision & Curriculum Development.
Yeager, D., & Dweck, C. (2012). Mindsets that promote resilience: When students believe that personal characteristics can be developed. *Educational Psychologist, 47*(4), 302–314.

Further Reading

Skemp, R. (1971). *The psychology of learning mathematics*. London: Penguin.

SECTION IIIC

LEARNERS OUT OF THE SCHOOL

CHAPTER

11

The Emotions Experienced While Learning Mathematics at Home

Janet Goodall, Sue Johnston-Wilder†, Rosemary Russell‡*

***University of Bath, Bath, United Kingdom, †University of Warwick, Coventry, United Kingdom, ‡AR & RR Education Ltd, Poole, United Kingdom**

INTRODUCTION

There can be few subjects in the school curriculum that provoke in learners such extreme emotions and inspire more love and hatred than mathematics (Hersh & John-Steiner, 1993). It is also well known that parental engagement in support of children can have a large and positive impact on children's learning in general (Desforges & Abouchaar, 2003; Harris & Goodall, 2007). In particular, parents and caregivers have a significant influence on their children's attitudes and self-concepts regarding mathematics (Cain-Caston, 1993; Dickens & Cornell, 1990; McMullen & de Abreu, 2011; Parsons, Adler, & Kaczala, 1982), and subsequently on their mathematics achievement. Supporting children's math education can also serve as motivation for parents, caregivers, and other family members to improve their own skills (McMullen & de Abreu, 2011; Skyrme, Gay, & Ratcheva, 2014).

However, we must also take into account that many parents themselves have had negative and sometimes traumatic experiences with mathematics. Asking such parents to support their child with math can be counterproductive. This has been vividly demonstrated by researchers such as Sian Beilock and her colleagues (Maloney, Ramirez, Gunderson, Levine, & Beilock, 2015) who found that if parents with math anxiety helped their children in first and second grades with mathematics at home, this had a

http://dx.doi.org/10.1016/B978-0-12-802218-4.00011-X

negative impact on progress. In the rest of this chapter we seek to extend the findings of Maloney et al. (2015), and other evidence of good and difficult experiences at home. Our work shows that parental math anxiety not only needs not be communicated to children, but it also can be overcome by parents, leading to positive experiences for both the parent and child.

To explain our reframing of the parental role, we rely on the prior work of Dweck (2000) regarding fixed-growth mindsets, Bandura (1977, 2000) regarding resilient self-efficacy and Lee and Johnston-Wilder (2014) regarding the growth zone model. In this chapter we will contrast TIRED maths (tedious, isolated, rote, elitist, de-personalised) (Nardi & Steward, 2003) with the maths that many young people can experience at school or at home: ALIVE (accessible, linked, inclusive, valued, empowering.) (Johnston-Wilder, Lee, Brindley, & Garton, 2015)

We consider what can go well and what can go badly, and then we look at the transformation of parents that is possible with a short intervention. People have asked us, "Is it really that easy?" We suggest that when the problems of math anxiety, exclusion, and trauma are factored in, it can really be that easy to help well-intentioned parents to be effectively supportive of their child's developing mathematical thinking.

We argue that this may not always result in better school grades, due to the difference in some cultures between education and modern, state-directed schooling, and the parent's need to work for the best interest of their individual child. It is in the child's best interest to make as much progress with math as possible and to develop an understanding of life-long learning, a growth mindset, inclusion, personal agency. It is also beneficial to understand the personal value of math, both as a useful set of tools for solving problems they may encounter and also as cultural capital, recognized as having worth in the adult community, to trade for interesting and valuable engagement in the wider community.

We begin our analysis by considering the distinction between mathematics education and math as experienced in school. By this we mean that, although the two are often thought to be the same, they may in fact be very different.

EDUCATION VS SCHOOLING

Learning is a natural and eventually well-honed experience. From a very young age, infants learn to avoid "laboring in vain" (Gerken, Balcomb, & Minton, 2011) and to assess what is learnable and what is not, seeking to achieve a maximal rate of learning. By learning, we mean deeper level learning in which the learner effectively compresses experiences into generalizations, such as "a certain smell is associated with food" or "the past tense in English is created by adding—ed" (Chomsky, 2014). Natural learning is an enjoyable experience of curious exploration and sense-making, learning from mistakes and guided by adults; it may be understood as sense-making activated by surprise (Dewey, 1897, 1916). Children's natural mathematical powers have been described by Mason, Burton, and Stacey (1982) as including the following dyads: noticing what is similar and different, discerning and classifying, stressing and ignoring, generalizing and specializing, and imagining and expressing. Such powers, and the development of associated mental imagery, can be encouraged in children by the adults and older learners around them. This happens first inactively, then iconically, and ultimately symbolically as the children develop (Bruner, 1966; Tall, 2013).

Thus we argue that mathematical thinking is already established at a young age; as a child experiences education and schooling, this ability develops further. If a child does not experience their natural mathematical powers being valued—if they experience their schooling as a process of exclusion and harm, rather than elucidation and mastery—these innate powers can be underdeveloped. Thus "being able to use mathematical thinking is a stated goal of schooling" (Stacey, 2006) but it proves elusive.

It is helpful here and throughout the chapter to distinguish between the process of mathematical thinking and learning new mathematics on the one hand and specific content on the other; parents are naturally capable of the former, though they may have lost sight of those skills through their own schooling.

Natural learning is worth doing. Based originally on Locke's (1880) concept of the naturalness of learning (admittedly in the well-to-do boy-child, but in a more modern view, extensible to all children), natural learning relates to those experiences of learning children have in their early years, which leave indelible impressions upon them. This learning involves positive emotion, engagement, meaning, positive relationships, success, getting stuck and unstuck, and balancing challenges and resources (Dodge, Daly, Huyton, & Sanders, 2012). It is fundamentally about making sense of experiences. Importantly, even from the very early days of the concept in Locke's writings, it is clear that these experiences can be positive or negative (Gregoriou & Papastephanou, 2013).

Education

Ideally, education is about "leading out" the growth potential in all of us. At all times this process should be guided by salutogenesis, meaning what keeps us well, and developing a personal "sense of coherence" (Antonovsky, 1993) that involves comprehensibility, manageability, and meaningfulness for the learners, rather than focusing on risks which must be avoided. In relation to mathematics per se, learners encounter math when exploring pattern, change, size, quantity, shape, and uncertainty. Mathematical activity can include pattern seeking, experimenting, describing, tinkering, inventing, visualizing, conjecturing, or guessing (Cuoco, Goldenberg, & Mark, 1996). Math education is concerned with increasing awareness of these processes (Gattegno, 1970) and their role in modern life.

Developing mathematical thinking is about developing habits of mind: defining, systematizing, abstracting, making connections, developing new ways to describe situations and make predictions, creating, inventing, conjecturing, and experimenting (Cuoco et al., 1996). Developing mathematical thinking may also be understood to include imagining and expressing, specializing and generalizing, conjecturing and convincing, focusing and defocusing, sense-making, looking back, getting unstuck (Mason et al., 1982), and articulating (Lee, 2006). This focus on describing mathematical thinking in terms of processes rather than content leads to the concept of mathematical literacy.

"Mathematical literacy is an individual's capacity to identify and understand the role that mathematical thinking plays in the world, to make well-founded judgements and to use and engage with mathematics in ways that meet the needs of that individual's life as a constructive, concerned and reflective citizen" (OECD, 2002). Further, an educated individual is always willing and able to learn more knowledge and skills as needed in life. They are resilient. As learning progresses and the individual becomes increasingly independent, educators may assume more of a coach role: listening, exploring options, supporting goal setting, and reviewing (Egan, 2013).

Schooling

"I honestly hadn't realized the pressure teachers and school leaders were putting children under" PERSONAL COMMUNICATION from PGCE STUDENT.

There is often a disconnection between education and what happens in schools. UK schools in particular are dominated by league tables, testing, and policy. Schooling, as a formalized, generally state-sponsored subset of education and learning, can become a process of "teaching to the test" (Wallace, 2009), seeking success by criteria that may not support the learning of individual children.

Perhaps one of the processes by which schooling becomes most harmful is in overusing the age-related findings of Piaget (Piaget, 1997; Piaget, Cook, & Norton, 1952; Piaget & Inhelder, 1969) as opposed to the enactive-iconic-symbolic (EIS) development described by Bruner (1966). Piaget taught that children began to develop symbolic thinking from age 11. In schools in the United Kingdom this has tended to mean that mathematical content can be introduced symbolically from the age of 11.

In fact, for a large majority of learners, most ideas are more accessible when met enactively (or kinesthetically). Once understood, iconic versions have meaning and can be explored. Only then can symbols take on the connected meaning needed for effective learning of mathematics. As an example, consider the teaching of algebra. Experienced teachers of many years standing have taught this topic largely symbolically and lamented about the poor responses of the learners. They find themselves surprised when learners who have previously been excluded become engaged and enabled by an experimental approach that is initially enactive, then iconic. Although children aged 11+ are "capable" of using symbolic thinking, Bruner gives us a ladder "EIS" to enable more learners to "access" symbolic thinking. Bruner's ladder of accessibility applies throughout the life span, so we can also use it with parents. For example, to understand what you are doing when you solve "simultaneous" equations, a powerful enactive example is "getting a meal on the table at a certain time." (For further exploration of this concept, see Lugalia, 2015; Lugalia, Johnston-Wilder, & Goodall, 2015.)

Elsewhere, (Johnston-Wilder & Lee, 2010a) we have described the process of systematic exclusion from mathematics, for example, by premature use of symbolic representation, as mathematical abuse. Some analysts have described this process as symbolic violence (Bourdieu, 1991, 1996) or structural violence (Galtung & Höivik, 1971). Bourdieu introduced the concept "symbolic power" to describe domination and exclusion. Symbolic power accounts for discipline used against another to confirm that individual's placement in a social hierarchy, at times in individual relations but most basically through system institutions, in particular education. Examples of this might include the use of mathematical symbols before children understand the basic concepts and not making links to existing understanding and knowledge, which often leads to feelings of exclusion and inability to grow or cope with the material. Structural violence refers to a form of violence wherein a social structure or social institution may harm people by preventing them from meeting their basic needs. According to Galtung, structural violence is an "avoidable impairment of fundamental human needs" (Galtung, 1969). In other words, in a math lesson, children are entitled to be treated with respect and to be safe, autonomous, and to have their curiosity supported and fostered, and their mental health should be safeguarded. (As an example, see Lugalia (2015), in which a complete year group of pupils went from being disheartened in algebra lessons to enjoying their lessons, to the point of running to arrive promptly.) This applies to the way schools deal with parents, as well, in their attempts to support parents' engagement with children's learning.

In our view, this applies to the worst cases of mathematical schooling (as distinguished from mathematics education or even more widely, learning). As with much structural violence, such as racism and sexism, mathematics exclusion is so common in many societies that it is accepted as inevitable. This also returns us to the difference between mathematics education (which would ideally lead to learning and mastery experiences) and mathematics as many experience it during schooling, as a process that leads instead to harm, loss of confidence, and exclusion and works to instill learned helplessness.

We argue that the most important role of parents and schools in this area is teaching learners both to safeguard themselves from such harm and to develop mathematical resilience, which includes a "can do if…" approach to mathematics. This also involves recognizing when harm may happen, even when that harm is caused by the well-intentioned others around the learner. However, many of the parents, teachers, and other adults with whom a learner comes into contact are also victims of the structural violence we call mathematics abuse, and they suffer from mathematics anxiety, helplessness and avoidance. While parental engagement supports learning in general, unaddressed anxiety does not (Maloney et al., 2015).

Learned Helplessness

Pekrun (2006) suggests that people's achievement emotions are in some ways determined by their feelings of control over situations; this would lead to the view that a feeling of not being in control, or not having control, would precipitate negative feelings toward particular activities (and outcomes). Coming from a different direction, the work on learned helplessness (Abramson, Seligman, & Teasdale, 1978; Hiroto, 1974; Maier & Seligman, 1976; Overmier, 2002) comes to a similar conclusion. People who have learned to be helpless in the face of particular experiences or circumstances feel unable to alter outcomes of situations, even if in fact they would be able to do so (Miller, 1984). Such individuals have internalized the idea that they cannot affect events and that results do not come from actions (Drath et al., 2008); in the case of learned helplessness around math, individuals become convinced that they are "no good" at math and no effort on their part will change this situation. This understanding of self in relation to a subject may be termed one's "identity," in this case, one's mathematical identity. This can be related to many people's images of mathematics which is often characterized negatively, for example, as "difficult, cold, abstract, and in many cultures, largely masculine" (Ernest, 1996, p. 802) or as "fixed, immutable, external, intractable and uncreative" (Buxton, 1981, p. 115). Importantly, however, McMullen and de Abreu (2011) suggest that these identities are not fixed, but rather they can be changed, altered by the individual, in reaction to experiences.

We suggest that in an ideal world parents and caregivers would feel enabled to encourage and support mathematical learning at home, while avoiding the harmful aspects of mathematical schooling and avoiding the inculcation of learned helplessness.

EXPERIENCES AT HOME

Research by Russell (2002) in the United Kingdom illuminated the previously unresearched "hidden help" parents give their children with mathematics. Russell defined hidden help as help that is initiated by parents, without prompting from school or researchers. Help of this kind is behind closed doors, in the privacy of the home, away from the view of schools and researchers (Russell, 2009).The research established that the practice exists; without prompting from school or researchers, parents do help their children with mathematics, and the practice is more widespread than had previously been acknowledged.

Russell found that though parents were very interested in helping their children with mathematics, they were encountering several difficulties and floundering in many areas when they attempted to do so; the following is an example:

> 'The parents felt their efforts to help did not achieve much, so this was an area of mixed emotions for them. These included concern, feelings of failure and resignation. They had difficulties understanding the schools' methods'. ***(Russell, 2002, p. 81)***
>
> There was a very strained relationship between father and child. There were reports of the child running out of the room and the parent losing patience with the child.' ***(Russell, 2002, p. 75)***

This aligns to research findings around parents' engagement with their children's learning overall. For example, Desforges has claimed that full parental engagement can increase the overall attainment of the child by 15%, "no matter what the social background of the family" (Desforges & Abouchaar, 2003). Parental engagement with children's learning is understood to be one of the best levers for raising achievement, with a good deal of literature to support this claim (Fan, Williams, & Wolters, 2012; Goodall, 2012, 2014a, 2014b; Goodall & Vorhaus, 2011; Gorard, See, & Davies, 2012; Harris & Goodall, 2007, 2008, 2009; Jeynes, 2012; See & Gorard, 2014, 2015). According to Sylva and colleagues (Sylva, Melhuish, Sammons, Siraj, & Taggart, 2008; Sylva, Melhuish, Sammons, Siraj-Blatchford, & Taggart, 2004), although levels of parental engagement are linked to socioeconomic status, what parents do with their children is far more important than who they are, in relation to children's progress and achievement. Even where families live in poverty children can achieve if their parents are involved and committed to their child's education (Sylva et al., 2004).

Psychologist Tanya Byron is quoted on the website of the UK Family Maths Toolkit (http://www.familymathstoolkit.org.uk/evidence) as saying, "Perhaps the single most important thing that parents can do to help their children with maths is to pass on a positive attitude." This is mirrored by the PISA finding that students' motivations to succeed in math is related to how their parents view math and what materials they are able to provide (OECD, 2013) (see also McMullen & de Abreu, 2011). In other words, some parents help their children to be resilient in the face of math, and they help to prevent children from learning to be helpless.

The National Numeracy Parent Survey (National Numeracy, 2015) provided a number of examples of positive experiences relating to home mathematics. One aspect of the positivity is parents coming to understand the practical nature of the mathematical activities and experiencing the relevance of mathematics to situations that children meet in everyday life. Examples of familiar family activity that can be seen to involve mathematical thinking include decorating, exercise, nutrition, and planning a meal. Such examples help to demonstrate to children that school mathematics can be linked to activities that they recognize as significant; children can be encouraged to make connections or links with existing knowledge and experience of everyday life. Parents can act as role models, as more knowledgeable others (Vygotsky, 1987) in applying mathematical thinking to the everyday task.

Adults can develop awareness to enable them to see school mathematics everywhere, and they can then share this awareness with children. Knitting, patchwork, and recipes are examples that often include algorithms that involve an unknown quantity. Phone and electricity tariffs are examples of $y=mx+c$; the path of a thrown ball or the shape of torch beam projected on a wall can provide examples of a parabola ($y=x^2$); slides and skate board parks can provide examples of changing gradient. An understanding of ways in which mathematics is used out-of-school can help improve children's attitudes towards mathematics and help children understand the value and relevance of mathematics in a variety of contexts (Jay, Rose, & Simmons, 2013). Ultimately the learner comes to understand that mathematical thinking can play a valuable role in achieving those goals.

Yet, this situation of parental confidence in math does not yet exist in our society. Maloney et al. (2015) showed that children learned less mathematics over the school year and were more likely to be math-anxious if parents, who are anxious about math themselves, provided frequent help on the child's mathematics homework. In 2014, an estimated 29% of Welsh parents reported making negative comments about mathematics in front of their child, comments such as "Don't ask me to do maths, I'm rubbish at it" (BBC, 2014). The Welsh government launched a campaign, "What You Say Counts," in an attempt to raise parental awareness that such negative attitudes to mathematics can adversely affect children (Welsh Government, 2014). If a parent reveals they are not good at mathematics, this can influence the way in which their child views mathematics (Yee & Eccles, 1988).

The work of Maloney et al. (2015) suggests that where parents suffer from mathematics anxiety, it is essential to provide an intervention before expecting such parents to support their children with mathematics at home. We would call this intervention "developing mathematical resilience."

We illustrate the extent of mathematics negativity at home in the United Kingdom with two recent surveys.

As part of their Parental Engagement Project, National Numeracy carried out a pilot project in 10 schools across the South East, South Wales, North East, and Midlands. At the end of the pilot project, there were indications the pilot had helped, for example, parents/caregivers to engage more with their children through everyday math. However, there were still barriers to engagement. Of the parents, 29% experienced conflict with their child about mathematics, 37% experienced lack of time to engage in mathematics with their child, 24% experienced a lack of confidence, and 40% felt that the new teaching methods were unfamiliar (National Numeracy, 2015).

One primary school in England, which took part in National Numeracy's Parental Engagement Project, carried out its own survey of parents. The school found that 59% of parents don't always enjoy helping child with math and 30% of parents find that doing math makes parents unhappy. Furthermore, 26% think children don't need math; this belief is an indictment of how they or their child were taught school mathematics. Almost a quarter of children think that math is a natural talent (harking back to the point above about actions not having an effect on outcomes, e.g., that one cannot learn to "be better" at math) (National Numeracy, 2015, p. 15).

In England, for the Department for Education, Goodall and Vorhaus (2011) acknowledge that parents' negative experiences of mathematics at school can have a large impact on their willingness to engage and, furthermore, some parents/caregivers are deterred from engaging in mathematics by other parents.

Some learners are not comfortable with the idea of their parents/caregivers getting involved and the most common reason for this was discussions ending up in arguments; for example, "doing mathematics alone is more peaceful—parents just take over." Children also mentioned differing methods as an issue.

Secondary school students gave a more mixed set of responses. Only a small number of students (all girls) were prepared to say that they engaged with their parents with mathematics. The majority of participants in these groups did not want to share mathematics with their parents. Again, it was seen as a source of confrontation: "when parents help, it just ends up in a massive fight." Students linked this increase in conflict to the difference in methods and also to a general change in relationships with parents at that age. They stated that it was easier to talk to parents when in primary school. Others lacked confidence that their parents could help them with their mathematics. Only one boy was able to say that his parent learns when he learns (p. 64).

In the next section, we discuss the results of an intervention on a mother/child dyad who were both experiencing mathematics anxiety and consequent arguing and negativity. The author's interviews with the mother in 2012 are reported; with the mother's permission, the following includes extracts of the mother's own account of the dyad's interactions with mathematics.

Heather's Bad Experiences

This section is based upon a journal kept by a young single mother after she had asked for an intervention to help her daughter at home with school mathematics. At the time of writing her journal, Heather was 28 and the mother of an only child, Rachel, who had just entered her final term in primary school.

Heather wrote that her own experience of mathematics in school had been negative. For example, having worked out an alternative method to answer a math question, she "got a red cross." Her teacher said "*it was because my working out was wrong and made me feel stupid in front of my peers.*" Up to that point, Heather had enjoyed "*the complex methods of long division and ... experimenting with a method of my own.*"

Heather left school without qualifications. She became a mother at the age of 17. Heather involved herself fully in supporting her daughter's education as best she could, and she attended all parental consultation evenings at the school. As her daughter grew, Heather expressed concern to school staff about Rachel's progress in math, but she felt that her concern was not taken seriously by school staff; she was "*told not to worry, she'll catch up, it will be fine, even if you did have a GCSE then you wouldn't be able to help.*" Heather requested school staff to send math homework for her daughter, but this did not materialize. Heather reported very negative interactions around math with her daughter at home, describing it as "*...a combination of falling out shouting, tears, avoidance feeling stupid on both parts.*"

In her writing, Heather showed that she had personal insight into some of the findings made by Beilock and her team (Maloney et al., 2015) because she felt that she was harming both her daughter's progress and their relationship at home. Heather also reported that "*... it broke my heart about a year ago when she came home upset and frustrated that [the teacher] had crossly said she wasn't trying hard enough, the tears pricking her eyes and the face that spoke a thousand words told me she had tried but just couldn't .....*" Heather's account indicates a common failure by teachers to distinguish reliably between effort and attainment.

As is common in those dealing with learned helplessness, Heather became adept at avoiding the mathematics and hence the anxiety that went with it. She wrote *"…whenever Rachel had shown an interest in day to day life and asked me a sum, or anything math related, I've always used avoidance techniques … if we were in the car, I could chat about anything, but if she asked about maths then I would regularly say 'hold on I've got to concentrate on driving at the moment ….'"* This also led to anger. Heather wrote: *"I suppose I got cross because I couldn't accept where I was mathematically and was ashamed as well as feeling guilty …."*

Both mother and daughter experienced anxiety. Heather wrote that *"… if Rachel told me she had maths homework on a Friday, we probably wouldn't approach it until Sunday but it would play on my mind and I'd spend my whole weekend tense, bracing myself for the onslaught of the dreaded maths homework…."*

By the time Rachel was in the final year of primary school, her teacher informed Heather that Rachel was too far behind in math to catch up to the expected level before entering secondary school. Heather reported *"…feeling like I'd just taken a blow to the stomach, complete shock , followed later by guilt for not doing more, anger for being let down so much myself as a child and now my daughter too, and working constantly to not let that cause a domino effect with Rachel."*

Rachel came home with practice papers for important math exams, and working together on the papers, Heather and Rachel fell into a familiar pattern, *"…well to say I damaged her confidence more and we fell out again big time was an understatement."*

Heather felt she had one last person she could ask for help to support her daughter: *"…during the last year or two I'd been working for a lady called Brenda … I knew her work was something to do with maths …."* Heather approached Brenda with some trepidation but fully determined to overcome her fears and embarrassment, in order to gain support for Rachel.

TRANSFORMATION

Heather's journal continues with a vivid description of her first consultation with Brenda about the mathematics. *"Well, I got to Brenda's house and took a deep breath and went through that door, and to say I was stunned at the response I got just does not describe it …. Brenda was so calm and understanding, it was as if I was asking for directions somewhere. She was just not fazed by my issues with maths at all in the slightest and took the paper and went through it with me" …. "at one point I thought, isn't this the part where I feel really stupid and frustrated as it is a year 6 paper, … but I don't think it mattered if it was a degree paper or reception class, there was just this huge feeling of acceptance and no expectation to be anywhere but where I was …"*

Heather and Rachel met with Brenda and the result was transformational, "*...we left with a better understanding ... and with a new sense of pride and respect for ourselves and for each other...*"

Heather described how Brenda helped her to overcome anxiety and supported her in helping Rachel overcome hers. Heather and Rachel were helped by Brenda to see examples of math in everyday life and to recognize ways in which they were already using math. For example, Brenda offered Rachel a piece of cake. When Rachel said "yes please," Brenda asked Rachel how many degrees of cake she wanted. To Heather, "*this delivered such a big message ... I suddenly saw a way I could relate maths to everyday life.*"

Brenda also gave Heather and Rachel a resource in the form of a math dictionary. Heather reported that: "*Rachel was so overjoyed she asked if she could take it in to her teacher, and her teacher asked if she could keep it for the weekend and that she had never seen one before. This filled Rachel with so much confidence.*" Heather noted that Rachel was greatly encouraged by this experience and that she was subsequently empowered to begin to overcome her previous lack of confidence.

Heather and Rachel's relationship around math subsequently became much more positive and productive, as Heather reported, "*Me and Rachel are now learning together ... we are both so much happier and actually enjoy the subject now, we have a laugh and work things out in our own way.*" Heather felt able to help her daughter and they were both greatly encouraged by the clear evidence that Rachel had made progress: "*I have managed to help Rachel to improve no end: her SAT result was a Level 3 , but her teacher assessment was a Level 4. ... She can now go up to senior school much better equipped for learning and a lot keener and happier in maths. I'm so proud. It feels like a real achievement considering that her class teacher (deputy head) didn't believe she could do it ....*"

This transformation also led to Heather and Rachel overcoming their joint learned helplessness in relation to maths; they are now able to deal with issues rather than avoid them, "*... now when she asks me a question and if I don't know we both find a way to figure it out, before she used to give up at that part.*"

Through their work with Brenda, Heather and Rachel experienced a change with their relationship with math that can validly be termed a transformation: from avoidance and anxiety, they moved through to resilience and progression.

Several important aspects of this transformation can be identified. These include the following:

- Brenda helped Heather and Rachel to accept their current situation, with positive regard (Rogers, 1951).
- Brenda listened closely to Heather and Rachel's accounts, explored their understanding, and included them in building understanding

and in setting goals (Egan, 2013). She also helped Heather to support Rachel in the same way.
- Brenda's support for Heather and Rachel emphasized the possibilities for change in order to address the ways in which they felt themselves to be dominated by the educational system (critical theory).
- Brenda's approach used a person-centered tool to support Heather and Rachel's development of awareness of their current feelings in relation to a particular mathematical challenge.
- Brenda's support was focused on introducing Heather and Rachel to supportive and effective processes, and not content.

According to Silver (1994), "*In symbolic politics, the power to name a social problem has vast implications for the policies considered suitable to address it… the discourse of exclusion may serve as a window through which to view political cultures.*" In the case of mathematics, the social problem is mathematics anxiety, avoidance, and helplessness. We have offered Heather's story as an example of how to change the culture of math exclusion to one of inclusion, engagement, and resilience.

Russell's Maths for Parents Course and Research

Since the autumn of 2013, Russell has been running Maths for Parents courses (Russell, 2014). The 4-week course addresses parents' anxieties about math, shows them how math is taught in schools currently, and gives them strategies for helping and encouraging their children. Her book (Russell, 2007) is also used as a resource. The course was developed based on more than 20 years of helping parents to support their children's math (Russell, 1996). It is resonant with research findings (Russell, 2002), Johnston-Wilder and Lee's (2010b) work on mathematical resilience, Dweck's work on the growth mindset (Dweck, 2000), Hughes' work on respecting and not dismissing what children are offering (Hughes, 1986), and Skemp's work on reconstructing schema (Skemp, 1971). Parents' past experience of school is often a source of negativity towards mathematics (Skyrme et al., 2014); thus the course has been run in people's homes, away from any pressure and unhappy memories of school.

Participants who start the course reporting negativity towards maths, for example, that they "hated maths" and "it was their least favorite subject," show a more positive attitude to mathematics after the course. For example, in response to the postcourse question, "*What are your feelings about maths after this course?*" one parent, Claire, wrote: "Feel happy. This course has changed my negative view" (Russell, 2014). Claire was able to recognize math in everyday life more easily after being on the course. She could see math as part of shopping, banking, and singing nursery rhymes, to name a few areas (Russell, 2016).

THE EVERYDAY MATHS PROJECT

The Everyday Maths Project (Jay et al., 2013) was based in primary schools in South West England. The aim of the project was to help parents to recognize their use of mathematics in everyday life and, by sharing this with their children, support their children's learning of mathematics.

There were two stages to the project. In the first stage, focus groups were formed with parents in order to find out their feelings about mathematics and helping their children with mathematics. The second stage involved running workshops for parents.

Through the focus groups, the researchers found that parents wanted information about what math their children were learning in school and how it was taught so that they could help at home because they were struggling to help and there were feelings of frustration (Jay et al., 2013). Many parents did not recognize methods used to teach mathematics. There was talk about fear of mathematics, with hardly any reference to enjoying math with their children; some parents said they felt upset if they were asked for help.

Though their own research indicated that parents and children are already engaging in many everyday activities that can be described as involving mathematics (Jay & Xolocotzin, 2012), it became clear too that parents did not value or recognize the wealth of mathematics that takes place in the home.

Through the ensuing four separate hour-long workshops, which were not about school math but about helping parents see the potential for conversations about math with their children through everyday activities such as finding a rope swing or playing Jenga, parents became more positive about discussing math with their children and were inspired to incorporate mathematics in talking with their children about everyday life.

Goldman and Booker (2009) discuss an alternative form of parental involvement that is less dependent on schoolwork and resourced by household activities. Alternatively, Goldman and Booker (2009) claim that parents often unknowingly face circumstances of problem solving that require mathematical knowledge and practice.

Winter, Salway, Ching Yee, and Hughes (2004) believe that early attempts to practice mathematics in everyday situations offer a context rich opportunity for children to learn and apply different forms of mathematics. All things considered, they argue parents do not need extensive mathematical knowledge to support their children's learning. Talking about ways that mathematics can help with everyday activities can be more useful than knowing a correct procedure or answer.

We would argue that the onus for removing math anxiety from the classroom rests with schools, and it will require significant input in training (including initial and continuing teacher training) as well as changes to curricula. Considering the points made above in relation to mathematical abuse, we would further argue that those with a responsibility for safeguarding become involved in this process. Additionally, this is an area, among many others, where clear collaboration between parents, schools, and communities is likely to bear fruit (Epstein & Sheldon, 2006; Goodall, 2014a, 2014b).

THE WAY FORWARD

Across the divide that exists between teachers and parents, based often on misconceptions of how each side views the other (Harris & Goodall, 2007), it is important to build bridges so that mutual respect and understanding can flourish and the child can gain the best from both worlds. All those involved in the process—school staff, other agencies and families—must realize that learning and education are far larger concepts than "schooling" and that learning is not restricted to the narrow confines and relatively short hours of the school experience.

To do this, it will be important that emotions are recognized and valued for what they are, and that challenges experienced by learners fit within the "yellow" area of the growth zone. In doing this, the factors that protect learners will be highlighted, and learned helplessness—"I can't do this, I'm no good at maths"—will be avoided because there will be no need to develop these defenses. This will require, as we have suggested above, that some learners move from the TIRED (Nardi & Steward, 2003) experience of the teaching of mathematics to an enabling, ALIVE experience (Johnston-Wilder et al., 2015). This will ensure that the current generation of children now learning math can avoid the pitfalls that hampered so many of their parents.

For those who have already developed a fixed mentality, however, more support will be needed, because they need to unlearn many entrenched habits of avoidance and flight. Parents need to be supported to gain confidence in relation to their own abilities in math (many of which they may not realize they possess), and then to support their children. Encouraging parents to work together with peers as well as with their children will help end the isolation that math anxiety can cause (the feeling that one is the only person struggling with something that comes easily to others). This will support them toward remaking their identities in relation to mathematics, moving from an ineffective, unable identity to one that is capable of achievement.

Postscript: Emancipation

Heather and Rachel demonstrated their resilience when Rachel was in year 10. Rachel, being off school, was using a math tutor who undermined her confidence and did not engage with Heather at all. Heather and Rachel discussed the situation with Brenda and refused further services from this tutor because they were aware of the damage being done not only to Rachel's learning, but to her confidence.

References

Abramson, L. Y., Seligman, M. E., & Teasdale, J. D. (1978). Learned helplessness in humans: Critique and reformulation. *Journal of Abnormal Psychology, 87*(1), 49.

Antonovsky, A. (1993). The implications of salutogenesis: An outsider's view. In A. P. Turnbull, J. M. Patterson, S. K. Behr, D. L. Murphy, J. G. Marquis, & M. J. Blue-Banning (Eds.), *Cognitive coping, families, and disability* (pp. 111–122). Baltimore, MD: Paul H. Brookes Publishing.

Bandura, A. (1977). Self-efficacy: Toward a unifying theory of behavioral change. *Psychological Review, 84*(2), 191.

Bandura, A. (2000). Exercise of human agency through collective efficacy. *Current Directions in Psychological Science, 9*(3), 75–78.

BBC. (2014). *Parents 'must be positive about maths', says Huw Lewis*. Retrieved from http://www.bbc.co.uk/news/uk-wales-26405765 (Accessed 28 February 2016).

Bourdieu, P. (1991). *Language and symbolic power*. Cambridge: Polity Press.

Bourdieu, P. (1996). *The state nobility: Elite schools in the field of power*. Cambridge: Polity Press.

Bruner, J. S. (1966). *Toward a theory of instruction*. Cambridge, MA: Harvard University Press.

Buxton, L. (1981). *Do you panic about maths? Coping with maths anxiety*. London: Heinemann.

Cain-Caston, M. (1993). Parent and student attitudes toward mathematics as they relate to third grade mathematics achievement. *Journal of Instructional Psychology, 20*(2), 96–101.

Chomsky, N. (2014). *Aspects of the theory of syntax*. Cambridge, MA: MIT Press.

Cuoco, A., Goldenberg, E. P., & Mark, J. (1996). Habits of mind: An organizing principle for mathematics curricula. *Journal of Mathematical Behavior, 15*, 375–402.

Desforges, C., & Abouchaar, A. (2003). *The impact of parental involvement, parental support and family education on pupil achievement and adjustment: A literature review*. London: Department of Education and Skills.

Dewey, J. (1897). My pedagogic creed. *School Journal, 54*(1), 77–80. Retrieved from http://dewey.pragmatism.org/creed.htm (Accessed 28 February 2016).

Dewey, J. (1916). *Democracy and education*. Copyright © 1916 The Macmillan Company. Retrieved from http://www.gutenberg.org/files/852/852-h/852-h.htm (Accessed 28 February 2016).

Dickens, M. N., & Cornell, D. G. (1990). *Parental influences on the mathematics self-concept of high-achieving adolescent girls*. AEL minigrant report no. 20.

Dodge, R., Daly, A. P., Huyton, J., & Sanders, L. D. (2012). The challenge of defining wellbeing. *International Journal of Wellbeing, 2*(3), 222–235.

Drath, W. H., McCauley, C. D., Palus, C. J., Van Velsor, E., O'Connor, P. M., & McGuire, J. B. (2008). Direction, alignment, commitment: Toward a more integrative ontology of leadership. *The Leadership Quarterly, 19*(6), 635–653.

Dweck, C. S. (2000). *Self-theories: Their role in motivation, personality and development*. Philadelphia, PA: Taylor & Francis.

Egan, G. (2013). *The skilled helper: A problem-management and opportunity-development approach to helping*. London: Cengage Learning.

Epstein, J. L., & Sheldon, S. B. (2006). Moving forward: Ideas for research on school, family, and community partnerships. In C. F. Conrad & R. Stein (Eds.), *SAGE handbook for research in education: Engaging ideas and enriching inquiry* (pp. 117–138). London: Sage.

Ernest, P. (1996). Popularisation: Myths, massmedia and modernism. In A. Bishop (Ed.), *International handbook of research in mathematics education* (pp. 785–817). Dordrecht: Kluwer.

Fan, W., Williams, C. M., & Wolters, C. A. (2012). Parental involvement in predicting school motivation: Similar and differential effects across ethnic groups. *The Journal of Educational Research, 105*(1), 21–35.

Galtung, J. (1969). Violence, peace, and peace research. *Journal of Peace Research, 6*(3), 167–191.

Galtung, J., & Höivik, T. (1971). Structural and direct violence: A note on operationalization. *Journal of Peace Research, 8*(1), 73–76.

Gattegno, C. (1970). *What we owe children, the subordination of teaching to learning*. New York, NY: Outerbridge and Diensfrey.

Gerken, L. A., Balcomb, F. K., & Minton, J. L. (2011). Infants avoid labouring in vain by attending more to learnable than unlearnable linguistic patterns. *Developmental Science, 14*(5), 972–979.

Goldman, S., & Booker, A. (2009). Making math a definition of the situation: Families as sites for mathematical practices. *Anthropology & Education Quarterly, 40*(4), 369–387.

Goodall, J. (2012). Parental engagement to support children's learning: A six point model. *School Leadership & Management, 33*(2), 1–18.

Goodall, J. (2014a). Engaging parents to raise achievement. *School Leadership Today, 63*, 56–60.

Goodall, J. (2014b). Re-thinking engagement. In Schools of Tomorrow (Ed.), *Growing engagement: Re-imagining relationships between schools, families and communities: Vol. 2* (pp. 8–19). Peterborough: Schools for Tomorrow.

Goodall, J., & Vorhaus, J. (2011). *Review of best practice in parental engagement*. London: Department of Education.

Gorard, S., See, B. H., & Davies, P. (2012). *The impact of attitudes and aspirations on educational attainment and participation*. York: Joseph Rowntree Foundation. Available at: jrf.org.uk/publications/aspirations-educational-attainment-participation.

Gregoriou, Z., & Papastephanou, M. (2013). The utopianism of John Locke's natural learning. *Ethics and Education, 8*(1), 18–30.

Harris, A., & Goodall, J. 2007. *Engaging parents in raising achievement: Do parents know they matter?* London: Department for Children, Schools and Families.

Harris, A., & Goodall, J. (2008). Do parents know they matter? Engaging all parents in learning. *Educational Research, 50*(3), 277–289.

Harris, A., & Goodall, J. (2009). *Helping families support children's success at school*. London: Save the Children.

Hersh, R., & John-Steiner, V. (1993). A visit to Hungarian mathematics. *The Mathematical Intelligencer, 15*(2), 13–26.

Hiroto, D. S. (1974). Locus of control and learned helplessness. *Journal of Experimental Psychology, 102*(2), 187.

Hughes, M. (1986). *Children and number: Difficulties in learning mathematics*. Oxford: Wiley-Blackwell.

Jay, T., Rose, J., & Simmons, B. (2013). Why parents can't always get what they (think they) want. *Proceedings of the British Society for Research into Learning Mathematics, 33*(2), 31–36.

Jay, T., & Xolocotzin, U. (2012). Mathematics and economic activity in primary school children. In *36th conference of the International Group for the Psychology of Mathematics Education*.

Jeynes, W. (2012). A meta-analysis of the efficacy of different types of parental involvement programs for urban students. *Urban Education, 47*(4), 706–742.

Johnston-Wilder, S., & Lee, C. (2010a). Developing mathematical resilience. In *BERA annual conference 2010, 1–4 Sep 2010*: University of Warwick. http://www.beraconference.co.uk/2010/downloads/abstracts/pdf/BERA2010_0636.pdf.

Johnston-Wilder, S., & Lee, C. (2010b). Mathematical resilience. *Mathematics Teaching, 218*, 38–41.

Johnston-Wilder, S., Lee, C., Brindley, J., & Garton, E. (2015). Developing mathematical resilience in school-students who have experienced repeated failure. In *ICERI 2015: 8th international conference on education, research and innovation, 16–18 November 2015, Seville, Spain.*

Lee, C. (2006). *Language for learning mathematics: Assessment for learning in practice*. London: McGraw-Hill Education.

Lee, C., & Johnston-Wilder, S. (2014). Mathematical resilience: What is it and why is it important? In S. Chinn (Ed.), *The Routledge international handbook of dyscalculia and mathematical learning difficulties* (pp. 337–345). Abingdon: Routledge.

Locke, J. (1880). *Some thoughts concerning education*. London: National Society's Depository.

Lugalia, M. (2015). *Pupils learning algebra with ICT in key stage 3 mathematics classrooms* [Ph.D. thesis]. Coventry: University of Warwick.

Lugalia, M., Johnston-Wilder, S., & Goodall, J. (2015). Using ICT and dialogic teaching: Impact on mathematical resilience and attainment in algebra of a Kenyan school year group. In *EDULEARN15 Proceedings*.

Maier, S. F., & Seligman, M. E. (1976). Learned helplessness: Theory and evidence. *Journal of Experimental Psychology: General, 105*(1), 3.

Maloney, E. A., Ramirez, G., Gunderson, E. A., Levine, S. C., & Beilock, S. L. (2015). Intergenerational effects of parents' math anxiety on children's math achievement and anxiety. *Psychological Science*. http://dx.doi.org/10.1177/0956797615592630.

Mason, J., Burton, L., & Stacey, K. (1982). *Thinking mathematically*. London: Addison Wesley.

McMullen, R., & de Abreu, G. (2011). Mothers' experiences of their children's school mathematics at home: The impact of being a mother-teacher. *Research in Mathematics Education, 13*(1), 59–74.

Miller, D. K. (1984). *Conditions affecting the occurrence of causal attributions in response to a learned helplessness induction* [Doctor of Philosophy]. Lubbock, TX: Texas Tech University.

Nardi, E., & Steward, S. (2003). Is mathematics T.I.R.E.D.? A profile of quiet disaffection in the secondary mathematics classroom. *British Educational Research Journal, 29*(3), 345–366.

National Numeracy. (2015). *Final report: Parental engagement project*. Lewes: National Numeracy. http://www.nationalnumeracy.org.uk/sites/default/files/parental_engagement_-_final_report.pdf.

OECD. (2002). *Education at a glance*. Paris: OECD Publishing.

OECD. (2013). *PISA 2012 Results: Ready to learn—Students' engagement, drive and self-beliefs: Vol. III*. Paris: OECD Publishing.

Overmier, J. B. (2002). On learned helplessness. *Integrative Physiological & Behavioral Science, 37*(1), 4–8.

Parsons, J. E., Adler, T. F., & Kaczala, C. M. (1982). Socialization of achievement attitudes and beliefs: Parental influences. *Child Development, 53*(2), 310–321.

Pekrun, R. (2006). The control-value theory of achievement emotions: Assumptions, corollaries, and implications for educational research and practice. *Educational Psychology Review, 18*(4), 315–341.

Piaget, J. (1997). *The moral judgement of the child*. London: Simon and Schuster.

Piaget, J., Cook, M., & Norton, W. (1952). *The origins of intelligence in children*. New York, NY: International Universities Press.

Piaget, J., & Inhelder, B. (1969). *The psychology of the child*. London: Basic Books.

Rogers, C. (1951). *Client-centered therapy: Its current practice, implications and theory*. London: Constable.

Russell, R. (1996). *Maths for parents*. London: Piccadilly Press.

Russell, R. (2002). *Parents helping their children with mathematics*. Bristol: University of Bristol.

Russell, R. (2007). *Help your child with numeracy: Age range 3–7*. London: Continuum.

Russell, R. (2009). Why do parents help their children with mathematics? *Proceedings of the British Society for Research into Learning Mathematics, 29*(2), 70–75.

Russell, R. (2014). *'So chuffed'. Blog: Maths 4 mums and dads.* http://maths4mumsanddads.co.uk/mfpc/so-chuffed.

Russell, R. (2016). *Nobody told me that before!.* London: National Numeracy. http://www.familymathstoolkit.org.uk/nobody-told-me-that-before.

See, B. H., & Gorard, S. (2014). *What do rigorous evaluations tell us about the most promising parental involvement interventions? A critical review of what works for disadvantaged children in different age groups.* London: Nuffield Foundation.

See, B. H., & Gorard, S. (2015). The role of parents in young people's education—A critical review of the causal evidence. *Oxford Review of Education, 41*(3), 346–366.

Silver, H. (1994). Social exclusion and social solidarity: Three paradigms. *International Labour Review, 133*, 531.

Skemp, R. (1971). *The psychology of learning mathematics.* London: Penguin.

Skyrme, S., Gay, S.-J., & Ratcheva, V. (2014). It's a massive confidence boost having your mum or your dad there: Discovering attitudes and barriers to parental engagement in mathematics with school age students. *Proceedings of the British Society for Research into Learning Mathematics, 34*(3), 61–66.

Stacey, K. (2006). *What is mathematical thinking and why is it important?.* Progress report of the APEC project: Collaborative studies on innovations for teaching and learning mathematics in different cultures (II)—Lesson study focusing on mathematical thinking.

Sylva, K., Melhuish, E. C., Sammons, P., Siraj, I., & Taggart, B. (2008). *Final report from the primary phase: Pre-school, school and family influences on children's development during key stage 2 (7–11).* Nottingham: Department for Children, Schools and Families. DCSF research report.

Sylva, K., Melhuish, E. C., Sammons, P., Siraj-Blatchford, I., & Taggart, B. (2004). *The effective provision of pre-school education (EPPE) project: Technical paper 12—The final report: Effective pre-school education.* London: DfES/Institute of Education, University of London.

Tall, D. (2013). *How humans learn to think mathematically: Exploring the three worlds of mathematics.* Cambridge: Cambridge University Press.

Vygotsky, L. (1987). Zone of proximal development. *Mind in society: The development of higher psychological processes* (pp. 52–91). Cambridge, MA: Harvard University Press.

Wallace, S. (2009). *Teaching to the test. A dictionary of education.* Oxford: Oxford University Press.

Welsh Government. (2014). *What you say counts.* Retrieved from http://learning.gov.wales/news/sitenews/what-you-say-counts/?lang=en (Accessed 29 January 2016).

Winter, J., Salway, L., Ching Yee, W., & Hughes, M. (2004). Linking home and school mathematics: The home school knowledge exchange project. *Research in Mathematics Education, 6*(1), 59–75.

Yee, D. K., & Eccles, J. S. (1988). Parent perceptions and attributions for children's math achievement. *Sex Roles, 19*(5–6), 317–333.

CHAPTER

12

Parents' and Children's Mathematics Anxiety

Sophie Batchelor, Camilla Gilmore, Matthew Inglis
Loughborough University, Loughborough, United Kingdom

AN OVERVIEW OF MATHEMATICS ANXIETY RESEARCH

What Is Mathematics Anxiety?

The notion of mathematics anxiety dates back many years. Research in this area is often seen to have begun with reports from mathematics teachers in the 1950s. In one of the earliest papers on the topic, Gough (1954) observed several students reacting emotionally toward mathematics and labeled this reaction "mathemaphobia." A few years later, Dreger and Aiken (1957) conducted the first empirical study of mathematics anxiety in a sample of university college students. They introduced the term "number anxiety," which they defined as "the presence of a syndrome of emotional reactions to arithmetic and mathematics" (p. 344). They showed that number anxiety could be separated from general anxiety. Furthermore, they showed that number anxiety was associated with lower grades in mathematics but not lower levels of general intelligence.

In the decades that followed, mathematics anxiety became defined in several ways. Lang (1968) asserted that mathematics anxiety is a phobia that, like any other phobia, influences individuals on three dimensions: (i) physiological, e.g., increased heart rate; (ii) cognitive, e.g., worrisome thoughts; and (iii) behavioral, e.g., avoiding mathematics classes (see also Faust, 1992). Meanwhile other researchers emphasized that mathematics anxiety comprises a range of emotional responses, from mild states of apprehension to intense feelings of uneasiness and distress (e.g., Richardson & Suinn, 1972). These emotional responses can arise not just in the mathematics classroom but in everyday activities such as managing personal finances, working out change, and calculating percentages while shopping.

http://dx.doi.org/10.1016/B978-0-12-802218-4.00012-1

What Do We Know About Mathematics Anxiety so Far?

With the construction of a number of scales designed to assess mathematics anxiety (e.g., the Mathematics Anxiety Rating Scale (MARS) developed by Richardson and Suinn (1972))[1], researchers have probed the nature of the construct and the consequences it can have (see, Hembree, 1990; Ma, 1999 for reviews). Hembree (1990) conducted a metaanalysis of 151 studies and found that high mathematics-anxious college students take fewer mathematics courses, demonstrate lower mathematics achievement, and hold negative self-perceptions about their mathematics abilities, when compared to their low mathematics-anxious peers. A metaanalysis by Ma (1999) yielded similar results from studies with elementary school students. The correlates of mathematics anxiety reported in these metaanalyses are summarized in existing book chapters by Ashcraft and Ridley (2005) and Ashcraft, Krause, and Hopko (2007).

For many years, research into mathematics anxiety tended to focus on older children and adults. It was traditionally believed that mathematics anxiety arose in the context of complex mathematics, surfacing at the end of elementary school when more advanced mathematical concepts were introduced (Maloney & Beilock, 2012). In this tradition, the most dominant account to explain the relationship between mathematics anxiety and mathematics performance was the "working memory disruption" hypothesis (Ashcraft & Kirk, 2001; Ashcraft & Krause, 2007; Hopko, Ashcraft, Gute, Ruggiero, & Lewis, 1998). Ashcraft and colleagues proposed that mathematics anxiety results in worrisome thoughts and ruminations that consume vital working memory resources needed to solve complex mathematical problems. In support of this account, they demonstrated that performance differences between high mathematics-anxious and low mathematics-anxious individuals manifested under high working memory load conditions, such as dual-task experiments or arithmetic problems involving carry operations (Ashcraft & Kirk, 2001).

Two recent strands of research have challenged this traditional view. Firstly, studies conducted with adults have shown that mathematics anxiety is associated with impairments in basic, low-level, numerical processing as well as higher-level mathematics (Lindskog, Winman, & Poom, 2017; Maloney, Ansari, & Fugelsang, 2011; Maloney, Risko, Ansari, & Fugelsang, 2010; Núñez-Peña & Suárez-Pellicioni, 2014). Maloney et al. (2011), for example, gave high mathematics-anxious and low mathematics-anxious adults a symbolic numerical comparison task, typically used to measure the precision of an individual's magnitude representations. They found that high mathematics-anxious individuals showed a larger numerical

[1] See Table 1 in Eden, Heine, and Jacobs (2013) for a recent summary of mathematics anxiety measures.

distance effect than their low mathematics-anxious peers. This suggests that their numerical magnitude representations, and in particular the mapping between nonsymbolic and symbolic representations, may be less precise (although see Dietrich, Huber, Moeller, & Klein, 2015, for an alternative account).

In addition to these findings, studies have demonstrated that mathematics anxiety is present in children much younger than previously thought; in other words, children who are in the early stages of formal schooling (Jameson, 2013; Krinzinger, Kaufmann, & Willmes, 2009; Ramirez, Gunderson, Levine, & Beilock, 2013; Thomas & Dowker, 2000; Wu, Barth, Amin, Malcarne, & Menon, 2012; Young, Wu, & Menon, 2012). Young et al. (2012) showed that children can develop mathematics anxiety by the age of seven years and that this anxiety is associated with a distinct pattern of neural activity. High mathematics-anxious children showed hyperactivity in the right amygdala (a brain region that is important for the processing of negative emotions) and reduced activity in the posterior parietal cortex and the dorsolateral prefrontal cortex (two brain regions that have been implicated in mathematical reasoning), when compared to their low mathematics-anxious peers (see Artemenko, Daroczy, & Nuerk, 2015 for an overview of the neural correlates of mathematics anxiety).

Importantly, the emergence of some developmentally appropriate mathematics anxiety scales has allowed researchers to start exploring the effects of mathematics anxiety on younger children's numerical development. Thus far, the initial findings have been mixed (Eden et al., 2013). Some studies have found a relationship between primary school children's mathematics anxiety and their mathematics performance (Ramirez et al., 2013; Vukovic, Kieffer, Bailey, & Harari, 2013; Wu et al., 2012), while others have not (Haase et al., 2012; Krinzinger et al., 2009; Thomas & Dowker, 2000; see Dowker, Sarkar, & Looi, 2016 for a discussion and possible explanations for these conflicting results). Further investigations are needed to help us understand not just the consequences but the causes of mathematics anxiety in the early years. Why do some children develop a fear of mathematics?

The causes of mathematics anxiety are not yet well understood (Eden et al., 2013), but they are typically reported to have multiple origins (Godbey, 1997; Jain & Dowson, 2009; Norwood, 1994). These can be broadly classified into environmental, personal, and cognitive factors (Rubinsten & Tannock, 2010). Environmental factors may include negative experiences at home (Simpkins, Davis-Kean, & Eccles, 2005) or at school (Greenwood, 1984; Newstead, 1998; Stodolsky, 1985). Personal factors may include mathematics self-efficacy (Cooper & Robinson, 1991; Jameson, 2014; Meece, Wigfield, & Eccles, 1990) and mathematics self-concept (Ahmed, Minnaert, Kuyper, & van der Werf, 2012; Ferla, Valcke, & Cai, 2009; Jameson, 2014; Lee, 2009). And cognitive factors may include domain-general factors such as attention (Rubinsten, Eidlin, Wohl, & Akibli, 2015) and domain-specific factors such as numerical magnitude representations (Maloney et al., 2010, 2011; Núñez-Peña & Suárez-Pellicioni, 2014) and computational skills (Carey, Hill, Devine, & Szücs, 2016; Harari, Vukovic, & Bailey, 2013).

In this chapter we focus specifically on the environmental influence of parents; a possible causal factor that is often cited in the literature but has received little empirical investigation. Given the scarcity of studies within this area, we start by drawing on research from the general childhood anxiety literature.

WHERE DOES CHILDHOOD ANXIETY COME FROM?

Research has uncovered a number of risk factors for the development of childhood anxiety—an umbrella term that can be used to refer to both trait anxiety and specific anxiety symptoms and disorders in children (Wood, McLeod, Sigman, Hwang, & Chu, 2003). One of the primary risk factors identified is parental anxiety. There is strong evidence for a link between parents' and children's anxiety (e.g., Woodruff-Borden, Morrow, Bourland, & Cambron, 2002), particularly when the anxious parent is the mother (e.g., McClure, Brennan, Hammen, & Le Brocque, 2001). In the clinical literature it has been shown that children who have a parent with an anxiety disorder are more than seven times more likely to be diagnosed with an anxiety disorder when compared to control children without anxious parents (e.g., Turner, Beidel, & Costello, 1987).

The Link Between Parent and Child Anxiety

There are several possible mechanisms for the transmission of anxiety between parent and child. These include predispositions or inherited traits (e.g., temperament), stressful life events (e.g., divorce or separation), parenting behaviors (e.g., overcontrol and disengagement), attachment relationships (insecure vs secure), and learning mechanisms (e.g., modeling, information transfer, reinforcement) (see Fig. 1 in Murray, Creswell, & Cooper, 2009, for a diagram of some of the possible pathways from parent to child anxiety). The direction of these influences is typically examined from parent to child and studies have tended to focus on single factors or main effects, but the development of childhood anxiety is nevertheless seen as a complex process attributable to many interacting factors (e.g., Vasey & Dadds, 2001). In the sections below we summarize some of the findings relating to predispositions and parenting factors.

Predispositions

Predispositions or inherited traits such as temperament have been linked to childhood anxiety in a number of studies. Central to this is the construct of "behavioral inhibition" (BI)—a term used to describe temperamental patterns of fearfulness and withdrawal in unfamiliar situations (e.g., Kagan, 1989). Longitudinal research has shown that BI is a relatively stable trait from infancy into childhood (e.g., Kagan, Reznick, Snidman, Gibbons, & Johnson, 1988), one which relates to parental anxiety (e.g., Rosenbaum et al., 1988) and predicts later childhood anxiety (e.g., Biederman et al., 1993; Rosenbaum et al., 1991). Biederman et al. (2001) demonstrated that the association between BI and childhood anxiety is particularly strong for social anxieties (social phobia and avoidant disorder) compared to other anxieties such as overanxious disorder, panic disorder, and generalized anxiety disorder. Overall, research suggests that BI in infancy may predispose children to developing later anxiety, particularly if it co-occurs with other risk factors in the environment.

Parenting Factors

Parenting Behaviors

Studies examining the relationship between anxiety and childrearing practices have highlighted the role of two parenting behaviors: autonomy vs control and warmth vs rejection (see McLeod, Wood, & Weisz, 2007; Rapee, 1997; Wood et al., 2003, for reviews). Low levels of autonomy and warmth (a parenting style labeled "affectionless control") have been associated with high levels of parental anxiety (e.g., Woodruff-Borden et al., 2002) and high levels of child anxiety (e.g., Moore, Whaley, & Sigman, 2004). A number of prospective and retrospective investigations have been conducted, with arguably the most compelling evidence coming from observational studies—as is the case in the examples given below.

Woodruff-Borden et al. (2002) observed 51 child-parent dyads (25 dyads with an anxious parent) completing two interactive tasks: an unsolvable anagrams task and a speech preparation and delivery task. They found that anxious parents were more withdrawn and disengaged from the tasks than nonanxious parents. There were no overall group differences in terms of parental control, but there was a trend toward anxious parents exerting more control in response to children's negative affect. In another study focusing on anxious children, Moore et al. (2004) observed 68 child-mother dyads during two conversational tasks: a conflict conversation (a discussion of something they argue about) and an anxiety conversation (a discussion of something the child is anxious about). Here they found that mothers of anxious children showed less warmth and more control than mothers of nonanxious children. This aligns with earlier observational findings (e.g., Hudson & Rapee, 2001), some of which found a link between mothers' control and children's anxiety, but only in daughters (e.g., Krohne & Hock, 1991).

Attachment Relationships

The link between attachment and anxiety dates back to the work of Bowlby (1973), who proposed that the bond (or attachment) between mother and child plays an important role in a child's personal and emotional development. According to Bowlby's theory, attachment relationships can be defined as secure or insecure, with securely-attached children hypothesized to experience less anxiety than insecurely-attached children. As the theory has developed, researchers have started to explore specific forms of insecure attachments and their individual associations with anxiety (Brumariu & Kerns, 2010). Ambivalent attachment (characterized by inconsistent and unpredictable interactions) has been compared to avoidant attachment (characterized by unavailability and unresponsiveness) and disorganized attachment (characterized by confusing and erratic behavior). While all insecure attachment relationships have been linked to anxiety, disorganized attachment is perhaps the most consistently related; children with disorganized attachment relationships have been found to show more catastrophizing and fewer adaptive coping strategies than their securely-attached peers (e.g., Brumariu, Kerns, & Seibert, 2012).

Learning Mechanisms

Another way in which parents may play a role in the development of child anxiety is through learning experiences (see Fisak & Grills-Taquechel, 2007 for a review). As well as learning through direct classically conditioned responses to stressful life events, children learn indirectly through the behaviors and instructions of others (e.g., Rachman, 1977). Rachman highlighted the role of two indirect learning mechanisms: modeling (or vicarious learning) and information transfer (or instructional learning). More recently, researchers have also emphasized the role of reinforcement processes (e.g., Rapee, 2002). Here, we briefly describe how each of these indirect learning mechanisms may operate in the context of parent-to-child transmitted anxiety.

Modeling. This is a process widely acknowledged through the social learning theory of Bandura (1986), who proposed that children may learn vicariously through parental modeling of anxious or avoidant behavior. As summarized by Fisak and Grills-Taquechel (2007), children may observe their parents responding fearfully (e.g., breathing heavily, verbalizing their worries) when confronted with certain stimuli and the child may then replicate this behavior when faced with a similar stimulus on a separate occasion. Numerous studies have demonstrated this phenomenon, both with animal (e.g., Mineka, Davidson, Cook, & Keir, 1984) and human samples (e.g., Gerull & Rapee, 2002). Mineka et al. (1984) demonstrated rhesus monkeys passing on fear of snakes to their offspring. Similarly, Gerull and Rapee (2002) showed that toddlers who observed their mothers responding fearfully to two novel toys (a snake and a spider) were more fearful and avoidant when later presented with those toys, compared to toddlers who observed their parents responding positively. In this particular study there was evidence for gender effects: The same pattern was observed for boys and girls but the effect was stronger in girls.

Information transfer. As well as transmitting anxiety responses through modeling, parents may communicate anxious messages or instructions to their child. This learning mechanism has received less attention in the anxiety literature (see Table 4 in Fisak & Grills-Taquechel, 2007), but there is some evidence to suggest that anxious children receive more fearful instructions from their parents (e.g., "be careful on ..." or "stay away from ...") than nonanxious children (Beidel & Turner, 1998). A series of studies by Field and colleagues have shown that when children are given negative information about a stimulus (e.g., an animal) they show greater fear of that stimulus, both in terms of their self-reports (e.g., Field, Argyris, & Knowles, 2001) and actual behavior (e.g., Field & Lawson, 2003). These effects were found when the information was provided by an adult (either a teacher or stranger) but not when it came from a peer. The experiments did not look specifically at parents but as Fisak and Grills-Taquechel (2007) remark, "it seems likely that similar, or perhaps even more robust, findings would be revealed for parents given that they are often children's most common source of information (both negative and positive) about the environment" (p. 223).

Reinforcement. The process of reinforcing or supporting a particular behavior can also act as a powerful learning mechanism. Rapee (2002) proposed that parents may reinforce children's anxieties in ways such as removing a child from a situation that makes them anxious or rewarding a child when they show anxiety or distress. In support of this hypothesis, studies have shown that parents may increase avoidant behaviors in children who are socially anxious, a phenomenon labeled the FEAR (Family Enhancement of Avoidant Responses) effect (e.g., Barrett, Rapee, Dadds, & Ryan, 1996). A number of retrospective self-report studies have confirmed these findings (e.g., Watt, Stewart, & Cox, 1998), but the precise mechanisms involved need further investigation. Researchers have emphasized the need for more longitudinal studies looking at how the different learning mechanisms may work together (Fisak & Grills-Taquechel, 2007).

Summary

In sum, research from the general anxiety literature has identified parents as an important risk factor for the development of childhood anxiety, with an abundance of studies investigating the potential mechanisms through which anxiety might be transmitted. There are a number of parental factors to consider, together with parent and child gender, which has sometimes been shown to moderate effects (e.g., Gerull & Rapee, 2002; Krohne & Hock, 1991). In the sections that follow we return to the literature on mathematics anxiety, focusing specifically on parental influences that can be plausibly linked to the development of mathematics anxiety.

DO PARENTAL INFLUENCES PLAY A ROLE IN THE DEVELOPMENT OF MATHEMATICS ANXIETY?

Parents are often cited as a possible cause of children's mathematics anxiety, yet few studies have directly tested this hypothesis. Gunderson, Ramirez, Levine, and Beilock (2012) suggest that there are two key ways in which parents may influence their children's developing attitudes (such as anxiety, self-efficacy, and self-concept) toward mathematics. One way is through their expectations and beliefs about their child's competence in mathematics. Another way is through their own attitudes toward mathematics. In other words, parents may function both as "expectancy socializers" and as "role models" (Parsons, Adler, & Kaczala, 1982). This draws on the literature on indirect learning mechanisms (discussed in the *Learning Mechanisms* section).

Parents as Expectancy Socializers

In support of the expectancy socializers account, Parsons et al. (1982) showed that parents' expectations of their child's mathematics ability predicted children's expectations and self-concepts, more so than children's previous mathematics performance (see also Jacobs, 1991; Jayaratne, 1987). More recently, in a study with younger children, Vukovic, Roberts, and Green Wright (2013) demonstrated a relationship between parents' expectations (and levels of home support) and children's mathematics anxiety. Interestingly, they found that parents' expectations had a positive effect on children's mathematical problem solving performance by reducing children's levels of anxiety.

Within this body of research there has been a large emphasis on gender issues. Studies have shown that parents' expectations of children's mathematics ability are often higher for boys than they are for girls, even when boys' and girls' levels of mathematics achievement do not differ (Eccles, Jacobs, & Harold, 1990; Yee & Eccles, 1988). These gender-biased expectations have been found for preschoolers as well as school-aged children (Blevins-Knabe & Musun-Miller, 1991), and they are greater for parents who hold stronger "mathematics = male" stereotypes (Jacobs, 1991). Note that they are still documented today, and not just in the older research literature (Stoet, Bailey, Moore, & Geary, 2016). In a recent analysis of data from the Programme for International Student Assessment (PISA) Study, Stoet et al. (2016) found that parents valued mathematical competence more in their sons than their daughters.

Parents as Role Models

In terms of the role models account, there is little research examining the relationship between parents' attitudes toward mathematics and their children's attitudes toward mathematics (Gunderson et al., 2012). Thus far, studies looking at the adult-child transmission of mathematics attitudes have tended to focus on teachers (e.g., Beilock, Gunderson, Ramirez, & Levine, 2010; Midgley, Feldlaufer, & Eccles, 1989), and from the recent studies that have focused on parents, the evidence is mixed.

Jameson (2014) looked at environmental factors relating to mathematics anxiety in second grade children (aged 7–9 years) and found no significant association between parents' mathematics anxiety and their children's mathematics anxiety or engagement in mathematics-related activities. In contrast, Maloney, Ramirez, Gunderson, Levine, and Beilock (2015) demonstrated that parents' mathematics anxiety predicted children's mathematics achievement and mathematics anxiety over the course of a school year, but only for children whose parents were frequently involved in helping with mathematics homework. The first- and second-grade children in this study were of a similar age to those in Jameson's study ($M = 7.21$ years, $SD = 0.62$) so the differences found cannot be accounted for by age or schooling. They may be accounted for by the mediating effect of parental homework involvement, or perhaps even the gender distribution of the samples. In an earlier study of 5th- to 12th-grade students, Jayaratne found that parents' attitudes toward mathematics were positively correlated with children's attitudes, but these correlations were only significant between mothers and daughters. Similar results have been observed in the general anxiety literature; for example, Adams and Sarason (1963) found that the strongest association between parent-child anxiety was for girls and their mothers.

These gender differences in the transmission of attitudes and anxieties have also been observed in studies looking at the influence of teachers. Beilock et al. (2010) demonstrated that female teachers' mathematics anxiety was related to girls' (but not boys') mathematics achievement and endorsement of mathematics—gender stereotypes. This may reflect the fact that children are more likely to emulate same-sex models than opposite-sex ones (Bussey & Bandura, 1984; Perry & Bussey, 1979). Further research is needed with male teachers and fathers to determine whether this is the case.

Other Parental Influences

Aside from acting as expectancy socializers and role models, parents may influence their children's attitudes toward mathematics through other behaviors (e.g., overcontrol when helping with mathematics homework), learning mechanisms (e.g., positively reinforcing mathematics avoidance), or attachment patterns. Drawing on research from the general anxiety literature, Bosmans and De Smedt (2015) investigated the relationship between parent-child attachment patterns and children's mathematics anxiety. The results of this study revealed a link between insecure attachment and mathematics anxiety: children's attachment anxiety and attachment avoidance were both positively correlated with their mathematics anxiety. This was the first study to examine the role of attachment patterns in the development of mathematics anxiety, and indeed the results indicate that it may be an important factor to consider.

Summary

Although only a small number of studies have investigated parental influences on children's mathematics anxiety, there is reason to propose that they may play an important role. There is some evidence to suggest that parents act as "expectancy socializers," while less is known about their function as "role models." We sought to address this gap by conducting a cross-sectional study of the relationship between parents' and children's mathematics anxiety. Given previous reports of gender differences in this domain, we ran separate correlations for boys and girls.

AN INVESTIGATION OF PARENTS' AND CHILDREN'S MATHEMATICS ANXIETY

The aim of the study was to investigate the relationship between mathematics anxiety in primary school children and their parents. To achieve this aim, we asked children (aged 6–9 years) and parents to individually complete a self-report measure of mathematics anxiety.

Method

Thirty-eight child-parent dyads took part in the study. Children (26 girls) were aged 6.0–9.8 years ($M = 7.6$ years, SD = 1.1 years) and they came from a range of socioeconomic backgrounds, as measured by UK postcode. The gender distribution of child-parent dyads is presented in Table 1.

TABLE 1 The Gender Distribution (Frequency and Percentage) of the Child-Parent Dyads.

	Daughter	Son
Mother	25 (65%)	11 (29%)
Father	1 (3%)	1 (3%)

Child Mathematics Anxiety Questionnaire

Children were asked to complete the Child Mathematics Anxiety Questionnaire (CMAQ) developed by Ramirez et al. (2013) for children aged 5–9 years. Two adaptations were made: all American words were replaced with British equivalents, and a 5-point scale was used instead of a 16-point sliding scale. For each of eight items children were asked to rate how they feel (or would feel) in various mathematics-related situations, on a five-point smiley-face scale. Four of the items involved specific mathematical calculations: for example, "How do you feel when you have to solve 34–17?" The other four items involved specific scenarios in which a child might be confronted with mathematics: for example, "How do you feel when getting your mathematics book and seeing all of the numbers in it?" An example item is presented in Fig. 1 and the full questionnaire can be found in Appendix C of Batchelor (2014).

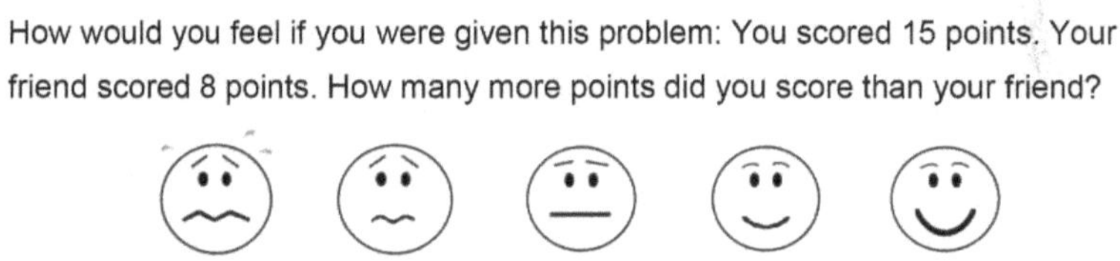

FIG. 1 An example item from the Child Mathematics Anxiety Questionnaire (CMAQ). Adapted from Ramirez, G., Gunderson, E. A., Levine, S. C., & Beilock, S. L. (2013). Math anxiety, working memory, and math achievement in early elementary school. *Journal of Cognition and Development, 14,* 187–202. doi: 10.1080/15248372.2012.664593.

For each item children received a score from 1 to 5 (where 1 = not anxious and 5 = very anxious). Total mathematics anxiety scores were computed by averaging across all items. The questionnaire showed good internal reliability, Cronbach's alpha = 0.81.

Parent Mathematics Anxiety Questionnaire

Parents were presented with the Mathematics Anxiety Scale-UK (MAS-UK) developed by Hunt, Clark-Carter, and Sheffield (2011). Because this 23-item questionnaire was originally designed and validated on British undergraduate students, some of the items were not relevant to this parent population. Therefore five items were removed from the scale.

For each of 18 items, parents were asked to rate how anxious they would feel in various mathematics-related situations on a 5-point scale, ranging from not at all to very much. There were items measuring the following three factors: mathematics evaluation anxiety (e.g., "having someone watch you multiply 12 × 23 on paper"), everyday/social mathematics anxiety (e.g., "working out how much your shopping bill comes to"), and mathematics observation anxiety (e.g., "listening to someone talk about mathematics"). The full questionnaire can be found in Appendix D of Batchelor (2014).

For each item parents received a score from 1 to 5 (where 1 = not anxious and 5 = very anxious). As with the child measure, total mathematics anxiety scores were computed by averaging across all items. The questionnaire showed good internal reliability overall, Cronbach's alpha = 0.89, and the following were scores for each of the three subscales: mathematics evaluation anxiety, Cronbach's alpha = 0.89; everyday/social mathematics anxiety, Cronbach's alpha = 0.81; and mathematics observation anxiety, Cronbach's alpha = 0.77.

Results

Overall, children showed individual differences in their reports of mathematics anxiety on the CMAQ ($M = 2.45$, SD = 0.87). Likewise, parents showed individual differences in their mathematics anxiety on the MAS-UK ($M = 1.78$, SD = 0.59). These individual differences are indicated by the means and standard deviations of scores on each of the five-point mathematics anxiety scales (the larger the standard deviation, the more the scores are dispersed around the mean, thus the greater the individual differences). The distributions of scores can be seen in the histograms in Fig. 2.

A one-way ANOVA was conducted to examine the effect of age on children's mathematics anxiety. This revealed no significant differences between the 6-, 7-, 8-, and 9-year-olds, $F < 1$. Further to this, an independent samples t-test was carried out to see whether there were any gender differences in children's mathematics anxiety. This demonstrated no significant

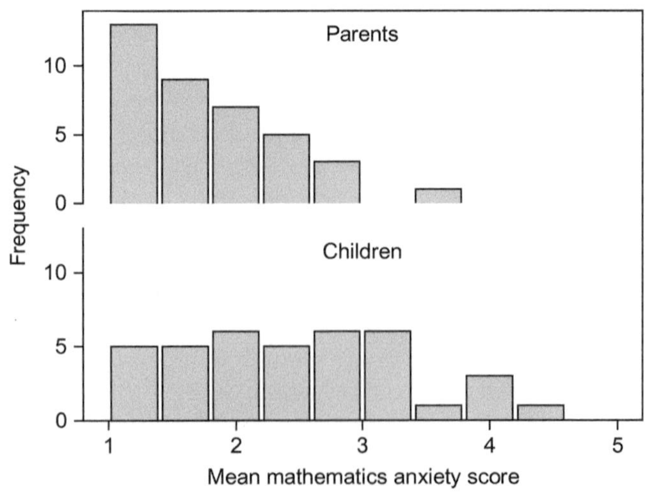

FIG. 2 Distribution of children's and parents' mathematics anxiety scores.

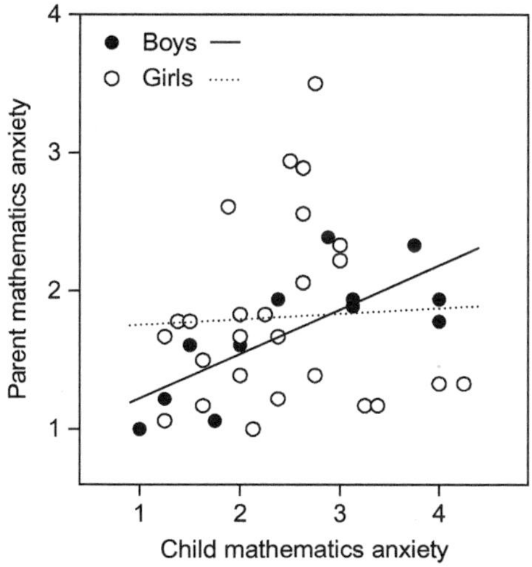

FIG. 3 Scatterplot depicting the relationship between children's and parents' mathematics anxiety scores.

differences between the boys' ($M=2.56$, SD=1.07) and girls' ($M=2.40$, SD=0.78) scores, $t(36)=0.53$, $P=0.598$. Given the lack of fathers who took part in the study ($N=2$), it was not possible to test for any gender differences in parents' mathematics anxiety.

Pearson correlations were calculated between parents' mathematics anxiety and sons' and daughters' mathematics anxiety. These revealed a significant positive association between the anxiety levels of parents and sons ($r=0.76$, $P=0.004$). In contrast there was no association between the anxiety levels of parents and daughters ($r=0.05$, $P=0.811$). As depicted in Fig. 3, the higher the mathematics anxiety in parents the higher the mathematics anxiety in sons, but not daughters. A Fisher's z test revealed that these correlations were significantly different, $z=2.39$, $P=0.017$. The 95% confidence intervals were [0.33, 0.93] and [−0.34, 0.43] for sons and daughters, respectively. We were unable to test for differences in terms of parents' gender; however, if the two fathers are excluded from the analyses then the results remained similar.[2]

Discussion

This study yielded three main results. First, in line with recent research, it was shown that mathematics anxiety is present in primary school

[2] There was a significant positive correlation between the anxiety levels of mothers and sons ($r=0.69$, $P=0.020$), but not mothers and daughters ($r=-0.02$, $P=0.915$). These correlations were significantly different, $z=2.09$, $P=0.037$.

children as young as 6 years (Ramirez et al., 2013; Wu et al., 2012; Young et al., 2012). The mean mathematics anxiety score was located around the midpoint of the five-point scale, with a wide spread of data points around the mean. Second, it was demonstrated that boys and girls show no significant differences in their anxiety. This adds to the mixed findings of gender differences in the literature (Devine, Fawcett, Szucs, & Dowker, 2012). Third, it was revealed that parents' mathematics anxiety is related to sons' mathematics anxiety but not daughters' mathematics anxiety. Below we interpret this result in view of the existing literature, considering possible reasons for this gender difference.

Only a handful studies have examined the adult-child transmission of mathematics anxiety and so far, the findings have been mixed. Jameson (2014) found no significant association between parents' and children's mathematics anxiety, whereas Maloney et al. (2015) found a significant association, but only for those children whose parents were frequently involved in helping with mathematics homework. Our finding that there was a gender difference in the association may not be at odds with these recent results: It is possible that we found a significant association for boys but not for girls because parents are more likely to be involved in mathematics-related activities with their sons than their daughters.

Note, however, that some studies—conducted with female teachers and mothers with older children—have found a gender difference in the opposite direction. In other words, it has been found that role modeling processes were stronger in girls (Beilock et al., 2010; Jayaratne, 1987). This contrasts with our finding, and given that the majority of parents in our study were mothers, the seemingly conflicting results cannot be reconciled by the fact that children are more apt to imitate same-sex models than opposite-sex ones (Bussey & Bandura, 1984).

How then can we account for these results? One possibility is that initially (in the early preschool and primary school years) parents engage in more mathematics-related activities with their sons than with their daughters. As a result, young boys may have more opportunities to pick up on their parents' mathematics anxiety than young girls. Later, as children get older and start to learn more complex mathematics in school, parents may become involved with girls' mathematics (e.g., helping with homework) as well as boys' mathematics, and the opportunity to pass on anxiety to daughters may arise.

In accordance with this hypothesis, there is evidence for varying gender differences in the provision of parents' mathematics-related activities for children of different ages (Chang, Sandhofer, & Brown, 2011; Jacobs & Bleeker, 2004; Simpkins et al., 2005). Chang et al. (2011) observed that mothers' number talk with their preschool children was significantly greater for boys than it was for girls; for example, mothers used number words with boys in approximately 10% of utterances compared to 5% of utterances with girls. In a study with older elementary school children, Jacobs and Bleeker (2004) found that while parents provided more mathematics-related toys to their sons, they were more involved with their daughters' mathematics homework.

As an alternative to this account, the gender differences in our study may stem from the way in which mathematics anxiety was measured. It is possible that boys and girls respond differently to self-report measures of mathematics anxiety: Girls may have been responding based on how they thought they should feel (e.g., in line with societal "mathematics = male" stereotypes), while boys were responding based on how they actually feel. Arguably, this would add more noise to the girls' mathematics anxiety scores, thus weakening the correlation between girls' mathematics anxiety and their parent's mathematics anxiety. In relation to this hypothesis, a recent study found that girls showed a discrepancy between their self-reported trait mathematics anxiety and their real-time state mathematics anxiety (Goetz, Bieg, Lüdtke, Pekrun, & Hall, 2013). Further studies employing real-time, state-based measures of mathematics anxiety may help to explain the gender differences observed in our study. More generally, state-based measures may provide a useful tool for investigating the causal mechanisms of children's mathematics anxiety.

Summary of Findings

This study investigated the role of parents in the development of children's anxiety toward mathematics. Interestingly, the findings indicate that gender impacts parental influence. Parents' mathematics anxiety was significantly associated with sons' mathematics anxiety, but not daughters' mathematics anxiety. This gender difference may reflect the fact that parents do more mathematics activities with boys and thus are more likely to transmit their mathematics anxiety to their sons than their daughters. Alternatively, it may be a result of measurement issues. Further investigations are needed to distinguish between these possibilities.

GENERAL DISCUSSION: KEY FINDINGS AND EMERGING QUESTIONS

Mathematics anxiety is a widely experienced phenomenon, estimated to affect about one in five people as they carry out numerical and mathematical tasks, both in school and in day-to-day life (Ashcraft et al., 2007). So far, research has uncovered a considerable amount about the consequences of mathematics anxiety, although it has only recently started to be investigated in young children. As more and more studies show the onset of mathematics anxiety in the early childhood years, there is greater emphasis on uncovering its origins or developmental roots: Why do some children become mathematics-anxious and others not?

Despite few empirical studies, there have been many anecdotal suggestions on the causes of mathematics anxiety—often highlighting environmental influences such as parents and teachers. In the current chapter we focused specifically on the possible role of parents. We drew on research from the general childhood anxiety literature before reviewing the empirical evidence on the relationship between parents' and children's mathematics anxiety. We conclude that there is some evidence for parental role modeling processes: parents' mathematics anxiety is related to children's mathematics anxiety, but this relationship may be moderated by gender and parental involvement.

This raises several interesting questions. First, why do we see these gender effects? Is it because parents are more involved in doing mathematics with their sons? Or, is it a result of the way girls respond to mathematics anxiety questionnaires? Further studies employing home numeracy assessments and different measures of mathematics anxiety would help us to dissect these possibilities. Second, it is important to ask how we might address the issue of parent-child transmission of mathematics anxiety. Are there any specific interventions that might help parents reduce the transmission of their mathematics anxiety? For example, should mathematics-anxious parents be encouraged to do less mathematics with their children? Can other adult role models (e.g., teachers) who are not mathematics-anxious provide a buffer against mathematics-anxious parents? It is likely that many environmental influences interact, but, to our knowledge, studies have not yet looked at these interactions. What happens if a child has a mathematics-anxious parent and a mathematics-anxious teacher? Are their outcomes multiplicatively worse? Finally, how do environmental influences interact with personal factors (e.g., lack of confidence) and cognitive dimensions (e.g., poor working memory)?

Overall, there is some evidence to support the anecdotal suggestions that parents play a role in shaping their children's attitudes and anxieties toward mathematics. These suggestions fit with the general anxiety literature that has identified parents as a key risk factor. In this general anxiety research, studies demonstrate a range of parenting influences, including parenting styles, attachment relationships, role modeling, information transfer, and reinforcement. It is possible that these findings can also be applied to the mathematics anxiety domain. Indeed, it may be useful to draw further on this research as we endeavor to uncover the developmental roots of children's mathematics anxiety.

References

Adams, E. B., & Sarason, I. G. (1963). Relation between anxiety in children and their parents. *Child Development*, *34*, 237–246. http://dx.doi.org/10.2307/1126844.

Ahmed, W., Minnaert, A., Kuyper, H., & van der Werf, G. (2012). Reciprocal relationships between math self-concept and math anxiety. *Learning and Individual Differences*, *22*, 385–389. http://dx.doi.org/10.1016/j.lindif.2011.12.004.

Artemenko, C., Daroczy, G., & Nuerk, H.-C. (2015). Neural correlates of math anxiety—An overview and implications. *Frontiers in Psychology*, *6*, 1333. http://dx.doi.org/10.3389/fpsyg.2015.01333.

Ashcraft, M. H., & Kirk, E. P. (2001). The relationships among working memory, math anxiety, and performance. *Journal of Experimental Psychology: General*, *130*, 224–237. http://dx.doi.org/10.1037/0096-3445.130.2.224.

Ashcraft, M. H., & Krause, J. A. (2007). Working memory, math performance, and math anxiety. *Psychonomic Bulletin & Review*, *14*, 243–248. http://dx.doi.org/10.3758/BF03194059.

Ashcraft, M. H., Krause, J. A., & Hopko, D. R. (2007). Is math anxiety a mathematical learning disability? In D. B. Berch & M. M. Mazzocco (Eds.), *Why is math so hard for some children. The nature and origins of mathematical learning difficulties and disabilities* (pp. 329–348). MD: Paul H. Brookes Publishing Co. Baltimore.

Ashcraft, M. H., & Ridley, K. S. (2005). Math anxiety and its cognitive consequences: A tutorial review. In J. I. D. Campbell (Ed.), *Handbook of mathematical cognition* (pp. 315–326). New York, NY: Psychology Press.

Bandura, A. (1986). *Social foundations of thought and action: A social cognitive theory*. Englewood Cliffs, NJ: Prentice-Hall.

Barrett, P. M., Rapee, R. M., Dadds, M. M., & Ryan, S. M. (1996). Family enhancement of cognitive style in anxious and aggressive children: Threat bias and the FEAR effect. *Journal of Abnormal Child Psychology*, *24*, 187–203. http://dx.doi.org/10.1007/BF01441484.

Batchelor, S. (2014). *Dispositional factors affecting children's early numerical development* [doctoral thesis]. Loughborough: Loughborough University. Retrieved from https://dspace.lboro.ac.uk/2134/17474.

Beidel, D. C., & Turner, S. M. (1998). *Shy children, phobic adults: Nature and treatment of social phobia*. Washington, DC: American Psychological Association.

Beilock, S. L., Gunderson, E. A., Ramirez, G., & Levine, S. C. (2010). Female teachers' math anxiety affects girls' math achievement. *Proceedings of the National Academy of Sciences of the United States of America*, *107*, 1860–1863. http://dx.doi.org/10.1073/pnas.0910967107.

Biederman, J., Hirshfeld-Becker, D. R., Rosenbaum, J. F., Hérot, C., Friedman, D., Snidman, N., et al. (2001). Further evidence of association between behavioral inhibition and social anxiety in children. *The American Journal of Psychiatry*, *158*, 1673–1679. http://dx.doi.org/10.1176/appi.ajp.158.10.1673.

Biederman, J., Rosenbaum, J. F., Bolduc-Murphy, E. A., Faraone, S. V., Chaloff, J., Hirshfeld, D. R., et al. (1993). A 3-year follow-up of children with and without behavioral inhibition. *Journal of the American Academy of Child & Adolescent Psychiatry*, *32*, 814–821. http://dx.doi.org/10.1097/00004583-199307000-00016.

Blevins-Knabe, B., & Musun-Miller, L. (1991). Parental beliefs about the development of preschool children's number skills. In *Paper presented at the biennial meeting of the society for research in child development, Seattle*.

Bosmans, G., & De Smedt, B. (2015). Insecure attachment is associated with math anxiety in middle childhood. *Frontiers in Psychology*, *6*, 1596. http://dx.doi.org/10.3389/fpsyg.2015.01596.

Bowlby, J. (1973). *Attachment and loss: Separation, anxiety and anger: Vol. 2*. New York, NY: Basic Books.

Brumariu, L. E., & Kerns, K. A. (2010). Mother–child attachment patterns and different types of anxiety symptoms: Is there specificity of relations? *Child Psychiatry & Human Development*, *41*, 663–674. http://dx.doi.org/10.1007/s10578-010-0195-0.

Brumariu, L. E., Kerns, K. A., & Seibert, A. (2012). Mother–child attachment, emotion regulation, and anxiety symptoms in middle childhood. *Personal Relationships*, *19*, 569–585. http://dx.doi.org/10.1111/j.1475-6811.2011.01379.x.

Bussey, K., & Bandura, A. (1984). Influence of gender constancy and social power on sex-linked modeling. *Journal of Personality and Social Psychology*, *47*, 1292–1302. http://dx.doi.org/10.1037/0022-3514.47.6.1292.

Carey, E., Hill, F., Devine, A., & Szücs, D. (2016). The chicken or the egg? The direction of the relationship between mathematics anxiety and mathematics performance. *Frontiers in Psychology*, *6*, 1987. http://dx.doi.org/10.3389/fpsyg.2015.01987.

Chang, A., Sandhofer, C. M., & Brown, C. S. (2011). Gender biases in early number exposure to preschool-aged children. *Journal of Language and Social Psychology*, *30*, 440–450. http://dx.doi.org/10.1177/0261927X11416207.

Cooper, S. E., & Robinson, D. A. (1991). The relationship of mathematics self-efficacy beliefs to mathematics anxiety and performance. *Measurement and Evaluation in Counseling and Development*, *24*, 4–12.

Devine, A., Fawcett, K., Szucs, D., & Dowker, A. (2012). Gender differences in mathematics anxiety and the relation to mathematics performance while controlling for test anxiety. *Behavioral and Brain Functions*, *8*, 2–9. http://dx.doi.org/10.1186/1744-9081-8-33.

Dietrich, J. F., Huber, S., Moeller, K., & Klein, E. (2015). The influence of math anxiety on symbolic and non-symbolic magnitude processing. *Frontiers in Psychology*, *6*, 1621. http://dx.doi.org/10.3389/fpsyg.2015.01621.

Dowker, A., Sarkar, A., & Looi, C. Y. (2016). Mathematics anxiety: What have we learned in 60 years? *Frontiers in Psychology*, *7*, 1–16. http://dx.doi.org/10.3389/fpsyg.2016.00508.

Dreger, R. M., & Aiken, L. R. (1957). The identification of number anxiety in a college population. *Journal of Educational Psychology*, *48*, 344–351. http://dx.doi.org/10.1037/h0045894.

Eccles, J. S., Jacobs, J. E., & Harold, R. D. (1990). Gender role stereotypes, expectancy effects, and parents' socialization of gender differences. *Journal of Social Issues*, *46*, 183–201. http://dx.doi.org/10.1111/j.1540-4560.1990.tb01929.x.

Eden, C., Heine, A., & Jacobs, A. M. (2013). Mathematics anxiety and its development in the course of formal schooling—A review. *Psychology*, *4*, 27–35. http://dx.doi.org/10.4236/psych.2013.46A2005.

Faust, M. W. (1992). *Analysis of physiological reactivity in mathematics anxiety* [unpublished doctoral dissertation]. Ohio: Bowling Green State University.

Ferla, J., Valcke, M., & Cai, Y. (2009). Academic self-efficacy and academic self-concept: Reconsidering structural relationships. *Learning and Individual Differences*, *19*, 499–505. http://dx.doi.org/10.1016/j.lindif.2009.05.004.

Field, A. P., Argyris, N. G., & Knowles, K. A. (2001). Who's afraid of the big bad wolf: A prospective paradigm to test Rachman's indirect pathways in children. *Behaviour Research and Therapy*, *39*, 1259–1276. http://dx.doi.org/10.1016/S0005-7967(00)00080-2.

Field, A. P., & Lawson, S. J. (2003). Fear information and the development of fears during childhood: Effects on implicit fear responses and behavioral avoidance. *Behaviour Research and Therapy*, *41*, 1277–1293. http://dx.doi.org/10.1016/S0005-7967(03)00034-2.

Fisak, B., Jr., & Grills-Taquechel, A. E. (2007). Parental modeling, reinforcement, and information transfer: Risk factors in the development of child anxiety? *Clinical Child and Family Psychology Review*, *10*, 213–231. http://dx.doi.org/10.1007/s10567-007-0020-x.

Gerull, F. C., & Rapee, R. M. (2002). Mother knows best: Effects of maternal modeling on the acquisition of fear and avoidance behavior in toddlers. *Behaviour Research and Therapy*, *40*, 279–287. http://dx.doi.org/10.1016/S0005-7967(01)00013-4.

Godbey, C. (1997). *Mathematics anxiety and the underprepared student* [Tech. rep. no. JC 990 063]. Murfreesboro, TN: Middle Tennessee State University.

Goetz, T., Bieg, M., Lüdtke, O., Pekrun, R., & Hall, N. C. (2013). Do girls really experience more anxiety in mathematics? *Psychological Science, 24,* 2079–2087. http://dx.doi.org/10.1177/0956797613486989.

Gough, M. F. (1954). Mathemaphobia: Causes and treatments. *Clearing House, 28,* 290–294.

Greenwood, J. (1984). Soundoff: My anxieties about math anxiety. *Mathematics Teacher, 77,* 662–663.

Gunderson, E. A., Ramirez, G., Levine, S. C., & Beilock, S. L. (2012). The role of parents and teachers in the development of gender-related math attitudes. *Sex Roles, 66,* 153–166. http://dx.doi.org/10.1007/s11199-011-9996-2.

Haase, V. G., Júlio-Costa, A., Pinheiro-Chagas, P., Oliveira, L.D.F.S., Micheli, L. R., & Wood, G. (2012). Math self-assessment, but not negative feelings, predicts mathematics performance of elementary school children. *Child Development Research, 2012,* 982672. http://dx.doi.org/10.1155/2012/982672.

Harari, R. R., Vukovic, R. K., & Bailey, S. P. (2013). Mathematics anxiety in young children: An exploratory study. *The Journal of Experimental Education, 81,* 538–555. http://dx.doi.org/10.1080/00220973.2012.727888.

Hembree, R. (1990). The nature, effects, and relief of mathematics anxiety. *Journal for Research in Mathematics Education, 21,* 33–46. http://dx.doi.org/10.1111/1467-8721.00196.

Hopko, D. R., Ashcraft, M. H., Gute, J., Ruggiero, K. J., & Lewis, C. (1998). Mathematics anxiety and working memory: Support for the existence of a deficient inhibition mechanism. *Journal of Anxiety Disorders, 12,* 343–355. http://dx.doi.org/10.1016/S0887-6185(98)00019-X.

Hudson, J. L., & Rapee, R. M. (2001). Parent–child interactions and anxiety disorders: An observational study. *Behaviour Research and Therapy, 39,* 1411–1427. http://dx.doi.org/10.1016/S0005-7967(00)00107-8.

Hunt, T. E., Clark-Carter, D., & Sheffield, D. (2011). The development and part validation of a UK scale for mathematics anxiety. *Journal of Psychoeducational Assessment, 29,* 455–466. http://dx.doi.org/10.1177/0734282910392892.

Jacobs, J. E. (1991). Influence of gender stereotypes on parent and child mathematics attitudes. *Journal of Educational Psychology, 83,* 518–527. http://dx.doi.org/10.1037/0022-0663.83.4.518.

Jacobs, J. E., & Bleeker, M. M. (2004). Girls' and boys' developing interests in math and science: Do parents matter? *New Directions for Child and Adolescent Development, 106,* 5–21. http://dx.doi.org/10.1002/cd.113.

Jain, S., & Dowson, M. (2009). Mathematics anxiety as a function of multidimensional self-regulation and self-efficacy. *Contemporary Educational Psychology, 34,* 240–249. http://dx.doi.org/10.1016/j.cedpsych.2009.05.004.

Jameson, M. M. (2013). The development and validation of the children's anxiety in math scale. *Journal of Psychoeducational Assessment, 3,* 391–395. http://dx.doi.org/10.1177/0734282912470131.

Jameson, M. M. (2014). Contextual factors related to math anxiety in second-grade children. *The Journal of Experimental Education, 82,* 518–536. http://dx.doi.org/10.1080/00220973.2013.813367.

Jayaratne, T. E. (1987). The impact of mothers' math experiences on their daughters' attitudes toward math. In *Paper presented at the biennial meeting of the society for research in child development, Baltimore.*

Kagan, J. (1989). Temperamental contributions to social behavior. *American Psychologist, 44,* 668–674. http://dx.doi.org/10.1037/0003-066X.44.4.668.

Kagan, J., Reznick, J. S., Snidman, N., Gibbons, J., & Johnson, M. O. (1988). Childhood derivatives of inhibition and lack of inhibition to the unfamiliar. *Child Development, 59,* 1580–1589. http://dx.doi.org/10.2307/1130672.

Krinzinger, H., Kaufmann, L., & Willmes, K. (2009). Math anxiety and math ability in early primary school years. *Journal of Psychoeducational Assessment, 27*, 206–225. http://dx.doi.org/10.1177/0734282908330583.

Krohne, H. W., & Hock, M. (1991). Relationships between restrictive mother-child interactions and anxiety of the child. *Anxiety Research, 4*, 109–124. http://dx.doi.org/10.1080/08917779108248768.

Lang, P. J. (1968). Fear reduction and fear behavior: Problems in treating a construct. In J. M. Schlien (Ed.), *Research in psychotherapy. III* (pp. 90–102). Washington, DC: American Psychological Association.

Lee, J. (2009). Universals and specifics of math self-concept, math self-efficacy, and math anxiety across 41 PISA 2003 participating countries. *Learning and Individual Differences, 19*, 355–365. http://dx.doi.org/10.1016/j.lindif.2008.10.009.

Lindskog, M., Winman, A., & Poom, L. (2017). Individual differences in nonverbal number skills predict math anxiety. *Cognition, 159*, 156–162. http://dx.doi.org/10.1016/j.cognition.2016.11.014.

Ma, X. (1999). A meta-analysis of the relationship between anxiety towards mathematics and achievement in mathematics. *Journal for Research in Mathematics Education, 30*, 520–540. http://dx.doi.org/10.2307/749772.

Maloney, E. A., Ansari, D., & Fugelsang, J. A. (2011). The effect of mathematics anxiety on the processing of numerical magnitude. *The Quarterly Journal of Experimental Psychology, 64*, 10–16. http://dx.doi.org/10.1080/17470218.2010.533278.

Maloney, E. A., & Beilock, S. L. (2012). Math anxiety: Who has it, why it develops, and how to guard against it. *Trends in Cognitive Sciences, 16*, 404–406. http://dx.doi.org/10.1016/j.tics.2012.06.008.

Maloney, E. A., Ramirez, G., Gunderson, E. A., Levine, S. C., & Beilock, S. L. (2015). Intergenerational effects of parents' math anxiety on children's math achievement and anxiety. *Psychological Science, 26*, 1480–1488. http://dx.doi.org/10.1177/0956797615592630.

Maloney, E. A., Risko, E. F., Ansari, D., & Fugelsang, J. A. (2010). Mathematics anxiety affects counting but not subitizing during visual enumeration. *Cognition, 114*, 293–297. http://dx.doi.org/10.1016/j.cognition.2009.09.013.

McClure, E. B., Brennan, P. A., Hammen, C., & Le Brocque, R. M. (2001). Parental anxiety disorders, child anxiety disorders, and the perceived parent–child relationship in an Australian high-risk sample. *Journal of Abnormal Child Psychology, 29*, 1–10. http://dx.doi.org/10.1023/A:1005260311313.

McLeod, B. D., Wood, J. J., & Weisz, J. R. (2007). Examining the association between parenting and childhood anxiety: A meta-analysis. *Clinical Psychology Review, 27*, 155–172. http://dx.doi.org/10.1016/j.cpr.2006.09.002.

Meece, J. L., Wigfield, A., & Eccles, J. S. (1990). Predictors of math anxiety and its influence on young adolescents' course enrollment intentions and performance in mathematics. *Journal of Educational Psychology, 82*, 60–70. http://dx.doi.org/10.1037/0022-0663.82.1.60.

Midgley, C., Feldlaufer, H., & Eccles, J. S. (1989). Change in teacher efficacy and student self- and task-related beliefs in mathematics during the transition to junior high school. *Journal of Educational Psychology, 81*, 247–258. http://dx.doi.org/10.1037/0022-0663.81.2.247.

Mineka, S., Davidson, M., Cook, M., & Keir, R. (1984). Observational conditioning of snake fear in rhesus monkeys. *Journal of Abnormal Psychology, 93*, 355–372. http://dx.doi.org/10.1037/0021-843X.93.4.355.

Moore, P. S., Whaley, S. E., & Sigman, M. (2004). Interactions between mothers and children: Impacts of maternal and child anxiety. *Journal of Abnormal Psychology, 113*, 471–476. http://dx.doi.org/10.1037/0021-843X.113.3.471.

Murray, L., Creswell, C., & Cooper, P. J. (2009). The development of anxiety disorders in childhood: An integrative review. *Psychological Medicine, 39*, 1413–1423. http://dx.doi.org/10.1017/S0033291709005157.

Newstead, K. (1998). Aspects of children's mathematics anxiety. *Educational Studies in Mathematics, 36*, 53–71. http://dx.doi.org/10.1023/A:1003177809664.

Norwood, K. S. (1994). The effect of instructional approach on mathematics anxiety and achievement. *School Science and Mathematics, 94*, 248–254. http://dx.doi.org/10.1111/j.1949-8594.1994.tb15665.x.

Núñez-Peña, M. I., & Suárez-Pellicioni, M. (2014). Less precise representation of numerical magnitude in high math-anxious individuals: An ERP study of the size and distance effects. *Biological Psychology, 103*, 176–183. http://dx.doi.org/10.1016/j.biopsycho.2014.09.004.

Parsons, J. E., Adler, T. F., & Kaczala, C. M. (1982). Socialization of achievement attitudes and beliefs: Parental influences. *Child Development, 53*, 310–321. http://dx.doi.org/10.2307/1128973.

Perry, D. G., & Bussey, K. (1979). The social learning theory of sex differences: Imitation is alive and well. *Journal of Personality and Social Psychology, 37*, 1699–1712. http://dx.doi.org/10.1037/0022-3514.37.10.

Rachman, S. (1977). The conditioning theory of fear-acquisition: A critical examination. *Behaviour Research and Therapy, 15*, 375–387. http://dx.doi.org/10.1016/0005-7967(77)90041-9.

Ramirez, G., Gunderson, E. A., Levine, S. C., & Beilock, S. L. (2013). Math anxiety, working memory, and math achievement in early elementary school. *Journal of Cognition and Development, 14*, 187–202. http://dx.doi.org/10.1080/15248372.2012.664593.

Rapee, R. M. (1997). Potential role of childrearing practices in the development of anxiety and depression. *Clinical Psychology Review, 17*, 47–67. http://dx.doi.org/10.1016/S0272-7358(96)00040-2.

Rapee, R. M. (2002). The development and modification of temperamental risk for anxiety disorders: Prevention of a lifetime of anxiety? *Biological Psychiatry, 52*, 947–957. http://dx.doi.org/10.1016/S0006-3223(02)01572-X.

Richardson, F. C., & Suinn, R. M. (1972). The mathematics anxiety rating scale: Psychometric data. *Journal of Counseling Psychology, 19*, 551–554. http://dx.doi.org/10.1037/h0033456.

Rosenbaum, J. F., Biederman, J., Gersten, M., Hirshfeld, D. R., Meminger, S. R., Herman, J. B., et al. (1988). Behavioral inhibition in children of parents with panic disorder and agoraphobia: A controlled study. *Archives of General Psychiatry, 45*, 463–470. http://dx.doi.org/10.1001/archpsyc.1988.01800290083010.

Rosenbaum, J. F., Biederman, J., Hirshfeld, D. R., Bolduc, E. A., Faraone, S. V., Kagan, J., et al. (1991). Further evidence of an association between behavioral inhibition and anxiety disorders: Results from a family study of children from a nonclinical sample. *Journal of Psychiatric Research, 25*, 49–65. http://dx.doi.org/10.1016/0022-3956(91)90015-3.

Rubinsten, O., Eidlin, H., Wohl, H., & Akibli, O. (2015). Attentional bias in math anxiety. *Frontiers in Psychology, 6*, 1539. http://dx.doi.org/10.3389/fpsyg.2015.01539.

Rubinsten, O., & Tannock, R. (2010). Mathematics anxiety in children with developmental dyscalculia. *Behavioral and Brain Functions, 6*, 46. http://dx.doi.org/10.1186/1744-9081-6-46.

Simpkins, S. D., Davis-Kean, P. E., & Eccles, J. S. (2005). Parents' socializing behavior and children's participation in math, science, and computer out-of-school activities. *Applied Developmental Science, 9*, 14–30. http://dx.doi.org/10.1207/s1532480xads0901_3.

Stodolsky, S. S. (1985). Telling math: Origins of math aversion and anxiety. *Educational Psychologist, 20*, 125–133. http://dx.doi.org/10.1207/s15326985ep2003_2.

Stoet, G., Bailey, D. H., Moore, A. M., & Geary, D. C. (2016). Countries with higher levels of gender equality show larger national sex differences in mathematics anxiety and relatively lower parental mathematics valuation for girls. *PloS one, 11*. http://dx.doi.org/10.1371/journal.pone.0153857.

Thomas, G., & Dowker, A. (2000). Mathematics anxiety and related factors in young children. In *Paper presented at the british psychological society developmental section conference, Bristol*.

Turner, S. M., Beidel, D. C., & Costello, A. (1987). Psychopathology in the offspring of anxiety disorders patients. *Journal of Consulting and Clinical Psychology, 55*, 229–235. http://dx.doi.org/10.1037/0022-006X.55.2.229.

Vasey, M. W., & Dadds, M. R. (2001). An introduction to the developmental psychopathology of anxiety. In M. W. Vasey & M. R. Dadds (Eds.), *The developmental psychopathology of anxiety* (pp. 3–26). Oxford: Oxford University Press.

Vukovic, R. K., Kieffer, M. J., Bailey, S. P., & Harari, R. R. (2013). Mathematics anxiety in young children: Concurrent and longitudinal associations with mathematical performance. *Contemporary Educational Psychology, 38,* 1–10. http://dx.doi.org/10.1016/j.cedpsych.2012.09.001.

Vukovic, R. K., Roberts, S. O., & Green Wright, L. (2013). From parental involvement to children's mathematical performance: The role of mathematics anxiety. *Early Education and Development, 24,* 446–467. http://dx.doi.org/10.1080/10409289.2012.693430.

Watt, M. C., Stewart, S. H., & Cox, B. J. (1998). A retrospective study of the learning history origins of anxiety sensitivity. *Behaviour Research and Therapy, 36,* 505–525. http://dx.doi.org/10.1007/BF01441484.

Wood, J. J., McLeod, B. D., Sigman, M., Hwang, W. C., & Chu, B. C. (2003). Parenting and childhood anxiety: Theory, empirical findings, and future directions. *Journal of Child Psychology and Psychiatry, 44,* 134–151. http://dx.doi.org/10.1111/1469-7610.00106.

Woodruff-Borden, J., Morrow, C., Bourland, S., & Cambron, S. (2002). The behavior of anxious parents: Examining mechanisms of transmission of anxiety from parent to child. *Journal of Clinical Child and Adolescent Psychology, 31,* 364–374. http://dx.doi.org/10.1207/S15374424JCCP3103_08.

Wu, S. S., Barth, M., Amin, H., Malcarne, V., & Menon, V. (2012). Math anxiety in second and third graders and its relation to mathematics achievement. *Frontiers in Psychology, 3,* 162. http://dx.doi.org/10.3389/fpsyg.2012.00162.

Yee, D. K., & Eccles, J. S. (1988). Parent perceptions and attributions for children's math achievement. *Sex Roles, 19,* 317–333. http://dx.doi.org/10.1007/BF00289840.

Young, C. B., Wu, S. S., & Menon, V. (2012). The neurodevelopmental basis of math anxiety. *Psychological Science, 23,* 492–501. http://dx.doi.org/10.1177/0956797611429134.

SECTION IIID

MATHEMATICS TEACHERS

CHAPTER

13

"I Hate Maths": Changing Primary School Teachers' Relationship With Mathematics

Mike Askew[*,†], *Hamsa Venkat*[*]

[*]University of the Witwatersrand, Johannesburg, South Africa
[†]Monash University, Melbourne, VIC, Australia

We were taught with expertise, patience and no humiliation.

This quote comes from a teacher in reply to being asked in an evaluation of the Wits Maths Connect-Primary (WMC-P) project (a 10 year research and development project to improve primary mathematics) what it was that they would never forget. Although not addressed so overtly, many other comments from the evaluation alluded to being treated well, not subjected to being humiliated, and to moving from hating mathematics and its teaching to enjoying, and—in many cases—even loving mathematics and teaching it.

The quote above, and similar responses, linked with our observations of mathematics lessons, suggests a widespread sense (in South Africa at least) of teachers having experienced being taught mathematics in ways that involved belittling them, and with such histories carrying through into further belittling learners, thus perpetuating experiences of learning mathematics as involving being the receiver of humiliation. Our language is intentionally strong here because the emotional tone of such lessons is a strong one, but even when primary teachers' emotions relating to mathematics are less strongly felt or stated, studies have noted that there is little joy in the engagement with mathematics (Hodgen & Askew, 2007). As a discipline, mathematics is frequently described as a disengaged and disembodied body of knowledge to be memorized (see, e.g., Ernest, 1999), and, particularly in the context of South Africa, rarely is there a sense of

http://dx.doi.org/10.1016/B978-0-12-802218-4.00013-3

teachers being mathematical practitioners who guide and encourage learners into becoming mathematicians—little sense of mathematics teachers and learners as people who "do" mathematics in the way, for example, that music teachers perform and play music, thus inculcating learners into what it means to be musicians. In short, there is scant evidence of either teachers or learners being in any kind of intimate relationship with mathematics. Rather, the converse holds—the knower most often establishes an alienated relationship with mathematics. The teachers we work with have inherited, and through their teaching practices sustain, those forms of mathematics teaching and learning that contribute to such alienated relationships, an alienation that in turn creates and maintains of feelings of inadequacy, in both teachers and learners.

Our starting position is to point out multiple dichotomies inherent in the opening quote and brief discussion above in relation to the ground of studies in mathematics education. Talk of alienated relationships with mathematics points to a separation of the knower from knowledge, or more importantly, from knowing; in this instance the separation of the teacher from the mathematics—an idea that we flesh out in detail later in this chapter. Adding to this is allusion to the experience of teaching and learning occurring in a situated sociocultural context in which "expertise, patience, and no humiliation" are rarities in comparison to prior experiences of working with mathematics. Inherent in the quote is also the idea of the intellectual endeavors of mathematics teaching and learning as inseparably bound up with the emotional aspects of these endeavors—contrasting with the frequent separation of these aspects in studies of teaching and learning.

In this chapter we use ideas drawn from theories of constructionism (Harel & Papert, 1991; Harré, 1986) to argue that some dissolution of these dichotomies of knower and knowledge (knowing), cognition and emotion, and individual and sociocultural has been important within our attempts to change primary teachers' relationships with mathematics.

THE DICHOTOMIES: SOME LITERATURE AND THEORY

Knower-Knowledge Separation

As noted, our observations of many South African lessons and the ways that teachers enact mathematics within those lessons frequently suggest that teachers' relationships with mathematics are based in alienation. This is revealed in teaching forms and practices that position mathematics as "out there," as something external and distinct from the doer or learner of mathematics, rather than something that can be related to internally.

Exemplifying this notion of "separateness" are pedagogies that establish engagement with mathematics as working with "physical tools" rather than "psychological tools," in the sense used by Vygotsky (1978). For instance, evidence in South Africa points to classroom work in early arithmetic where teachers continually encourage learners to rely on counting objects—first fingers, then counters, and then tally marks (Schollar, 2008). Even when learners have begun to move from counting to using known or derived facts to calculate, we, and others, have witnessed many teachers directing these learners back into using counting strategies (Ensor et al., 2009). In later arithmetic, learners are introduced to the tool of a vertical algorithm, but learners' written records frequently show that this algorithm too is acted upon as a physical tool to the extent that the subcalculations required in working with digits in columns are still carried out by relying on tally marks, rather than being done mentally with learners displaying little sense of what the calculations might mean. A strong bifurcation is enacted and created that keeps the learner separate from the mathematics and—by implication—the knower separate from the knowledge, from becoming knowing (Holzman, 2009).

On the theoretical side, this dichotomy has been critiqued within both constructionist and new materialist conceptions. Harel and Papert (1991) in their development of the theory of constructionism noted that traditional epistemological approaches have tended to treat knowledge as "abstract, impersonal, and detached from the knower," arguing instead for "more personal, less detached knowledge" (p. 10). A more radical reworking of this position is exemplified by the embodied position taken up by de Freitas and Sinclair (2014) in their argument for the "materiality" of mathematical concepts in which the mathematical "body" is viewed as

> an assemblage of human and non-human mathematical concepts. The new materialism we propose aims to embrace the "body" of mathematics as that which forms an assemblage with the body of the mathematician, as well as the body of her tools/symbols/diagrams, in the "dance of agency" that makes up mathematical activity (p. 454)

Individual: Socio-Cultural Separation

Constructivist (as distinct from constructionist) positions frequently suggest a view of learning mathematics as an individual and solo endeavor (see, e.g., von Glasersfeld, 1995). Our position on doing and learning mathematics is that these are not solitary activities; they essentially involve acting within relationships, both with other persons but also with mathematics itself. We follow the distinction noted by Prawat and Peterson (1999) that while constructivist views attend to an individual's constructions, the "constructionist" position focuses on the cultural and social construction of knowledge and emotions. With the constructionist position, mathematics gains meaning and significance from the rules, processes, and expectations created and sustained by *the community of mathematicians* and in doing mathematics one enters into that community. Even if physically alone when doing mathematics, from a sociocultural perspective, the individual's engagement is with an on-going stream of activity that has been shaped by the ongoing community. In this view, the endeavor of mathematical activity is only possible because of the products and processes that have been historically shaped by the broader community of mathematicians. Tools are important products of this history, and meaningful problem solving involves working with tools, rather than through tools. To do this, one has to establish, even if not in conscious awareness, a relation with mathematics and its tools. In looking at mathematical activity it is thus not sufficient to look separately at the actions of a problem solver and at the mathematical tools being acted upon—a key third element is the relationship between the user and the tool, which, by implication reinforces the dissolving of the knower-knowledge dichotomy, the relationship between the knower and the mathematics.

Dissolving these two dichotomies—knower-knowledge and individual-sociocultural— continues, however, largely to leave aside the emotional aspects of teaching and learning. In the next section, we pay more extensive attention to the emotional-intellectual dichotomy, using literature and constructionist theory to point to what we see as shortcomings of more separatist views.

Emotional–Intellectual

Much work on teacher development focuses, rightly, on the intellectual, on the cognitive, knowledge-based side of mathematics teaching. However, just as Zembylas (2004) has argued in the case of science teachers that emotions play an important part in the nature of a teacher's self-esteem (toward teaching), so too we argue that primary mathematics teachers' emotions toward the discipline have to be considered as intrinsic to professional development. This is especially important in the context of South Africa, where, thanks both to reports of international comparisons and to results of national tests, teachers are constantly reminded of how poorly their learners are performing. Given the limitations identified in learners' levels of attainment in mathematics in South African primary schools (Department for Basic Education, 2011) and the evidence of the need to improve teachers' mathematical content, and pedagogic content, knowledge (Carnoy & Chisholm, 2008; Taylor & Taylor, 2013) it is tempting to use the precious time made available for professional development to focus on the knowledge and cognitive projects. These elements have certainly been critical and important within our work, but broader evidence in the South African context has noted that beyond issues of what teachers are not able to (or "can't")do, there is also evidence of a more emotional reluctance, and perhaps even rejection, relating to some aspects of teaching (the "won't" do) (NEEDU, 2013, pp. 20–30). In our work, therefore, we include explicit attention to the emotional aspects of engaging with mathematics. As part of the WMC-P project, we built in a series of workshops—I Hate Maths (IHM)—with the deliberate intention of foregrounding the emotional aspects of engaging with mathematics.

Emotional engagement with mathematics, for us, involves entanglement, and entanglement in turn, necessitates a rejection of more dualist views of emotion—in which emotions are separate from the rational "knowing" of mathematics. Taken as separate, emotions may impact on the knowing, but they have to be addressed and managed in ways that keep emotions distinct from the "real" work of coming to know, since emotions are not seen as contributing to the work of coming to know; they only act in ways that inhibit or support it. This view popularly plays out in the often expressed desire of teachers to find ways to make mathematics "fun," ways that are most often realized by "wrapping" the (supposedly) unpleasant pill of mathematics inside some variant of a "spoonful of sugar" that makes it more palatable, and perhaps even hides the fact that mathematics is being worked on at all. Disguising mathematics will not, however, change participants' emotional relationships with the discipline because negativity is largely assumed in this kind of response.

Harré's (1986) social constructionism attends to emotions in ways that build on the dissolutions of the two earlier dichotomies discussed: knower/knowledge, individual/sociocultural. For Harré, emotions are culturally and social determined, through the different discourses that groups use to talk about emotions—a position suggesting that if these discourses can change over time and place, then so too can the associated emotions. This constructionist view of emotions points to a double signification. First, there is an improvizational aspect to emotions, in that they arise from interpretations of situations. Just as a Wittgensteinian view of discourse (Wittgenstein, 1958) positions talk as functioning within specific language games—that words attain meaning through their use—so too a social constructionist view of emotions posits emotions as arising through taking part in a culture's "emotional games." Second, emotions are not simply genetically determined; that is, they are not merely the responses of a physical body positioned out of the social and the cultural; the physical, social, and cultural are intertwined.

One implication of a constructionist stance on emotions is that some emotions acquire more significance than others within particular cultures through being deemed as "acceptable" when publicly articulated. With respect to mathematics, for example, it is commonplace in some cultures (e.g., in South Africa and the United Kingdom), for primary teachers to talk about how they have never liked mathematics, thus contributing to establishing and maintaining the view that disliking mathematics is "normal." Emotions are thus as much part of the local culture because they are the "property" of individuals.

Griffith (1998) identifies two models for the social construction of emotions: the social concept and the social role. The social concept model (Solomon, 2003) argues that manifesting an emotion is dependent on making judgments about situations, and the emotional categories into which situations are classified are culturally determined. So while a fear of snakes could be seen as an emotional response to a natural object, fear of mathematics is an emotional response to a cultural object. This is exemplified in adults' responses to mathematics. Many adults claim to be no good (which is as much an emotional response as a cognitive claim) at mathematics but when pressed reveal a considerable degree of mathematical competency. When this competency is pointed out them, they will argue that what they can do is not "real" mathematics, that real mathematics is all the stuff that they cannot do. It is not mathematics per se that is addressed here but the socially constructed categories of "my mathematics" and "real mathematics." Armon-Jones states that "emotions are constituted by nonnatural attitudes, these being acquired in, and explicable by reference to, specifically sociocultural contexts ... such attitudes and their external referents are either irreducibly, or significantly sociocultural in nature" (Armon-Jones, 1986, pp. 36–37).

While the social concept theory of emotions addresses the construction of categories of events that provoke emotions, the social role theory goes beyond examining how categories of events provoke emotions, and it theorizes how events set up the manifesting of emotions. Averill (1980, p. 312) points out that social constructionists taking up the social role position argue that emotion arises as a result of transitory social roles that include an individual's appraisal of the situation, which is interpreted as a passion rather than as an action. So while a social role is a culturally proscribed "normed" pattern of behavior, such as mother or manager, a transitory social role is taken up and enacted on an intermittent basis within particular social situations—for example, a strict teacher or kind stranger. Averill argues that, for example, "being loving" or "being afraid" are roles taken up in particular situations, roles that reflect a cultural ascribing of what is acceptable or appropriate in that situation. Rather than love or fear "naturally" or spontaneously arising, with the concomitant assumption that such emotions thus cannot easily be changed, according to Averill, emotions can be adopted, can be acted out: "having" an emotional response arises from the transitory social roles we take on. If a social role is taken on often enough, then this becomes a habituated acting-out of emotions that in turn results in the sense that that emotions have origins that are internal and individual, rather than social. Over time, individuals do not consciously produce an emotional response, but the public provides reinforcement for particular responses.

From this position, encouraging teachers to take on a different social role vis-a-vis their relationship with mathematics can support them enacting a different emotional response to the discipline; over time, through acting in accordance with these new social role expectations, the production of changed emotions becomes relatively automatic.

Shame and Shaming

This two-way movement between the social and the individual is particularly relevant to the notion of shame as an emotion that has been written about both generally, and with specific reference to primary mathematics teaching. From a constructionist perspective there is a dialectical relation here— shame is not simply provoked by being excluded from a group; the social experience of being excluded is what culturally comes to define, and subsequently provoke, shame. With respect to mathematics we argue that the experience of shame often gets transmuted into one of dislike, if not hate, for the discipline. If being exposed as inadequate is provoked through the context of doing mathematics, then it is emotionally safer to take mathematics is the root of the humiliation, not the cultural and social milieu that has set up the expectation that it is shameful either to get mathematics wrong or to not immediately understand it. Scheff (1994) notes there are "two movements that disguise emotion: denial of inner feeling and projection of it onto the outer world" (p. 43). Hence "I got it wrong, but it was the mathematics that caused it."

If, as Harré (1986) argues, emotions are identified by how an individual uses them in particular contexts, then shame, an emotion particularly relevant in discussing primary teachers' relationships with mathematics, is an emotion provoked by an individual's interpretation that the context is going to expose them in some way, whether or not that might be the case. Several researchers have argued that shame provokes both alienation and marginalizes individuals (for an overview of this see Bibby, 2002). Zembylas goes as far as to argue that "shame may be a mark of powerlessness in a pervasive sense of personal inadequacy" (2004), echoing Bibby's argument that shame is a key aspect of teachers' anxiety toward and fear of mathematics.

Bibby argues that three things principally engender feelings of shame vis-a-vis mathematics: talk and the nature of mathematical vocabulary, unsupportive and judgmental contexts, and the sorts of mathematical questions asked (Bibby, 2002, p. 713). Contrasting each of these with everyday contexts highlights key differences. Mathematical vocabulary (and in particular a perceived lack thereof) may be interpreted as a barrier to conversation, but contrast this with talking about problems in nonmathematical contexts where one is more likely to be met with empathy and support, rather than the judgmental reactions experienced in many mathematics lessons. Mathematical questions also pose threat in the way that they differ from everyday questions: in the mathematics classroom the inquirer usually knows the answer to the questions—they are in effect pseudo questions. Even compared to other "school" questions, mathematical questions make different, potentially shame-provoking, demands. As Bibby points out, asking a question like "what is 18 times 45" differs from a question like "what is the capital of France?" Getting the latter wrong is a failure of memory (assuming one once knew the answer), but in the mathematical case more than memory is being called into play: in carrying out the calculation "understanding and performance, and hence intelligence and status are also being questioned" (Bibby, 2002, p. 713).

THE IHM WORKSHOPS: EMOTIONS AT THE HEART OF PRINCIPLES AND PRACTICES

Our awareness of the dangers of the dualisms and dichotomies discussed above led to a conscious choice to establish a public workshop series entitled "I Hate Maths." This naming was not simply ironic, aimed at raising a wry grin. We wanted to openly acknowledge the strength of feelings of antipathy toward the discipline, both so that participants could be open about their feelings but also to help them recognize that they were far from alone in having those feelings and, through such public acknowledgement, work with emotions.

Our theoretical starting point in designing the IHM workshops was accepting and enacting Harré's argument that

> In the case of emotions, the overlay of cultural and linguistic factors on biology is so great that the physiological aspect of some emotional states has had to be relegated to a secondary status, as one among the effects of the more basic sociocultural phenomena. ***(Harré, 1986, p. 4)***

From a constructionist perspective, discourse is a primary influence on emotions and attending to emotion words and to available emotion vocabularies was central to the design of the workshops. By focusing on the local language made available within the workshops, together with what Harré refers to as the "local moral order"—in this case creating a safe environment where getting things wrong did not carry any moral implications—the workshops invoked and explicitly worked with the "two social matters that impinge heavily on the personal experience of emotion" (p. 9).

In the workshops we incorporated attention to issues of vocabulary, judgement, and mathematical questions by using tasks and associated activities that were deliberately framed in nontechnical language, where all offerings from the participants were accepted as valid forms of communicating. Activities were chosen so as not to pose questions with single correct or incorrect responses but with a focus on playing with ideas foregrounded. In particular, the workshop leaders worked at adopting a stance of not being at the center of judging what was right or wrong. In early workshops participants were frequently surprised that asking the facilitators if what they had done was correct was met with "what do you think?" Any questions raised were collectively shared and put back to the participants, rather than being judged by the facilitators.

The setup of the IHM workshops was designed deliberately to go further than simply providing experiences for the teachers that might unsettle their historical relationship with mathematics—experiences that might begin to dissolve the boundaries between the teacher as the knowing subject and mathematics as the knowledge object. Over and above this, the workshops were designed to explicitly introduce into the collective space the language of emotions and to encourage a different way of talking about relationships with mathematics that might, ultimately, change the nature of those relationships. Rather than treating emotions as a "reaction to" mathematics—a position that would have maintained the separation of mathematics as a trigger for emotions from the teacher who responds to the trigger—the discourse of the workshops was laden with the language of emotions and relationships. The emotional "tone" of the workshops was explicitly talked about, in contrast to the more usual focus on the "rational" processes of mathematics, such as justifying, generalizing, conjecturing, and so forth (which were not absent from the workshop, but were not allowed to crowd out attention on the emotional aspects).

Importantly, the style of the workshops was based around dialog rather than discussion (Bohm, 2004). A shift toward dialog extends beyond changing the nature of one's relationship with the content of the talk. It also engenders a change toward the "other," a shift from me and you to "I and thou" in Bohm's terms, in contrast to the emotional and moral tone of discussion that can often involve holding fast to one's opinion and is geared to "win over" the other rather than engage with them. As Robin Alexander points out, dialog is at the heart of developing caring teachers and learners:

> Dialogue requires willingness and skill to engage with minds, ideas and ways of thinking other than our own; it involves the ability to question, listen, reflect, reason, explain, speculate and explore ideas; to analyse problems, frame hypotheses, and develop solutions; ... Dialogue within the classroom lays the foundations not just of successful learning, but also of social cohesion, active citizenship and the good society. ***(Alexander, 2008, p. 122)***

While Alexander focuses on the relationships established with other participants in and through dialog, from our constructionist position, we consider dialog to also be central to changing relationships with mathematics as a means to invoking Harré's two social matters affecting the personal experience of emotion—the language used and the local moral order. The former was addressed in discussing the activities, initially focusing on how the participants felt about what they had done rather than thoughts on what they had learned. The latter was addressed by not only working with the language of positive emotions, but also explicitly acknowledging that no activity is going to be universally liked and that it was acceptable to voice dislike of an activity without fear of censure. We also followed de Freitas and Sinclair (2014) in taking a more dialogic approach to the discipline involving engaging with it in ways that are more organic. By this, we mean treating mathematics as emerging and growing out of human embodied activity, so that the learner is authoring the mathematics—the authority is local and personal, rather than mathematics as a platonic collection of immutable truths to be passed on by an external authority. One's relationship with mathematics can, through dialog, change, and move toward experiences similar to how mathematicians report their relationship with the discipline, although perhaps not as dramatically. Gauss, for example, reported being "seized" by mathematics and Bertrand Russell went as far as claiming "I had not imagined there was anything so delicious in the world. From that moment until I was 38, mathematics was my chief interest and my chief source of happiness" (Russell, 2000).

Lortie (1975) talked of the "psychic rewards" of teaching—the joys and satisfactions arising from caring for and working with learners, but alongside this we argue for the need to recognize the psychic rewards of engagement with mathematics.

The workshops were designed around broad themes such as games, generality, or pattern. The workshop would start with a brief introduction and overview but quickly would move into an interactive style of working. Even with audiences of up to 200 seating theater-style the emphasis was on paired and small group work. Simple resources were provided where necessary. For example, in the games workshop participants were provided with dice, counters, and digit cards. Individuals were given a handout to record some of outcomes of the games played, but even this was incorporated into an activity as the handout could be cut, folded, and turned into a booklet. Activities were chosen to be "low threshold" and "high ceiling"—that is, the mathematical knowledge needed to begin to engage with an activity was kept to a minimum, but the activities also had to have the potential to develop the mathematics by extending into structure and generality. One such example was the game: "say 10": working in pairs and starting from 1, players took it in turns to count on 1 or 2 more each time, the winner being the player to say 10: "1," "2," "3, 4," "5, 6," "7," "8," "9, 10, I win." From this low threshold start, raising the ceiling included looking at winning strategies, changing the rules ("Say 15"; count on 1, 2, or 3), and extending the game to involve fractional counts. Inevitably as the ceiling was raised, some participants began to feel uncomfortable, but rather than defuse the emotional tone, by facilitators "helping" participants to reengage with the mathematics, the dialog deliberately turned to explicitly noting the emotional tone and talking about why it had changed and what that might mean.

In order to further provoke a range of emotional responses, many of the IHM workshop activities had their origins in games originally designed for the drama classroom (see, e.g., Spolin, 1986)—games that we put a mathematical "spin" on. In some senses a drama lesson could be seen as the antithesis of a typical mathematics lesson in that drama is, necessarily, embodied, emotional, and engaging. But it was precisely these sorts of reactions that we sought to engender, in contrast to the stereotypical view of mathematics as disembodied, rational, and dull.

One activity, for example, involved working with a partner and reciting, in unison, a multiplication table and at the same time engaging with the partner in a pat-a-cake clapping action. The demand of enacting both the verbal and the physical actions simultaneously means that one or other of the tasks—reciting or clapping—soon breaks down. This "failure" is, however, accompanied not by a feeling of distress at getting it wrong but joyous laughter at the silliness of the two tasks, each simple on their own, undertaken together becoming so much more difficult than the sum of their parts. Space is thus opened up to talk about what has just happened, why a sense of failure in mathematics is more often accompanied by a sense of shame than one of exhilaration, and what the lessons are from an experience where failure is responded to differently. Again, this is a deliberate working with emotion, bringing into the open the two constructionist tenets of emotion language and local moral order. An important dialogic stance was to accept, honor, and work with negative as well as positive emotions, to help participants appreciate that whatever they felt and know they were not alone in having those feelings, and that it was okay and safe to express feelings.

Thus the IHM workshops were set up in ways to provoke emotional responses through doing mathematics that was different from what these teachers might have previously experienced.

TEACHERS' RESPONSES TO THE WMC-P PROFESSIONAL DEVELOPMENT

The IHM approach was, for us, part of a broader raft of interventions, some focusing more centrally on teachers' ways of "knowing" and "doing" mathematics. Across all of them, although most strongly marked in the IHM workshops, was attention to teachers' relationships with mathematics, rather than separating this from "mathematical knowledge" or "mathematical practices."

An evaluation at the end of 5 years included distributing a questionnaire to participating teachers. The questions asked on this schedule included items inquiring into teachers' perceptions of change relating to the WMC-P project, whether they viewed any changes produced as important, what they would have changed about the project, what they had learned that they would never forget, and what they had shared from their learning with others.

Responses were received from 120 teachers across the ten partner schools, with emotion-laden terms imbuing several of these responses (the opening quote is one such response). Words rooted in the terms "love," "enjoy," "comfortable," and "confident" occurred frequently across the responses, most often linked to describing a change in feelings about mathematics. Moreover, this change of feeling was often described in ways that were bound up with changes in understanding, backing our sense of the need to dissolve the emotional-intellectual separation:

> 'Before I used to hate Maths with all my heart, after I attended this course I really love Maths. I thank Maths Connect because it helped me to understand maths (Grade 2 teacher)
>
> Learners and educators have developed love for maths. As a teacher it has changed my subject from I hate Maths to I love Maths. I am now connected to maths particularly from grade 1 to 3 (Grade 2 teacher)
>
> What have changed is that I now have love for math and I enjoy teaching now more than before because the different strategies of teaching math that I was taught by Wits maths connect have made everything so simple to me in such a way that I now don't have that bad attitude about maths (Grade 1 teacher)

Several responses indicated a sense of changes in emotional relations resulting in changes to practice and classroom relations and thus what is regarded as mathematical success:

> I have learned that you must have love for maths, then you will enjoy doing it and do it practically its (sic) a way of doing it correct (Grade 2 teacher)
>
> That learning maths can be such fun and enjoyable if you have love for it and also if you are confident about what you are teaching (Grade 1 teacher)
>
> I say wow to maths Connect because they made me to love working with numbers with my children (Grade 3 teacher)

There was also a sense of this changed relationship with mathematics feeding through into learners' emotional responses to mathematics in classrooms:

> Most of the educators who attended Wits Maths Connect have confidence when teaching Maths. They don't hate it anymore. (National Test) results have improved which means our learners are now enjoying and understand better than before. (Grade 2 teacher)

Importantly there was also a sense of a changed willingness to "forward share" this change of sentiment with learners, teachers, and their own children as parents:

> The performances of learners in the class are improved. I've improved in teaching maths in the classroom and I can help my colleagues in how to prepare lessons of maths and motivate them to love maths (Grade 4 teacher)

Rather than being passive "deliverers" of the curriculum, teachers' expressed working with mathematics with a sense of agency and investment in this working:

> Mentally and physically. I changed dramatically because now I know that if they say count on, where must I start teaching my learners?—that I must also use other materials to help my learners to understand more, e.g. If I use the beads—I must also use the abacus and the number grid/board, to emphasise or make them understand. If they say more or less—how must I do my work? Physically! Oh yes my confidence is more important when teaching my learners, I've gained a lot (Grade 1 teacher)

A focus group was set up to explore some of these issues further. The following dialog transcribed from the group discussion reveals the teachers' awareness of the importance of engaging with the emotional aspects in mathematics lessons, although the final offering suggests that there is still some way to go in creating supportive whole class environments. The discussion followed the teachers playing a card game in pairs that required rapid multiplication of two numbers, a game that had provoked much hilarity for those present in the group.

MA: It's not often in mathematics lessons that I hear people laugh. So what is it about that activity that makes you laugh?

T1: You are laughing at yourself making mistakes. And you are laughing with the others, at everyone's mistakes

T2: It was a little exciting, that's why we were laughing.

MA: What made it exciting?

T3: Everyone had something to do. If children were playing they would be having something to do all the time.

MA: How do you think children would feel?

T4: It would energise them.

T5: It provides plenty of change. They would not be bored of the same thing. Not just copying from the board but working for themselves.

T6: The educational value of this, it's very educational.

MA: Why is it educational? Some people would say the children are just playing a game.

T6: It would engage them in thinking about multiplication, about pairs of numbers that multiply together.

(*Continued*)

T7: And I'm thinking, that I wasn't quick with the calculations so I need to go back and work on that. So the children would be seeing what they can do, thinking about if they can do and going and working on it if they couldn't. It would bring the child to thinking, "ok I need to do something about this.

MA: What was it about that activity that made you want to figure out what was going on?

T7: You realize, ok, I don't know this thing and I'm not playing as well, as somehow I want to be like the others. I'd pull aside, I'd have to train myself and then come back in. It will help the children realize I've got to master this too.

MA: When doing something with the whole class, does that same sense of I've got to master this too', come through?

T7: Some children, some will come and ask you.

T8: But in the whole class, some are too shy to ask you. In small groups they are more likely to say, ma'am I don't understand what to do. In the whole class, they won't ask because they are frightened of being laughed at. When it's just the little groups they have got the freedom to say I don't understand it to someone next to them, but when it's the whole class, that embarrassing.

CONCLUDING COMMENTS

Our primary focus in this chapter has been to describe the rationales for developing an intervention in which teachers' emotional relationships with mathematics were explicitly in the foreground, as well as sharing the IHM workshops' orientations toward tasks and activities. In our explicit move away from mathematics as rational, abstract, and cold toward mathematics as emotional, embodied, and warm, we set out to generate a more human mathematics, a mathematics toward which the teachers might sense a shift from the division of "me versus mathematics" to "the two of us—me and mathematics." In doing this, we found the constructionist emphasis on developing a language of emotions and explicitly attending to the local moral order helping in guiding our actions. Teacher responses to our insertion of the usually marginalized aspects of emotion and bodily sensations have been positive and similarly emotion-laden. Importantly for us, in centering on the emotional, the intellectual project has been enhanced and supported (see, e.g., Venkat, 2013; Venkat & Askew, 2012), rather than sidelined.

Our claim here is not one of causality for the impact of the IHM workshops. We do not yet know if the attitudinal and emotional changes that teachers articulated are either deep or long lasting. We do, however, claim that the workshops offered the opportunity for teachers to engage in forms of dialog about mathematics that differed from the norm. It differed both in focusing on the affective responses to the experience and in promoting a discourse of joy, pleasure, and curiosity in contrast to shame, fear and disengagement. Teachers' responses to this orientation suggest that opportunities to change their relationships with mathematics have been both welcome and generative. Thus we view the workshops as having the potential for valuing emotional change as both a worthy and attainable goal for mathematics education.

References

Alexander, R. (2008). Essays on pedagogy. *London*. New York, NY: Routledge.

Armon-Jones, C. (1986). The thesis of constructionism. In R. Harré (Ed.), *The social construction of emotions*. Oxford: Blackwell.

Averill, J. R. (1980). A constructivist view of emotion. In R. Plutchik & H. Kellerman (Eds.), *Emotion: Theory, research, and experience* (pp. 305–339). New York, NY: Academic Press.

Bibby, T. (2002). Shame: An emotional response to doing mathematics as an adult and a teacher. *British Educational Research Journal*, *28*(5), 705–721.

Bohm, D. (2004). *On dialogue*. New York/Oxford: Routledge Classics.

Carnoy, M., & Chisholm, L. (2008). *Towards understanding student academic achievement in South Africa: A pilot study of grade 6 mathematics lessons*. Pretoria: Spencer Foundation.

de Freitas, E., & Sinclair, N. (2014). *Mathematics and the body. Material entanglements in the classroom*. Cambridge: Cambridge University Press.

Department for Basic Education (DBE). (2011). *Report on the annual national assessments of 2011*. Pretoria: DBE.

Ensor, P., Hoadley, U., Jacklin, H., Kuhne, C., Schmitt, E., & Lombard, A. (2009). Specialising pedagogic text and time in foundation phase numeracy classrooms. *Journal of Education*, *47*, 5–30.

Ernest, P. (1999). Forms of knowledge in mathematics and mathematics education: Philosophical and rhetorical perspectives. *Educational Studies in Mathematics*, *38*, 67–83.

von Glasersfeld, E. (1995). *Radical constructivism: A way of knowing and learning*. London: Falmer Press.

Griffith, P. E. (1998). *What emotions really are: The problem of psychological categories*. Chicago: University of Chicago Press.

Harel, I., & Papert, S. (1991). *Constructionism*. New York, NY: Ablex Publishing.

Harré, R. (1986). An outline of the social constructionist viewpoint. In R. Harré (Ed.), *The social construction of emotions* (pp. 2–14). Oxford: Blackwell.

Hodgen, J., & Askew, M. (2007). Emotion, identity and teacher learning: Becoming a primary mathematics teacher. *Oxford Review of Education*, *33*, 469–487.

Holzman, L. (2009). *Vygotsky at work and play*. New York/Hove: Routledge.

Lortie, D. C. (1975). *Schoolteacher*. Chicago, IL: University of Chicago Press.

NEEDU. (2013). *National report 2012: The state of literacy teaching and learning in the foundation phase*. Pretoria: National Education Evaluation & Development Unit.

Prawat, R., & Peterson, P. L. (1999). Social constructivist views of learning. In J. Murphy & K. S. Louis (Eds.), 2. *Handbook of research on educational administration. A project of the amercian educational research association* (pp. 203–226). San Francisco, CA: Jossey Bass.

Russell, B. (2000). *The autobiography of Bertrand Russell*. New York/London: Routledge.

Scheff, T. J. (1994). *Bloody revenge: Emotions, nationalism and war*. Boulder, CO: Westview press.

Schollar, E. (2008). *Final report: The primary mathematics research project 2004–2007—Towards evidence-based educational development in South Africa.* Johannesburg: Eric Schollar & Associates.

Solomon, R. C. (2003). *What is an emotion? Classic and contemporary readings.* New York, NY: OUP.

Spolin, V. (1986). *Theater games for the classroom: A teacher's handbook.* Evanston, IL: Northwestern University Press.

Taylor, N., & Taylor, S. (2013). Teacher knowledge and professional habitus. In N. Taylor, S. van der Berg, & T. Mabogoane (Eds.), *Creating effective schools.* Pearson: Cape Town.

Venkat, H. (2013). Using temporal range to theorize early number teaching in South Africa. *For the Learning of Mathematics, 33*(2), 31–37.

Venkat, H., & Askew, M. (2012). Mediating early number learning: Specialising across teacher talk and tools? *Journal of Education, 56*, 67–90.

Vygotsky, L. S. (1978). *Mind in society.* Cambridge, MA: Harvard University Press.

Wittgenstein, L. (1958). *Philosophical investigations.* Oxford: Blackwell.

Zembylas, M. (2004). Emotional issues in teaching science: A case study of a teacher's views. *Research in Science Education, 34*(4), 343–364.

CHAPTER

14

Using Students' Emotional Experiences to Guide Task Design in Mathematics Content Courses

Kelli M. Slaten, Sarah E. Ives*†

*Georgia Gwinnett College, Lawrenceville, GA, United States
†California State University, Sacramento, CA, United States

Recent research concerning emotional factors and the learning of mathematics is often focused on developing appropriate theoretical approaches that connect the affective domain and the cognitive domain (e.g., Hannula, Evans, Phillippou, & Zan, 2004; Zan, Brown, Evans, & Hannula, 2006). We think it is important to also address how educators can integrate students' emotions and attitudes about mathematics into task design in order to capitalize on students as individual learners. Additionally, elementary teachers need to possess a robust understanding of mathematics (Ma, 1999). The experiences within their preparation content courses should include tasks designed to positively influence their attitudes/identity as well as deepen their mathematical knowledge. This chapter describes a study involving preservice elementary teachers' emotional experiences with mathematics and how those experiences informed our uses of writing assignments in content courses for education majors.

We approach the study of preservice elementary teachers' emotions toward mathematics from their self-reported emotions and attitudes. This kind of self-reporting and reflection provides an accurate characterization of their identities as learners and future teachers of mathematics. Although self-reflection is not directly observable (von Glasersfeld, 1987), mathematical understanding can be inferred from observations of students engaged in mathematical tasks, such as writing about mathematics.

Understanding Emotions in Mathematical Thinking and Learning
http://dx.doi.org/10.1016/B978-0-12-802218-4.00014-5

The tasks we report here involve writing explanations about mathematical concepts as well as describing emotions and attitudes associated with doing mathematics.

The research reported in this chapter adds to the development of practical strategies for addressing emotional issues in the learning of mathematics. We believe those involved in the design of instruction for student teachers and practicing teachers of mathematics will find this chapter useful.

BACKGROUND LITERATURE

Several themes are relevant to this study. We present the background literature on these themes as follows: first, there is a discussion of the research on emotions in learning mathematics, followed by literature on using writing in mathematics—as a way to (1) communicate mathematics, (2) learn mathematics, and (3) assess learning of mathematics; and finally, there is a focus on task design in mathematics education, specifically written mathematical biographies or "mathographies."

Emotion in Learning Mathematics

The subject of mathematics tends to elicit strong emotion—both positive and negative—in many people. "For many students studying mathematics in school, the beliefs or feelings that they carry away *about* the subject are at least as important as the knowledge they learn *of* the subject" (Philipp, 2007, p. 257). Skemp (1987) offered an explanation for why this is the case. First, the final criterion of any piece of mathematics is consistency: "In mathematics perhaps more than any other subject the learning process depends on agreement, and this agreement rests on pure reason" (p. 85). Second, when students are asked to learn mathematics through a series of memorized rules, the rules are void of reason, and thus this insults the students' intelligence. The learner has to expand or restructure personal schemas to assimilate meaninglessness. Third, "Viewed in this light, one can begin to see why some learners acquire not just a lack of enthusiasm for mathematics but a positive revulsion" (p. 85).

In general, researchers determined that emotions have their own memory pathways in the brain and serve as a critical source of information for learning (Lazarus, 1991; LeDoux, 1994). In the process of problem solving, Belavkin (2001) claimed that emotion always accompanies and makes a positive contribution to the process. Friesen and Francis-Poscente (2009) argued, "experiences that are charged with mathematical emotion are not some kind of extraneous distraction or curious side effect, but they are at the very core of involvement with mathematics" (p. 166). Appropriate emotional conflict can positively affect psychological and learning functions. In fact, Thompson and Thompson (1989) claimed that frustration is required in developing an appreciation for problem solving.

Emotions can arise when students perform tasks involving writing about mathematics. Russek (1998) stated, "It [writing in mathematics] is used to open doors of communication with students who may have math anxiety or who have 'I hate math!' feelings, students who may have never really 'spoken' to their mathematics professor before" (p. 36). Writing gives students an opportunity to voice their emotions and possibly change from negative to positive, especially when writing tasks contribute to students' increased confidence in their own mathematical abilities.

Writing in Mathematics

The use of writing in mathematics is not new; however, the literature regarding this topic tends to be practitioner-based (e.g., Keith, 1990; Meier & Rishel, 1998; Russek, 1998). Several areas of research about writing in mathematics illustrate how it can be used in a variety of effective ways in the teaching and learning of mathematics. In this study, we focus on three ways: writing as communication, as a way to learn, and as a way to demonstrate knowledge.

Communication

Teachers of mathematics are required not only to understand mathematics, but also to be able to communicate it in ways students can understand. "Those who really understand mathematics are not common; those who can communicate it, less so; those who are also excellent group leaders, fewer still; while those who can also communicate this last ability are rare indeed" (Skemp, 1987, p. 91). The development of communication skills is receiving significant attention in the teaching and learning of mathematics (e.g., Common Core State Standards Initiative (CCSSI), 2010; NCTM, 2000; Paige, 2009). Writing is an active learning strategy that engages students in learning. When students are required to write as part of their mathematics education experiences, the goal is usually to help students learn to *communicate* about mathematics and not just *do* mathematics (Borasi & Rose, 1989; NCTM, 2000; Powell & Lopez, 1989; Pugalee, 2001). When students write to communicate mathematically, they must think about mathematics in a new way. Writing further contributes to student learning because it actively involves the student in making connections between what is being learned and what is already known (Mayher, Lester, & Pradl, 1983).

Writing to Learn

Another example of an approach to writing about mathematics and relevant to this study is the approach known as *writing to learn*. This approach to writing is designed to allow students to actively build cognitive connections between what they understand and what they are learning (Bean, 2011; Mayher et al., 1983). The process of writing about mathematics serves as an effective tool for facilitating student learning. Writing about mathematics encourages students to organize the mathematics in their own terms, thereby supporting conceptual learning and retention (Keith, 1990).

Powell and Lopez (1989) included written summaries within a developmental mathematics college course and found students were better able to negotiate meaning through the use of writing. In a study of writing in mathematics, Grossman, Smith, and Miller (1993) made the assertion that "a student's ability to explain concepts in writing is related to the ability to comprehend and apply mathematical concepts" (p. 4). Therefore, the use of writing in mathematics is not just a viable learning activity but provides a means for instructors to assess the levels of understanding of their students. When used in conjunction with an understanding of students' emotional dispositions toward mathematics, students have further opportunities to make cognitive connections using tasks based on writing about mathematics rather than memorization of procedures.

Assess Learning

When writing to demonstrate knowledge, students show what they have learned by synthesizing information and explaining or applying their understanding of concepts and ideas. The use of writing prompts can serve as a formative assessment of understanding. Russek (1998) stated, "Writing is a valuable assessment tool. It is used to assess attitudes and beliefs, mathematics ability, and ability to express ideas clearly" (p. 36). Writing about mathematics can be an effective means of assessment of individual student learning and provide insight into what students understand beyond memorization of procedural skills. Furthermore, using writing as a diagnostic tool can help educators learn common student errors, curriculum decisions, levels of student understanding, and students' attitudes and beliefs about mathematics (Drake & Amspaugh, 1999). Clearly, writing about mathematics provides educators with rich information about student learning. Therefore, given that students' emotional experiences with mathematics can be accessed through their writing, it is imperative educators focus on their goals for using writing about mathematics via task design.

Integrating Writing in Task Design

Task design in mathematics content courses for future elementary teachers is a multifaceted endeavor. Tasks need to be high in cognitive demand, strengthen mathematical knowledge, and develop pedagogical content knowledge while also building the mathematical knowledge needed for teaching (Ball, Thames, & Phelps, 2008; Shulman, 1986; Smith & Stein, 1998). Learning tasks designed to develop mathematical understanding involve two critical cognitive processes: reflection and communication (Hiebert et al., 1997). Writing is a task requiring both thought processes of reflection and communication. These thought processes help students build cognitive connections among mathematical concepts.

A primary goal of our mathematics content courses for preservice teachers is to deepen their understanding of the mathematics content they will teach. Shulman (1987) termed this particular type of teacher knowledge *subject matter knowledge* or SMK. Much of teachers' SMK is obtained before college and is shaped by previous experiences (Ball & McDiarmid, 1990). Moreover, although preservice mathematics teachers take a substantial number of mathematics courses, this does not guarantee depth of SMK. In this study, we aimed to explore how writing about mathematics helps develop preservice elementary teachers' SMK while accounting for their emotions based on their prior experiences with learning mathematics and their concerns about future teaching of mathematics. One way to do this is through the use of a mathematical autobiography.

Mathematical Autobiographies

The "mathematical autobiography," also referred to as a "mathography," has been in use for decades, though little research has been done on the effects of its use. What has been written is mostly practitioner-based. For example, Russek (1998) described how she used this assignment with her undergraduate students. She used the prompt, "Write a 'mathography' in which you describe your feelings about and experiences in mathematics, both in and out of school. Include the completion of the statement: What I like most (or least) about math is…" (p. 37). This task opens communication between the instructor and student and gives students the opportunity to write about their feelings concerning a topic that can often be, to students, threatening and mysterious.

Another use of mathographies is described by McCulloch, Marshall, DeCuir-Gunby, and Caldwell (2013) in their professional development project with in-service elementary teachers. The teachers each wrote a mathematical autobiography storytelling their identity as learners of mathematics. "Mathematics autobiographies have the potential to help teachers reflect on their identities as mathematics learners and to understand their role in the development of their students' mathematics identities" (p. 380).

Hauk (2005) used mathographies in her college liberal arts mathematics courses to examine university students' K-12 mathematical experiences. She analyzed college students' mathographies and found that, through the process of writing mathographies, students' mathematical self-regulation increased. "Instead of seeking to rid students of their reactions to mathematics, a mathematical autobiography allows students to acknowledge these responses and build reflective awareness of them" (p. 49). Students were able to make self-evaluations of their mathematical ability, efficacy, and potential.

The literature regarding emotions relating to mathematical experiences and the use of writing in the learning of mathematics support our assertion that these should be integrated components of task design in content courses for preservice elementary teachers. Through attention to student emotions and deliberate use of writing to develop SMK, educators can help students as individual learners. In order to better understand the preservice elementary teachers in this study as individual learners, we employed a conceptual framework to analyze their emotional experiences as learners of mathematics and the specific referents associated with their emotions.

CONCEPTUAL FRAMEWORK

We consider two frameworks to serve as the conceptual basis for analysis of students' self-reported emotions and attitudes about mathematics. First, McCulloch et al. (2013), in describing teachers' identities as learners, analyzed written mathematics autobiographies to characterize teachers' experiences as mathematics learners. McCulloch et al. (2013) provided a framework for characterizing teachers' mathematical stories. Second, in his effort to conceptualize student attitudes' toward mathematics based on emotion and cognition, Hannula (2002) analyzed student attitudes categorized according to evaluative processes used by the learner. He extended Mandler's (1989) theory of emotion by providing an analytic framework to better understand student attitudes about mathematics.

Mathographies

We coded the preservice teachers' written mathographies based upon the framework provided by McCulloch et al. (2013). Their framework characterized six categories of teachers' experiences as learners of mathematics:

1. *Consistently frustrated*: the student had early negative experiences that were seen as indicative of a lifelong weakness in learning mathematics;
2. *Roller coaster*: the student had both positive and negative experiences in learning mathematics;
3. *Minor setback*: the student had a couple of negative experiences in learning mathematics, but these experiences did not hinder personal growth;

4. *Smooth track*: the student did not experience any negative events in learning mathematics;
5. *Negative turning point*: the student had positive experiences with learning mathematics until a specific event resulted in a more negative perspective of mathematics;
6. *Positive turning point*: the student had negative experiences with learning mathematics until a specific event resulted in a more positive perspective of mathematics.

Emotional Evaluative Processes

The analytic framework provided by Hannula (2002) is based on four categories of evaluative processes used by students when considering their emotions in the learning of mathematics. These four categories of describing attitude and their changes are (1) emotional disposition, (2) associations (e.g., What comes to mind from the word "mathematics"?), (3) expectations (e.g., Are you looking forward to a mathematics class?), and (4) personal values (e.g., Would you study mathematics if it were optional?). We have adapted these categories as follows:

1. *Emotions in situations*: specific emotional reactions based on prior experiences,
2. *Emotions associated with the stimuli*: general emotional dispositions about mathematics,
3. *Expected consequences*: emotional expectations of consequences when imagining a specific mathematical situation,
4. *Relating situation to personal values*: emotions about math in relation to the personal goal of becoming a teacher.

Our goals for using this conceptual framework for analysis were two-fold. We wanted to understand how the experiences of these preservice elementary teachers had affected their recalled emotions toward learning mathematics. We also wanted to better understand how specific emotions from their experiences were perceived, or evaluated, as addressed in their self-reflections. That is, what types of referents were associated with their emotions?

METHODOLOGY

Context and Participants

This study is a large-scale qualitative analysis of preservice elementary teachers' ($n = 473$) self-reported emotions and attitudes about their past experiences in mathematics learning as well as concerns about their future teaching. Analysis of the collected data began with a detailed investigation of individual students' written mathographies. We sought to find common themes of emotional experiences among a large group of preservice elementary teachers.

The majority of mathographies were collected from a Hispanic-serving institution in the Southern United States; these were collected over a span of seven years (2007 13). Written reflections about experiences with writing in a mathematics content course were also collected from several preservice elementary teachers from this institution ($n=10$). The challenges facing Hispanic preservice elementary teachers (e.g., a high percentage are nonnative English speakers, many are not full-time students, and some have financial and familial responsibilities) might have an influence on individuals' emotional experiences. While this is not the focus of this manuscript, we acknowledge that contextual factors might be worth considering in future studies (see Tillman, An, & Boren, 2013).

Additionally, several mathographies ($n=14$) were collected from another midsized university in the Southeastern United States along with their written reflections about their experiences with writing activities in a mathematics content course. The preservice elementary teachers at both institutions were required to complete both mathematics content courses and teaching methods courses.

Data Collection

The mathographies were assigned at the beginning of the preservice teachers' first mathematics content course as the first homework assignment. The assignment, shown below in Fig. 1, asks for the preservice elementary teachers' emotions about their experiences with mathematics in the past and their concerns about teaching mathematics in their future.

In addition to the mathographies, we collected and analyzed written reflections ($n=24$) at the end of the semester to assess how writing in a content course contributed to the preservice teachers' understanding of mathematics. These reflections were completed as an in-class "Free Write" task asking the preservice elementary teachers to reflect on their writing experiences within the content course as shown in Fig. 2:

Data Analysis

We used a narrative analysis approach (Hatch, 2002; Huberman & Miles, 2002; Polkinghorne, 1995) and QSR International's NVivo 10 qualitative data analysis software (NVivo qualitative data analysis software, 2014), to analyze the mathographies, within the conceptual framework

Mathematical Me

This journal assignment will allow you to reflect on your own experiences as a mathematician. Just as everyone is not a concert pianist, you may not be a concert mathematician—but you can produce more than mathematical chopsticks!

Please share your emotions, attitudes, and beliefs about mathematics and what your mathematical experiences mean to you, both for **learning** and **teaching** mathematics. Some ideas to get you started are:

- your mathematical strengths and weaknesses
- your successes and frustrations
- any concerns about this class or your future teaching mathematics
- your philosophy of what mathematics is, how it should be taught, and how it is best learned

Your journal should be at least **three pages** long, double-spaced, with **1-inch margins**. Do NOT add extra information at the top—just your name and the title of the homework.

FIG. 1 Mathematical Me assignment.

> In the space below, explain the ways the written work in this class has helped (or hindered) your understanding of the mathematics. Take a few minutes to think about it before you begin writing.
>
> Write a minimum 2 paragraphs. Use complete sentences. Use mathematical language where appropriate.

FIG. 2 Free write assignment given at the end of content course.

based on the work of Hannula (2002) and McCulloch et al. (2013), as described above. The use of a narrative analysis approach is based on the perspective that learners make sense of their experiences through personal stories (Hatch, 2002; Huberman & Miles, 2002; Polkinghorne, 1995). As researchers and educators, we can better understand individuals' personal experiences through this type of narrative form. It provides a principle way learners make sense of their experiences and give them meaning

(Huberman & Miles, 2002). The use of previously developed frameworks or categories is often the first step in a narrative analysis, followed by another round of analysis to identify relationships between and among these established categories (Polkinghorne, 1995).

Analysis of the data was conducted in three phases. In the first phase, all 473 mathographies were uploaded into NVivo and coded within the software program. This first phase of analysis involved coding each mathography in its entirety based on the McCulloch et al. (2013) six categories of preservice elementary teachers' experiences as learners of mathematics: (1) consistently frustrated, (2) roller coaster, (3) minor setback (4) smooth track, (5) negative turning point, and (6) positive turning point. Each mathography was assigned exclusively to one category. In order to check coding validity, once the first researcher coded the entire set of mathographies, the second researcher then coded a random set of 10% ($n=45$) of the total mathographies. The small number of disparities ($n=10$) were discussed and resolved before continuing to the second phase of analysis.

The second phase of analysis involved conducting word searches across all mathographies for uses of words referring to the basic emotions: *enjoy, feel, like, frustrate, love,* and *hate.* Each explicit use of these words was then coded using Hannula's (2002) four categories of evaluative processes: (1) emotions in situations, (2) emotions associated with stimuli, (3) expected consequences, and (4) relating situation to personal values. The coding of these categories was based strictly on explicit uses of these emotion words referring to emotions and feelings about their experiences with mathematics and their concerns about their futures' as teachers. Uses of these emotion words were not coded if used in reference to experiences not directly relating to mathematics. For example, the statement "A good teacher, in my opinion, makes learning fun, and makes you want to show up to class and makes the overall experience of school enjoyable" was not coded as a use of the word "enjoy." This statement references this preservice elementary teacher's belief about how a teacher could make learning enjoyable, not in reference to her personal emotions toward mathematics. A table of the frequencies of these emotion words is presented below in Table 1:

In this phase, we also read the 24 free write reflections in order to assess any themes concerning how the written assignments helped them better understand the mathematics content they learned in the course. These assignments were read to find any emergent themes about their learning and were also used as a reference for guiding the interpretation of the findings of the mathography analysis.

The third phase of coding utilized queries between the first two phases of coding. We explored relationships within the coded documents and passages by using NVivo to create documents for each of the six categories of experiences as learners along with the coded evaluative processes

TABLE 1 Frequency of Emotion Words Within Categories of Evaluative Processes

	Emotions in situation	Emotional disposition	Expected consequences	Personal goals
Enjoy	177	84	15	26
Feel	122	68	32	77
Frustrate	209	24	23	31
Hate	47	19	3	5
Like	185	78	16	30
Love	112	50	6	34

within each category. A sample document of the query "Consistently Frustrated vs. Personal Goals" is included as an Appendix. Several themes emerged from the third phase of analysis: for example, emotions about math were related to previous teachers or specific concepts, emotions based on grades or assessments, and emotions about their personal perspectives of mathematics. We are particularly interested in the themes relating to emotions about learning specific mathematical concepts and practices. This theme is important for our study because these specific concepts could serve as a basis for task design incorporating writing. We also found an important theme based on concerns about future teaching and the desire to not share negative emotions about mathematics with their students. As teacher educators, we believe this theme needs to be addressed during preservice teachers' educational experience in order to help them overcome their fears of teaching mathematics. Each of these themes provide rich areas for development of writing tasks for use within content courses to help develop the content knowledge of these future elementary teachers.

FINDINGS

In this section, we provide excerpts from the analysis of the mathographies and free write reflections, specifically relating to the themes of emotions about content and future teaching and the activity of writing about mathematics. These excerpts support our assertion that emotions and attitudes are important components of task design. We noticed many participants commented about the communication of mathematics, both regarding past experiences and concerns about future teaching. Therefore, we incorporated written assignments with a focus on explaining specific mathematical concepts in our content courses for preservice elementary teachers.

Themes From Mathographies

Each coded category of preservice elementary teachers' experiences as learners of mathematics were compared to the coded evaluative processes within each category. Within the six categories of experiences as learners, 18% ($n=85$) were coded "consistently frustrated," 34% ($n=161$) were coded "roller coaster," 24% ($n=112$) were coded "minor setback," 7% ($n=34$) were coded "smooth track," 11% ($n=51$) were coded "negative turning point," and 6% ($n=30$) were coded "positive turning point." Here we present excerpts from mathographies of preservice elementary teachers who are "consistently frustrated." This choice is based on the notion that these are the preservice elementary teachers who would benefit the most from effective task design involving writing about mathematics to strengthen their SMK.

Emotions About Specific Mathematical Concepts

The following excerpts are from the mathographies of preservice elementary teachers coded within the categories of experiences as learners of mathematics as "consistently frustrated" and within the categories of evaluative processes as "emotions in situations." The excerpts relate to specific mathematical content commonly taught in elementary grades. Although positive emotions were reported about experiences with specific mathematical content, we provide excerpts where negative emotions were reported. The negative emotions expressed about these mathematical concepts reveal areas rich for written assignments within content courses designed for preservice elementary teachers.

> ...I think that I really started disliking math when we had to memorize the "times tables" later on in elementary. I know I didn't love long division either...
>
> ...I, to this day, cannot tell you every multiplication answer. You ask me what 8x6 is I could not tell you unless I worked it out on a piece of paper or punched it in the calculator. Not knowing a simple thing like that makes me feel stupid...
>
> ...Here are some of my frustrations with math such as critical thinking problems, dividing fractions, graphing equations, histograms, permutations and combinations...
>
> ...My mom helped me growing up but it wasn't in a fun way it was more like making a times table chart and placing it on my room door and making a tape with the multiplications that I would have to listen to over and over again. You're probably thinking didn't that help? It did, but it wasn't fun and it made he hate math even more...
>
> ...Of course, surprise surprise, my biggest frustration was the "F" word, FRACTIONS...
>
> ...I have always hated math word problems they are the worst...
>
> ...My other gripe about math is fractions. Fractions really terrify and frustrate me...
>
> ...I was soon introduced to fractions and I guess this is where I began to hate math and not really care about it...
>
> ...We would have to recite certain multiplication facts to our teacher, but if we didn't know them all we got sent back to our seats and we didn't receive a stickers on

our stars hanging on the wall. I was embarrassed and frustrated that my friends got sticker and I didn't. After that I just believed that I would never learn math…

From these excerpts, we find many common mathematical concepts taught at the elementary level are associated with negative emotions. Concepts such as multiplication, division, and fractions were a source of frustration for these preservice elementary teachers.

Concerns About Future Teaching

The following excerpts are also from the mathographies of preservice elementary teachers coded as "consistently frustrated," but these excepts were coded within the categories of evaluative processes as "personal goals." The excerpts reveal their concerns about their future teaching practice and their hopes of not sharing their negative emotions about mathematics with their students.

…I want to learn how to like math so that when I'm actually teaching my students have no idea that for the longest time math was my least favorite subject…

…My concern for teaching math in the classroom, is how I am going to be able to teach math when I have a hard time myself understanding how to complete it myself? I know that I will be teaching young children, but if I have deep frustrations with mathematics, how will that not come out when I am trying to teach them and they do not understand how to do it either?…

…I have disliked math for so long and for so many reasons, but deep down I have a feeling that this will be the one subject I will teach with all my authority to keep my students from feeling the way I did…

…I would like this semester to be a positive venture, and that I will learn something which will change my fears to feelings of achievement instead. I want to be able to explain to my students what I want them to learn with a clear and good understanding so that I may convey this to them…

..Seeing as I don't have many strengths in math, I'm nervous about teaching my future students math. I never enjoyed math in from grade school to college, so I'm anxious about the task of making students excited about math, when I don't even know how that feels in the first place…

…I want to feel comfortable when I teach the subject of math…

…So I hope that I can be successful in helping children love math and maybe even I could learn to love the subject too…

…I really feel that knowing the fundamentals of math is a key element to being able to be an effective teacher, and that more than anything is my goal…

…As a teacher, I cannot show my own frustrations in math or my fear of math. I hope that I will find a way to love math that way I will have a positive influence on my own students…

These excerpts show although the preservice elementary teachers have negative emotional associations with particular mathematical concepts, they do not want their future students to have those same negative emotions as a result of their teaching. Additionally, we see that although these preservice elementary teachers have anxiety and express nervousness about teaching mathematics, they are hopeful that they can be successful.

Themes From Free Writes

Below we present excerpts from the free write assignment regarding how written explanations of content contributed to the preservice elementary teachers' improved communication and understanding of the content.

> ...Being able to write in math helps me understand the material better and it will only help me explain to my future students how to do the problems...
>
> ...Through writing I have also been able to better explain math. When teaching others it is very important to be clear, and through writing out explanations and definitions I believe I understand and comprehend what we learned more...
>
> ...The main thing the writing in this class has helped me with is communicating math concepts and being able to explain them well...
>
> ...Within this course being able to explain a math problem helped me understand it a lot better...
>
> ...In many ways writing out my work explaining how I figured out the problem has helped my understanding of mathematics...
>
> ...I think written work in this class has helped me understand math a lot better because I've never had to think this deeply about mathematical connections...
>
> ...As a teacher, we need to know how to interpret their explanations and the only way we are going to learn is practicing on our own...

These excerpts reveal the potential benefits of using written explanations about mathematical concepts in content courses for preservice teachers. For these preservice elementary teachers, they not only felt their SMK improved, but also that the task of writing explanations will help them become better teachers.

DISCUSSION

The research presented here makes both theoretical and methodological contributions to the existing literature on the role of emotion when educating future teachers to teach mathematics. By requiring preservice elementary teachers to write about their own experiences and emotions in relation to mathematics, they were better able to reflect on and make sense of these experiences and emotions. In addition, writing allowed researchers to see how cognitive and emotional aspects integrate in very concrete situations, such as those described in the identified themes: emotion about specific mathematical concepts and concerns about future teaching.

The integration of the frameworks developed by McChulloch et al. and Hannula makes a theoretical contribution to studying emotion in mathematics. The three phases of our analysis allowed us to first categorize the prospective teachers' experiences as learners of mathematics, and then we cross-analyzed these categories with the emotional evaluative processes developed by Hannula. This integrated framework gave us a way to interpret learners' emotional experiences as seen in their writing. We gained an understanding of our learners in a way so as to capitalize on these findings when designing tasks that, ideally, not only improve their emotions but also their SMK.

The emotions reported by the preservice elementary teachers revealed how their previous learning experiences shaped their attitudes toward mathematics, which provided us with valuable information about each student as an individual learner. The data reported in this study illustrate how those preservice elementary teachers who had early negative experiences with mathematics associate their negative emotions with the learning of specific concepts such as multiplication and fractions. They often have negative emotions associated with the memorization of mathematics. The data also reveal the concerns these preservice elementary teachers have about teaching mathematics. They are often worried about preventing negative emotions about mathematics in their future students. We speculate that our findings from analyzing the free write assignment indicated the task of explaining mathematical ideas may help preservice elementary teachers gain confidence in their own content knowledge.

A fundamental difficulty in learning to teach is how teachers' learn to use their SMK (Ball, 2000). Wilson, Shulman, and Richert (1987) found that "...teachers need more than a personal understanding of the subject matter they are expected to teach. They must also possess a specialized understanding of the subject matter, one that permits them to foster understanding in most of their students" (p. 104). What teachers know about their subject is not, by itself, sufficient for effective teaching; they must also find ways to communicate their knowledge to their students. Our study supports the need for preservice elementary teachers to develop communication skills about mathematics through the practice of writing about mathematics.

In her study of elementary teachers' SMK, Ma (1999) states that SMK is a result of connecting teachers' mathematical competence and their concern about teaching mathematics. We have shown that for many preservice elementary teachers, negative emotions from their early experiences as learners of mathematics also connect to their worries about their future teaching and not instilling negative emotions in their students. The use of written explanations of mathematical concepts helps develop SMK by addressing mathematical competence and increasing confidence in their own knowledge and their concerns about teaching mathematics.

Giving students the opportunity to write during the learning process affords educators additional opportunities to contribute to the development of their students' understanding. We believe teacher educators can readily incorporate the use of mathographies at the beginning of an academic term in order to assess the emotional experiences of their students in their learning of mathematics. Instruction can be tailored based on student responses in an effort to retain positive emotional reactions and to turn negative emotions toward more positive ones. Emotions bias the ways students cognitively process the learning of mathematics; therefore, their consideration should be an integral element of instructional planning and monitoring. Future studies are needed to understand how students' emotions about mathematics can be used in a wider variety of learning tasks to help students gain both knowledge of mathematics and confidence in themselves as mathematics learners and future teachers.

APPENDIX: NVIVO CODING QUERY FOR "CONSISTENTLY FRUSTRATED VS. PERSONAL GOALS"

Internals\\Mathographies\\ - § 1 reference coded [4.42% Coverage]
Reference 1 - 4.42% Coverage

As a future educator, I fear that I will place the same doubt and instability among my students because I do not possess a strong background in mathematics.

Internals\\Mathographies\\ - § 2 references coded [13.31% Coverage]
Reference 1 - 9.51% Coverage

My frustrations are that even though I got to where I am now I don't think that I've really learned much after high school; that's why through this class I hope to get rid of the math fear that I have and hope to learn new and interesting ways to teach children that math can be fun, as well as teaching strategies that parents can use at home with their children without torturing them.

Reference 2 - 3.80% Coverage

I want to learn how to like math so that when I'm actually teaching my students have no idea that for the longest time math was my least favorite subject.

Internals\\Mathographies\\ - § 1 reference coded [7.46% Coverage]
Reference 1 - 7.46% Coverage

My concern for teaching math in the classroom is how I am going to be able to teach math when I have a hard time myself understanding how to complete it myself? I know that I will be teaching young children, but if I have deep frustrations with mathematics, how will that not come out when I am trying to teach them and they do not understand how to do it either?

Internals\\Mathographies\\ - § 2 references coded [10.75% Coverage]
Reference 1 - 3.98% Coverage

Although, I now see that not having a goal added to my learning frustrations, I also attribute some frustration to not seeing the purpose of math as it pertained to my life.

Reference 2 - 6.78% Coverage

I'm hopeful that I will feel more confident in myself as an educator as I progress throughout the semesters and pray that I do not lose hope along the way. I feel that the frustrations that I have had will only help me by being relatable to kids who have their own apprehensions about the subject.

Internals\\Mathographies\\ - § 2 references coded [4.82% Coverage]
Reference 1 - 1.36% Coverage

I do not want to be that person who hates math and everything about it anymore.

Reference 2 - 3.46% Coverage

I have disliked math for so long and for so many reasons, but deep down I have a feeling that this will be the one subject I will teach with all my authority to keep my students from feeling the way I did.

Internals\\Mathographies\\ - § 1 reference coded [6.72% Coverage]
Reference 1 - 6.72% Coverage

I would like this semester to be a positive venture, and that I will learn something which will change my fears to feelings of achievement instead. I want to be able to explain to my students what I want them to learn with a clear and good understanding so that I may convey this to them.

Internals\\Mathographies\\ - § 1 reference coded [3.32% Coverage]
Reference 1 - 3.32% Coverage

I know with my degree plan it covers K through sixth and right now I do not feel very confident in teaching sixth graders algebra.

Internals\\Mathographies\\ - § 1 reference coded [8.36% Coverage]
Reference 1 - 8.36% Coverage

Seeing as I don't have many strengths in math, I'm nervous about teaching my future students math. I never enjoyed math from grade school to college, so I'm anxious about the task of making students excited about math, when I don't even know how that feels in the first place.

Internals\\Mathographies\\ - § 1 reference coded [3.36% Coverage]
Reference 1 - 3.36% Coverage

I feel like I have to do lots of planning and refreshing myself before I'm able to present to my class. There again all my effort and time into one subject. It's so frustrating.

Internals\\Mathographies\\ - § 1 reference coded [6.62% Coverage]
Reference 1 - 6.62% Coverage

I really don't worry about math because that is not something I'm planning on teaching, but if I do have to teach math I know that I could handle it because it would be in the elementary school level.

Internals\\Mathographies\\ - § 2 references coded
[4.00% Coverage]
Reference 1 - 2.45% Coverage

Myself as a mathematician is not as strong as I would like to be, being that I want to be an elementary teacher.

Reference 2 - 1.55% Coverage

I also would like to be better at math than the students I am teaching.

Internals\\Mathographies\\ - § 1 reference coded
[2.83% Coverage]
Reference 1 - 2.83% Coverage

That feeling, if only for a moment, restored faith in my math skills, and more than anything made me realize that I could teach, and that was all I wanted to do.

Internals\\Mathographies\\ - § 1 reference coded [1.09% Coverage]
Reference 1 - 1.09% Coverage

I want to feel comfortable when I teach the subject of math

Internals\\Mathographies\\ - § 1 reference coded [4.19% Coverage]
Reference 1 - 4.19% Coverage

I am looking forward to learning how to teach math to children because I predict that my feelings will change. It is an indescribable feeling when you instill knowledge into a child.

Internals\\Mathographies\\ - § 1 reference coded [2.58% Coverage]
Reference 1 - 2.58% Coverage

I hope that I can be successful in helping children love math and maybe even I could learn to love the subject too.

Internals\\Mathographies\\ - § 1 reference coded [5.09% Coverage]
Reference 1 - 5.09% Coverage

I have had some unsuccessful situations in explaining math to students, and I feel so ashamed of myself.

Internals\\Mathographies\\ - § 1 reference coded [3.73% Coverage]
Reference 1 - 3.73% Coverage

As a teacher, I cannot show my own frustrations in math or my fear of math. I hope that I will find a way to love math that way I will have a positive influence on my own students.

Internals\\Mathographies\\ - § 1 reference coded [3.18% Coverage]
Reference 1 - 3.18% Coverage

I really feel that knowing the fundamentals of math is a key element to being able to be an effective teacher, and that more than anything is my goal.

Internals\\Mathographies\\ - § 1 reference coded [4.44% Coverage]
Reference 1 - 4.44% Coverage

I feel confident in the fact that I will be able to empathize with any students having trouble with their mathematical assignments and help them narrow in on how they can become concert mathematicians.

Internals\\Mathographies\\ - § 1 reference coded [7.81% Coverage]
Reference 1 - 7.81% Coverage

I feel like my skills in math are lacking so severally that I may not have any real strengths in the subject; instead, only weaknesses. It is that sad reality that makes me fear my ability to teach math to younger children and why I was afraid to take the required math courses for my degree.

References

Ball, D. L. (2000). Bridging practices: Intertwining content and pedagogy in teaching and learning to teach. *Journal of Teacher Education, 51*(3), 241–247.

Ball, D. L., & McDiarmid, G. W. (1990). The subject-matter preparation of teachers. In W. R. Houston (Ed.), *Handbook of research on teacher education* (pp. 437–449). New York, NY: Macmillan.

Ball, D. L., Thames, M. H., & Phelps, G. (2008). Content knowledge for teaching: What makes it special? *Journal of Teacher Education, 59*, 5389–5407. http://dx.doi.org/10.1177/0022487108324554.

Bean, J. C. (2011). *Engaging ideas: The professor's guide to integrating writing, critical thinking, and active learning in the classroom*. San Francisco, CA: Jossey-Bass.

Belavkin, R. V. (2001). The role of emotion in problem solving. In *The AISB'01 symposium on emotion, Cognition and affective computing proceedings, March 21–24, Heslington, England* (pp. 49–57).

Borasi, R., & Rose, B. J. (1989). Journal writing and mathematics instruction. *Educational Studies in Mathematics, 20*, 347–365.

Common Core State Standards Initiative (CCSSI). (2010). *Common core state standards for mathematics*. Washington, DC: National Governors Association Center for Best Practices and the Council of Chief State School Officers. http://www.corestandards.org/assets/CCSSI_Math%20Standards.pdf.

Drake, B., & Amspaugh, L. (1999). What writing reveals in mathematics. *Focus on Learning Problems in Mathematics, 16*(3), 43–50.

Friesen, N., & Francis-Poscente, K. (2009). The Tower of Hanoi and the experience of lived number. In P. Lang (Ed.), *Re-thinking e-learning research: Foundations, methods, and practices* (pp. 149–170). New York, NY: Peter Lang Inc.

Grossman, F. J., Smith, B., & Miller, C. (1993). Did you say "write" in mathematics class? *Journal of Developmental Education, 22*(4), 2–6.

Hannula, M. S. (2002). Attitude towards mathematics: Emotions, expectations and values. *Educational Studies in Mathematics, 49*, 25–46.

Hannula, M., Evans, J., Phillippou, G., & Zan, R. (2004). Affect in mathematics education—Exploring theoretical frameworks. In *Proceedings of the 28th conference of the International Group for the Psychology of Education, Vol. 1* (pp. 107–136).

Hatch, J. (2002). *Doing qualitative research in education settings*. Albany, NY: State University of New York.

Hauk, S. (2005). Mathematical autobiography among college learners in the United States. *Adults Learning Mathematics: An International Journal, 1*(1), 36–56.

Hiebert, J., Carpenter, T., Fennema, E., Fuson, K., Wearne, D., Murray, H., et al. (1997). *Making sense: Teaching and learning mathematics with understanding*. Portsmouth, NH: Heinemann.

Huberman, A., & Miles, M. (2002). *The qualitative researcher's companion*. Thousand Oaks, CA: Sage Publications.

Keith, S. (1990). Exploring mathematics in writing. In P. Connolly & T. Vilardi (Eds.), *Writing to learn mathematics and science* (pp. 134–146). New York, NY: Teachers College Press.

Lazarus, R. (1991). *Emotion and adaption*. New York, NY: Oxford University Press.

LeDoux, J. E. (1994). Emotion, memory and the brain. *Scientific American, 270*(6), 50–57.

Ma, L. (1999). *Knowing and teaching elementary mathematics*. Mahwah, NJ: Lawrence Erlbaum.

Mandler, G. (1989). Affect and learning: Causes and consequences of emotional interactions. In D. B. McLeod & V. M. Adams (Eds.), *Affect and mathematical problem solving: A new perspective* (pp. 3–19). New York, NY: Springer-Verlag.

Mayher, J. S., Lester, N., & Pradl, G. M. (1983). *Learning to write/writing to learn*. Upper Montclair, NJ: Boyton/Cook.

McCulloch, A., Marshall, P., DeCuir-Gunby, J., & Caldwell, T. (2013). Math autobiographies: A window into teachers' identities as mathematics learners. *School Science and Mathematics, 113*(8), 380–389.

Meier, J., & Rishel, T. (1998). *Writing in the teaching and learning of mathematics*. MAA notes, 48 Washington, DC: Mathematical Association of America.

NCTM. (2000). *Principles and standards for school mathematics*. Reston, VA: National Council of Teachers of Mathematics.

NVivo qualitative data analysis software (2014). QSR International Pty Ltd. Version 10.

Paige, J. (2009). The 21st century skills movement. *Educational Leadership, 67*(1), 11.

Philipp, R. A. (2007). Mathematics teachers' beliefs and affect. In F. K. Lester (Ed.), *Second handbook of research on mathematics teaching and learning* (pp. 257–315). Reston, VA: National Council of Teachers of Mathematics.

Polkinghorne, D. E. (1995). Narrative configuration in qualitative analysis. *International Journal of Qualitative Studies in Education, 8*(1), 5–23.

Powell, A. B., & Lopez, J. A. (1989). Writing as a vehicle to learn mathematics: A case study. In P. Connolly & T. Vilardi (Eds.), *Writing to learn mathematics and science* (pp. 157–177). New York, NY: Teachers College Press.

Pugalee, D. K. (2001). Writing, mathematics, and metacognition: Looking for connections through students' work in mathematical problem solving. *School Science and Mathematics, 101*(5), 236–245.

Russek, B. (1998). Writing to learn mathematics. *Writing Across the Curriculum, 9*, 36–45. http://wac.colostate.edu/journal/vol9/russek.pdf.

Shulman, L. S. (1986). Those who understand: Knowledge growth in teaching. *Educational Researcher, 15*, 4–14.

Shulman, L. S. (1987). Knowledge and teaching: Foundations of the new reform. *Harvard Educational Review, 57*, 1–22.

Skemp, R. R. (1987). *The psychology of learning mathematics*. Hillsdale, NJ: Lawrence Erlbaum.

Smith, M., & Stein, M. (1998). Selecting and creating mathematical tasks: From research to practice. *Mathematics Teaching in the Middle School*, *3*(5), 344–350.

Thompson, A. G., & Thompson, P. W. (1989). Affect and problem solving in an elementary school mathematics classroom. In D. B. McLeod & V. M. Adams (Eds.), *Affect and mathematics problem solving: A new perspective* (pp. 162–176). New York, NY: Springer-Verlag.

Tillman, D. A., An, S. A., & Boren, R. L. (2013). Hispanic female preservice elementary teachers' mathematics teaching self-efficacies, attitudes, and student outcome expectations. *Journal of Mathematics Education*, *6*(2), 27–47.

von Glasersfeld, E. (1987). Learning as a constructivist activity. In C. Janvier (Ed.), *Problems of representation in the learning and teaching of mathematics* (pp. 3–18). Hillsdale, NJ: Lawrence Erlbaum.

Wilson, S. M., Shulman, L. S., & Richert, A. E. (1987). 150 different ways of knowing: Representations of knowledge in teaching. In J. Calderhead (Ed.), *Exploring teachers' thinking* (pp. 104–124). London: Cassell Education.

Zan, R., Brown, L., Evans, J., & Hannula, M. (2006). Affect in mathematics education: An introduction. *Educational Studies in Mathematics*, *63*, 113–121.

SECTION IV

THEORETICAL ADVANCES

CHAPTER

15

Digging Beneath Dual Systems Theory and the Bicameral Brain: Abductions About the Human Psyche From Experience in Mathematical Problem Solving

John Mason[*,†], *Martina Metz*[‡]

[*]University of Oxford, Oxford, United Kingdom

[†]Open University, Milton Keynes, United Kingdom

[‡]University of Calgary, Calgary, AB, Canada

INTRODUCTION

In this chapter, we use a phenomenological approach to explore the behavior of students exploring mathematics. At the core of our approach is an ancient psychological description of the human psyche, consisting of enaction, affect, cognition, attention, will, and witness. The notion of attention as being multiply structured (Mason, 1998; van Hiele, 1986) is part of the proposed elaboration. The ideas draw upon and develop the notion of frames of mind (Hudson, 1968; Minsky, 1986), which align with the notion of polyphrenia, otherwise known as multiple selves (Bennett, 1964), and microidentities (Varela, 1999). The stance taken develops the notion of *adherences and coordinations* among enaction, affect, cognition, attention, and will, observed by the witness. This, it is proposed, can be used to make useful and informative sense of experience of mathematical thinking and of working with others to help them to develop their mathematical thinking (see also Mason, 1998; Metz & Simmt, 2015).

The milieu within which mathematical thinking is initiated, sustained, and promoted is of course highly significant, and much discussed

http://dx.doi.org/10.1016/B978-0-12-802218-4.00015-7

currently in a vast literature that promotes discussion. Often, however, this is done without attending to the practices and ethos that make that social immersion effective. For example, on the one hand there is the notion of apprenticeship, illustrated by Lave (1988) observing Liberian tailors; and Seeley-Brown, Collins, and Duguid (1989) promoting a modern day version of apprenticeship. On the other hand there is the notion of structured group work as promoted by Boaler (2010) among others. We suggest that the practices identified in these various approaches work most effectively when they take into account the full human psyche as well as drawing upon didactic specifics of learning and doing mathematics (Mason, 2002a).

Our whole approach is informed by the use of the *discipline of noticing* (Mason, 2002a), which provides philosophically well-founded methods for enriching the noticing of opportunities to act freshly in the moment, or in other words, to learn from experience.

The chapter will show how automaticities (habitual actions that require no explicit in-the-moment conscious cognitive input) often have affective as well as enactive and cognitive components, how they are structured by the structure of attention, how they enable or are enabled by will, and how this is all informed by inner witnessing. It will suggest ways in which emotional reactions can be worked at and around, through the ways in which students begin to trust both their own mathematical powers and the teacher. Habits of mind, emotion, action, and attention will be contrasted with intuition and with experiencing the ineffable in mathematics, including how the ineffable can be brought to expression through entry into the implicit (Gendlin, 1978, 2009).

Trying to appreciate students' attempts at learning how to solve mathematical problems, by which we mean sophisticated exploration rather than routine exercises, we return to the theme of McLeod and Adams (1989), who in their ground-breaking book drew attention to the central role of emotions during mathematical problem solving. Our purpose is not simply to re-emphasize the important role that emotions play in mathematical thinking, but to integrate this into a more sophisticated view of the human psyche.

The thrust of our study concerns the combining of dual process theory as adumbrated by Kahneman (2012) and Kahneman and Frederick (2002) with a more comprehensive view of the human psyche involving six interacting components: enaction, affect, and cognition, together with attention, will, and witness. We include along the way the bicamerality of the human brain (Jaynes, 1976; McGilchrist, 2009) and the notion of micro-identities (Varela, 1999) or *coordinated adherences*, which some describe using the term *multiple selves* (Bennett, 1964; Minsky, 1986). We consider it vital not to try to isolate emotions or affect, but rather to integrate them into a complex comprehensive view of the psyche. As with the Sufi story of the blind men and the elephant (Shah, 1970; see also Wikipedia 1, n.d.), dual process theory, emotions, and cognition are but components of a complex phenomenon known as human beings.

We are interested in lived experience, and so our approach to research is to begin phenomenologically. Because we wish to act consistently with our principles, our approach to reporting research is again phenomenological. The data we offer is what comes to you as you read brief-but-vivid descriptions of incidents and accounts-of phenomena, rather than transcripts of incidents observed by researchers accounting for what is observed. This is part of the *discipline of noticing* (Mason, 2002a, 2002b). Thus the chapter begins with descriptions of some phenomena that we hope readers will recognize. The intention is that that experience goes beyond mere recognition by summoning up images, triggering recollection of past experience, and perhaps even then resonating with, or triggering access to, more recent experiences with similar qualities. In our accounts we use the personal pronoun as the events recounted are personal.

We then elaborate on the theories that we are combining. This provides a complex framework for considering in detail people's response to mathematical tasks in order to highlight but not overstress the emotional component of experiences of thinking mathematically. The complexity of the human psyche is not clarified or appreciated by isolating different components and trying to study them in isolation. Our claim is that by sensitizing oneself to the complexity of the human psyche, and by striving to notice various components, it is possible to be more in tune with and more aware of the lived experience, and this in turn enriches the ability to notice what may be happening for students, both in terms of appreciating potential stumbling blocks they may experience in a particular mathematical context and in terms of noticing (and helping them to notice) the way their attention is structured. In this way, we may better choose effective pedagogic actions rather than reacting in habitual ways to students who need help.

For example, when a student gives an incorrect answer to a question, describing the student as "unable" to answer the question correctly is a habituated reaction that ignores or overlooks the many reasons why a student might not think to say what the teacher wants to hear; or if a student gives a muddled or inarticulate response to a question that is not what the teacher had in mind, thinking that the student does not understand and turning to another student in search of a clearer response can become a habituated action, rather than, for example, treating the response as an attempt to articulate an almost inchoate sense and working with that student in order to clarify and streamline their narrative.

The chapter concludes with an extended description of work on a particular mathematical problem concerning goldfish, making use of our elaborated theories to make sense of and to draw lessons from that experience.

PHENOMENA

The first three phenomena are offered as access to experiences in which emotional energy is manifested physically in action and cognitively in thought.

Phenomenon 1: Distraction

> You hear a sudden sharp sound. Your attention immediately shifts to the possible source manifested as head or body turning and looking. There may even be some adrenalin flow, the body and cognition both being alert, and some emotion may be triggered. All previous thoughts are for the moment obliterated, and it may be difficult later to re-enter that previous state.

It is well known physiologically that under duress the body is the first to act. Emotions then start flowing, while cognition is last, constructing a narrative about what may or may not have happened (Kahneman, 2012; Mandler, 1989; McGilchrist, 2009).

Phenomenon 2: Blurting

> During a conversation a thought arises as to something you might contribute, but you reject it for some reason. Moments later during a lull in the conversation you suddenly find yourself blurting out the beginnings of that thought.

A closely allied phenomenon is *oar-wetting*: during a meeting, once you have managed to speak, speaking again (and again) requires much less effort. Those who speak first tend to dominate the whole meeting: once you get your "oar wet" it tends to stay wet!

Phenomenon 3 Instant Lesson

Context: several times, while visiting a school, it suddenly became clear as my host and I approached the classroom that I was to conduct the lesson. In some cases I was told the topic as I entered the room; in another, I was told in the corridor as we approached. In another case, the teacher did not show up, so two of us, intending to be observers, took over the lesson.

> Suddenly faced with a class of children and a topic, what happens? In each case I went into automatic mode, enacting the first actions that came to me.

In one case the topic was to be factoring quadratics, so I immediately asked the learners to expand some brackets with me, then invited them to consider the reverse process on a particular quadratic and announced that they would be able to deal with it by the end of the lesson. I then invited them to construct some examples of pairs of factors, to expand them, and to use this as fodder for working out how to go backwards. I suggested that being systematic in their construction of examples would make life easier!

In another case when the topic was parallel lines, I worked on mental imagery of a line with a copy being translated while the original and a transversal remained fixed. Together with the class I built up a complex diagram with several sets of parallel lines and transversals and tasked the students with finding which angles were equal. Finally they constructed examples and explored how few angles they could specify so that all the other angles could be determined.

Comment

The accounts concentrate on behavior (what I did) and its accompanying cognition (thoughts). Under the surface there are emotional energies flowing associated with adrenalin. Overall the moment of entering the classroom activated a particular set of actions, emotions, and thoughts that were connected or coordinated; in short, there were adherences that permitted energy to flow in certain directions but blocked out others. The situation was not sufficiently daunting to block all action, but the actions enacted were coordinated with and by past experiences of the topic and of pedagogic preparation. The fact that they are typical of how I am under tension in a classroom suggests that the various components adhere to each other or are, in some structural manner, coordinated. Getting learners to construct the examples on which they work (Watson & Mason, 2002) is one aspect of, whenever possible, getting learners to make significant choices in mathematics in order to increase their commitment and to complexify their relationship with mathematical objects while expanding and enriching their example spaces. Their emotions are brought into play through their commitment (will) and attention, because it is their own construction.

Phenomenon 4: Emotional "Portcullis"

> You offer a mathematical task to some learners and some or all immediately say, "I can't do this" or something similar. Pens are put down and action stops.
>
> You are offered a mathematical task and immediately emotions well up inside you and you experience a state of, "I can't do this".

These are both examples of a lack of any possible action surfacing and/or a welling up of negativity. This can be associated with lack of success in the past, with a current emotional state, including disposition and particular patterns of will-directed behavior, and even a sense of cognitive stress in the form of tiredness or lack of concentration. The habitual aspect is a re-awakening of emotional patterns of energy that then activate a restricted collection of actions and thoughts. It is as if a portcullis drops, separating you from the task. Because this state, at least in its negative form, is often aligned with a rising feeling of alienation (Boaler, 2000; Duffin & Simpson, 1993), of feeling remote from the task and indeed from the situation as a whole, we choose to label it as *re-emoting*, both because it refers to a surfacing of past emotional states and because of the contraction *remoting*, which sums up the usual effects.

Phenomenon 5: Sensing and Feeling

> You are working on a problem but for some reason have to put it to one side for a time. One day, perhaps in the shower or while walking or doing something else, an idea pops into your head. You feel inspired. You sense that this new idea might work, which you describe to someone else as a gut-feeling. You find your spirits rising, your energy beginning to flow in fresh channels, and you find much of your attention is now directed toward following the leads offered by those new channels, despite the fact that you were in the middle of doing something else (that you may now find difficult to attend to or even to recall).

Notice how the word "feeling" is associated with the body (enaction), with emotion and the flow of energy, and with cognition. The use of a single term for all three suggests that in order to make sense of experience it is necessary to make as much use as possible of a full and rich structure for the human psyche.

Comment

We think that the accounts of phenomena 4 and 5 show ways in which emotions (note the etymological connection with *motion*) provide the energy that enables actions to be enacted, moderated perhaps by intentions and cognition.

THEORETICAL FRAME

We begin with the difference between *reacting* to some stimulus and *responding* to it (Mason, 2009a, 2009b, 2011, 2015). This distinction lies at the heart of the dual systems theory (also known as two type theory), which was initiated by Stanovich and West (Kahneman, 2012, p. 48) and exploited and developed as a way of explaining a great deal of human behavior by Kahneman (2012; see also Kahneman & Frederick, 2005). The notion of dual systems has also been used to account for phenomena in mathematics education by Leron and Hazzan (2006; see also Leron, n.d.).

In the dual systems theory, System S1 is seen as reactive, consisting of actions that arise automatically from habit. It is quick to act. Kahneman and Frederick (2005, p. 273) use the following task to illustrate S1.

> A baseball bat and ball cost together one dollar and 10 cents. The bat costs one dollar more than the ball. How much does the ball cost?

They found that many people yield to the immediate impulse to say 10 cents. This may be partly due to the neat split between $1 and $10 cents, which triggers a reaction before cognition considers that action. That action then influences subsequent cognition. We have also noticed some people freezing, which is an emotional blockage arising when there are no (re) actions available. Our aim in this chapter is to elaborate on these possibilities, augmenting the simplistic dual systems or dual types of Kahneman and others, so as to include the whole of the psyche. Of course, an impulse may serve as a useful starting conjecture requiring modification, but only if, in Kahneman's terms, S2 kicks in and delays the proposed S1 reaction, parking it while other options are considered.

By contrast with S1, S2 is seen as responsive and considered. It is interesting that Kahneman privileges S2 over S1 in his campaign to develop rational action based on statistical reasoning. He sees S2 as involving cognition, which weighs up available actions in light of the gap between current state and goal state(s), and the suitability of available resources and actions. Bennett (1993; see also Shantock Systematics Group, 1975) used these four terms as a tetrad to delineate activity (Fig. 1):

The diagram acts as a reminder to ensure balance, so that the resources are adequate for movement from current state to goal state, that the tasks are appropriate for such a movement, and that the tasks make use of the resources in a balanced way with respect to both current and goal state. It can assist in digging deeper than merely asking that cognition consider the proposals generated by S1.

S1 is seen as providing S2 with possible actions using assumptions based on whatever limited data is available, in line with past experience, a process characteristic of abduction (Eco, 1983) and consonant with the notion of "frames" that act as soon as all parameters have assigned values (Minsky, 1975, 1986). If all the parameters already have default values, the frame will fire without waiting for more data. Of course it is often

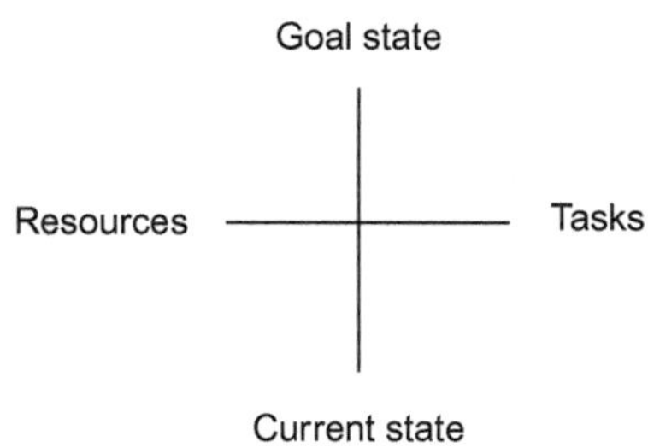

FIG. 1 The systematics tetrad of activity.

essential to act immediately, without taking time to consider pros and cons of various possible actions, especially in a classroom; but it is sometimes valuable, even essential, to park the first actions proposed and to respond rather than react.

Kahneman develops an elaborate narrative applied to a wide range of studies in which subjects are interpreted as being prey to S1 offering actions to S2, which, being somewhat lazy, often fails to evaluate them, so they simply "happen." Our suggestion is that, while offering a useful perspective, a dualist approach does not sufficiently account for the complexity of human behavior because it fails to involve the whole of the human psyche.

There is evidence that the body has to prepare muscles for an action to be carried out, which may or may not be enacted, and this has to happen before cognition is sufficiently informed to be able to observe a thought about it, much less actually initiate action (Norretranders, 1998). For example, if you consider the possibility of saying something and then try not to, it will often slip out unexpectedly: you literally "find yourself" saying it, as in Ph1 above. It seems that action and perception are not consequential but rather independent (Guillery, 2014), though what the philosophical implication of this might be seems to be highly contentious (Hacker & Guillery, 2014).

A similar dichotomy, the development of the bicameral brain, was used by Jaynes (1976) to account for a shift from universal holistic "knowing" to a reliance on shamans and subsequently priests to provide the "god-voice" or inner direction that was seen to be atrophying in the general population due to the evolution of the brain. He saw shamans as still in touch with the experience of a unified rather than split brain. The predominant tendency to react rather than respond is thus associated with the bicamerality of the now-split modern brain, in that the god-voice of right-brain holistic knowing can now be blocked out by the separated left-brain. Another similar dichotomy concerning left and right sides of the brain operating differently has been proposed by McGilchrist (2009). He develops a whole discourse around dialog between left and right sides of the brain, a distinction which was exploited and popularized by Edwards (1992) for drawing. McGilchrist applies it to all aspects of life. Although McGilchrist considers the role of attention and emotion because they function uniquely in each brain hemisphere, each of the dichotomies discussed here explains our tendencies to react or respond with a view of the human psyche that arises from the coordination of two distinct components.

Reacting and responding have been observed as basic to human behavior for a long time. For example, Shakespeare's *Much Ado About Nothing* can be seen as a study of the humorous consequences of reacting rather than responding in a considered manner, and this theme can be found in folktales of many different countries (Calvino, 1980). An expanded view

FIG. 2 Indian toy from the collection of John Mason.

of the human psyche can be found in the Upanishads and as an image in various Middle Eastern stories. The human psyche is likened to a horse (or horses), chariot, driver, and owner (see Fig. 2). Later versions change the chariot into a carriage, and then a hansom cab, as appropriate to the times. The owner hires a driver, who has the responsibilities of maintaining the chariot fabric and tackle as well as for looking after the horse(s). If the chariot body is allowed to decay, if the reins and harness go moldy or stiff, if the horses become hungry or mangy, or if the driver becomes dissolute, then the chariot cannot be used properly, and the owner will stay away (Gurdjieff, 1950, pp. 1193–1199).

If the horses are not guided, they will rear at the slightest movement seen out of the corner of their eyes. They will follow every whim of their own, ignoring desires of the driver, grazing at the roadside at every opportunity. If the carriage is not maintained, it will creak and crack under stress. If the driver is not awake to where the chariot is headed, then the route will be missed, and there may be crashes.

Reins and harness (mental imagery) enable the driver (intellect) to guide and direct the horses (emotions, senses), while the shafts (habits) enable the horses' energy to move the chariot (body, enaction). Note the etymological link between motivation, motive, and emotion. When crossing difficult terrain metaphorically, it matters where the horses place their feet and how they work together … how they direct their attention and their energy against the yoke. These are not controlled by the driver (let alone the owner). In short, the horses supply more than brute force that is directed by the driver. The owner can be thought of as will, since when present, it is the will which informs the driver's choices. Confusion or elision between will, attention, and action are very common. For example, Locke (1710: Introduction para 40) observed that "the will seldom orders any action, nor is there any voluntary action performed, without some desire accompanying it; which I think is the reason why the will and desire are so often confounded." The reins as links between the driver and the horses have something in common with the corpus callosum in the brain, resonant with McGilchrist's notions of *master* and *emissary*.

Enacting a habitual action without considering other possibilities is like a man scrabbling in the gutter under a streetlamp. When asked what he is doing, the man says, "I am looking for my door key." When asked, "Where did you lose it?," the man replies, "Over there," pointing across the street in the dark. "So why are you looking here?" "Because there is more light!" (Shah, 1983, p. 9; see also Webref, 2013).

Following an automatically generated action even when it is not apparently making progress is rather like the person searching for the key: it is easier to look, to act in familiar ways than it is to initiate an unfamiliar action. This observation can be used to make sense of a great deal of student behavior in classrooms, especially the behavior that so disappoints teachers. It also points to a need for teachers to help students develop and/or access a broader range of behaviors, which can inform their behavior outside of school as well as in class.

Missing entirely from Kahneman's dual process theory is a third state that might be described as System 3 (S3). This is a centered state in which S1 and S2 have been quietened so that there is access to "the still point of the turning world" (Eliot, 1969), to insight fed by creative energy (Bennett, 1964). It can be prepared for, as in Synectics (Gordon, 1961), in Mindfulness, and in "U Theory" (Scharmer, 2007), and as in the parable of the wise and foolish virgins. However it cannot be mechanically reproduced, as signaled by the Sufi story concerning someone who, while walking in a garden, stooped to pick up a stone and in that moment became enlightened, so hundreds of people visited that garden and stooped to pick up stones, hoping to be similarly enlightened, without effect. Eliot develops the notion in *The Four Quartets*, particularly *East Coker* and *Burnt Norton*, and it is experienced in various ways by mathematicians who await a critical insight, for example, appearing as a surprise, as Poincaré experienced while stepping on a bus (Hadamard, 1945).

As in "Theory U," approaching S3 involves at least suspending actions and thoughts that become available, letting go of emotions, particularly negative ones, and directing attention on a single "point," akin to contemplation and meditation, which is why it is usefully distinguished from S1 and S2, even if it cannot be intentionally initiated at will. The "Aha!" moment is a release of energy that has suddenly arrived but which is too strong to be held and transformed internally (Gurdjieff, 1950).

Extending the Chariot Image

In the chariot image there is no direct analog for attention, except perhaps in terms of what the driver (cognition) and the horses are attending to, and how. Furthermore there is no simile for a sixth component of the psyche, namely the *inner witness,* which observes but does not judge, does not comment, and does not engage in action. The inner witness, also referred to as the *executive* (Schoenfeld, 1985) is the "still calm voice" of the Quaker poem (Whittier, 1872) that observes without engaging and that asks questions such as "why are we doing this?" and "might there be a better way?" It is this inner voice, this "witness on the shoulder" that is so vital to develop as a problem solver, because as it becomes more and more experienced it can divert attention when progress is at a standstill or when effort is being expended without result. An early stanza of the Rg Veda provides a supplementary image in terms of two birds: (see Fig. 3.)

> Two birds, close-yoked companions,
> both clasp the self-same tree;
> one eats of the sweet fruit,
> the other looks on without eating.
> ***(quoted in and translated by Bennett, 1964, p. 108)***

It is through strengthening and refining the inner witness, monitor, or executive, that a rich repertoire of actions can be accumulated, accessed, and control exercised over flows of energy through the psyche.

Attention is often closely allied with will, as in the adage "you are where your attention is." However, James (1890) saw attention as a stream of consciousness; it is more helpful to see attention as structured on three levels. At the macro level, attention has *focus, locus,* and *scope.* Using the metaphor of one or more searchlights illuminating the night sky, there may be one or more foci simultaneously (multiple lights), which may be experienced as located in front of, beside, behind, or inside the body, and which may be broadly or narrowly focused. At the micro level, attention can consist of gazing (holding a whole), discerning details, recognizing relationships between discerned details, or

FIG. 3 From the collection of John Mason.

perceiving properties as being instantiated as relationships (Mason, 1998). These are closely allied to "levels" identified by van Hiele-Geldof (1957) and developed by van Hiele (1986), but they were independently developed from Bennett (1964). Finally, there is a meso level consisting of the current *absolute, axis,* or *center of gravity* of the individual. Thus adolescent attention tends to be oriented around sex and social relationships, whereas young adult attention tends to be oriented around career and homemaking, or adventure-seeking.

Learner engagement with tasks is in itself a field of research with a literature which is too large to summarize here (see, e.g., Henningsen & Stein, 1997; Skilling, 2013). The term *engagement* is of course an observer's construct to summarize a plethora of interconnections among components of the psyche. Engagement involves an activated will (to initiate action), a disposition (positive being more helpful than reluctant or negative), a repertoire of actions as a resource to call upon, and a flow of energy to enable specific actions to be undertaken (the terad of activity). Specific actions, emotions, thoughts, and particular forms of attention and initiative are all coordinated or interconnected so as to adhere to or trigger each other, creating the behavior observed by a teacher or researcher.

The terms *reacting, remoting* or *re-emoting,* and *responding* provide a useful framework for considering the nature of someone's utterance, whether your own or that of a student or teacher who has been asked a question or otherwise invited to speak (Molina & Mason, 2009; Scataglini-Belghitar & Mason, 2012). They align with the traditional Western constructs of enaction, affect, and cognition, but they also take into account intention (will). It makes a significant difference to any analysis, any attempt to account for observed behavior, to be clear on whether what is proposed as data is dominantly *reaction, remotion,* or *response,* especially if you are going to decide that your interpretation of their actions and words are what they truly think or believe or consider, and whether you consider these to be stable and robust over time and change of situation. It can prove fruitful to ask yourself, before analyzing data through other frameworks, what you would have to be attending to in order to be in a state in which you said and did what your subjects said and did. Being clear about what is being attended to, and how, sheds considerable light on what credence can be given to the use of other distinction-making analytic frameworks.

Polyphrenia: Adherences, Selves, and Micro-identities

It takes very little self-observation to reach the conclusion that we feel, think, and act differently at different times, and this has been noted many times in literature. Each collection of actions, emotions, and thoughts (what some people refer to as a self) is constructed to deal with specific situations (being a mathematics learner in an educational institution, a sports person or a computer expert, a son or daughter, sister or brother, or indeed any of these in a particular context in a particular moment in time). Each self is characterized by different flows of energy activating different aspects of the psyche (Bennett, 1964; De Geest, 2006). Not only are we different people with different practices when in the classroom, in a meeting, or at home, but we can be different at different moments of a lesson or a meeting or while working on a mathematical problem, depending on what we are currently attending to, what habitual reactions and re-emotions are drawn to the surface, and what energies are flowing.

One way to account for this multiplicity is through the notion of adherences between actions, feelings, and thoughts—between what we attend to and how we attend to it and the degree of will power to which we have access at any moment (Mason, 2009a). In any particular situation there are actions, feelings, and thoughts that arise or become available because of our current state and because of the situation. Energy flows through well-worn pathways, activating action, triggering emotions, and stimulating thoughts, as well as sensitivities to attend to certain things in certain ways and strength to exercise willpower. These follow particular patterns according to our past experience of similar situations, even when we aren't consciously aware (or are only vaguely aware) of those past experiences. A particular action may become available with an associated collection of feelings and thoughts; a particular emotional state may arise with associated actions and thoughts; a particular thought may arise, with associated actions and feelings. These might all be referred to as "adherences" between actions, feelings and thoughts, and states of attention and willpower.

Traditionally, sets of adherences have been called *selves* (Bennett, 1964; Roberts & Donahue, 1994), with associated *frames* of mind that are disposed to "fire" (Minsky, 1975, 1986) or come to the fore (Hudson, 1968). Varela (1999) referred to them as *micro-identities,* because for a moment they "become us": they dominate, generating our behavior and our experience. Houston (1997) used the term *polyphrenia* for "the orchestration of our many selves our extended health. We have a vast crew within."

Of particular interest are transitions between adherences, between selves. When one self is dominating, absorbing the flow of energies, there may be another self or collection of adherences gaining strength in the background. Sometimes the transition is abrupt due to changing environment (leaving one situation and entering another); sometimes the transition is smooth as new possibilities with their corresponding actions and attitudes gain strength; sometimes a transition occurs when one way of being proves inadequate and another takes over control of the energy flow. We suggest this is a matter for personal investigation through catching such transitions in oneself.

The metaphor of multiple-adherences or person-as-multiple-selves can be read into classic literature such as the *Iliad*, the *Odyssey*, the *Parliament of the Birds*, and the *Legend of King Arthur*, not to say novels, plays and soap operas, in which the various characters can be interpreted at a psychological level as aspects of a single individual, and at a sociological level as members of a social group. Modern forms of the same idea include the *society of mind* (Minsky, 1986), the *autopoetic construction of a self* (Maturana & Varela, 1972) and Varela's (1999) notion of *micro-identities*.

Adherences are, according to ancient psychology, constructed or formed in response to social and psychological forces, in order to defend core emptiness from exposure to the outside world, and to help us cope with the unexpected. Personality is the many layered shell which shields that emptiness from everyday awareness. Multiplicity evolves because each social context involves a collection of practices that define participation (*cf* Wenger, 1998).

Thus adherences can be thought of as comprising particular associations or coordinations between actions, feelings, and thoughts, together with both characteristic ways of attending and access to particular strengths of will power. Each "self" has characteristic ways of releasing affective energy, and characteristic ways of transforming that energy enactively, cognitively, attentively and willfully. Different adherences may vie for domination, or a group of them may act in concert to produce a particular "state." Without a strong inner witness, we may not be aware that the currently dominating adherence arises from many possibilities.

Polyphrenia helps account for different ways of dealing with unexpectedness. In any given situation, reactions and responses will depend on which collections of adherences are currently active or dominating. This explains why sometimes we notice opportunities to act, while at other times, in apparently similar situations, we do not. It also explains why sometimes the opportunities to act that we notice may have been appropriate in another similar situation but not in the current one. Sometimes we may feel that we are "not ourselves" when we are reacting to situations with one set of adherences when (especially in retrospect) others of our "selves," some "other us" would prefer to be leading. The notion of "true identity" can be challenged when someone realizes (makes real to "themselves") that each situation brings to the fore a particular collection of adherences, with characteristic ways of acting, feeling, thinking, attending, and taking initiative.

In the historical psychological literature, a person has been likened to a self-sufficient community, such as a ship or a feudal estate. The owner is absent, and the servants (selves, adherences) have become slack and undisciplined. Each vies with the next to act as gatekeeper, make choices, and issue instructions. However, most instructions are ignored or only halfheartedly and incompletely carried out. For example, Plato uses the metaphor of the "ship of state" (Republic Book VI 488e–489d: Hamilton & Cairns, 1961): Here, the philosopher king navigator, who alone has knowledge of star-gazing, is dismissed by bickering sailors vying for the attention of the ship owner so as to control the direction of the ship. In addition to taking Plato at face value, various spiritual traditions advocate interpreting characters in parables and stories as aspects of a single person (Bourgeault, 2008; Jaynes, 1976). This proves fruitful in many contexts, including making sense of Plato's stories.

A good example of the operation of—and work to change—resident selves occurs in the extensive research of Dweck (2000). She shows how diverting inner and outer discourse concerning "I can't" away from its derivatives "I won't" and "I don't" to "I can if I try harder or with appropriate support" can significantly impact how students respond to difficulty. Over time this can alter people's sense of believing that "thinking mathematically is basically genetic" so that failure is to be avoided, to "thinking mathematically is like a muscle which can be strengthened with exercise."

Ballard (1928) located differences between English and American teaching of arithmetic in different views of how behavior is generated. He distinguished, on the one hand, a predominantly English view that "arithmetic is logic," and on the other, a predominantly American view that "arithmetic is habit." The contrast is interesting and significant; but it is not new. It resembles, in fact, the antithesis between the Platonic and the Aristotelian views of virtue. To Plato virtue is knowledge; to Aristotle it is habit. To Plato it is an intellectual grasp of the consequences of our acts; to Aristotle it is the practice of choosing the mean between two extremes.

> These differences in emphasis and outlook are not of mere theoretical import: they vitally affect practice. They prescribe what we shall teach, how we shall teach it, and how we should test it. ***(Ballard, 1928, p. xii)***

For researchers, how they view the human psyche has profound implications for how they interpret and make sense of data based on observations of behavior. For teachers, how they view the human psyche has profound implications for the atmosphere they generate in the classroom, and hence learner relations with mathematics because of relationships with their teacher. For learners, how they view their own psyche has profound implications for how they engage with what is new or unfamiliar, as Dweck (2000) has shown so vividly.

THE GOLDFISH PROBLEM

The following is an account of work by one of us (MM) on a problem that at the time of writing had been circulating widely. In many ways it is similar to the baseball-and-bat task used by Kahneman and Frederick (2005, p. 273) mentioned earlier:

> If out of 200 specimens, 98% are of one type, how many of that type must be removed so that 96% are of that type?

Seriously working on this task before reading on will enable the reader to link their own experience to our comments. First we offer an initial comment, then offer an account of work on the task with commentary, making use of our theoretical frame alongside.

In reading this response, we invite the reader to attend to the italicized descriptors of sensing, feeling, knowing, noticing, and deciding. Here, *sensing* might be described as the vague, general awareness of intuition, a *potential* knowing that is as yet unarticulated but nonetheless largely cognitive, with some emotional support. There may perhaps be an expectation or a suspicion, but perhaps not that strong. *Feeling* is emotion, often in the form of satisfaction or dissatisfaction (of various shades) with part of a solution, but also the surprise that can occur when the intuitive elements that had conflicted with logical knowing (making the logical solution feel counter-intuitive) begin to be unpacked, bringing intuition and logic into greater agreement. Here, *knowing* is associated with logical certainty—a state in which actions are enacted with some confidence (even if they are counter-intuitive), and possibly with recourse to justification.

We then invite the reader to return to the account and to consider the bold-printed self-references (e.g., I, me, my) and to consider the various voices to which these refer: At times, these "I's" may be seen to be conversation with one another, while in other cases, "I" is the witness attempting to narrate. The accounts are of course cast in language, which implies alignment with enculturated forms of description. It also depends on sensitivities and energies in order to notice, to mark, and to record (Mason, 2002a, 2002b). The act of articulating, however, necessarily channels attention to those aspects of experience being described, and it can make it difficult to trap other aspects, which are in constant flux and even influenced by the acts of description. It can also create the illusion that particular adherences of senses, feelings, knowings, and states of attention are more stable than they actually are. Nonetheless, we suggest that, however imperfect, attention to the multiple adherences that form our selves in every moment is significant for deepening our understanding of the dynamic nature of our learning selves and can alter the way we attend to students. Our aim in offering an account of Martina's experience is to seek resonance with the experience of attending to multiple adherences in play, not necessarily with the goldfish problem itself but in other mathematical explorations.

Martina's Account & Some Analytical Comments

My immediate *gut reaction* to this problem (all the while with alarm bells alerting me to the fact that it *must be* wrong, if only because it made the problem trivially easy and **I** *knew* there was supposed to be something tricky about it), was this: Since 98% are Type A, then 2% or 4 must be Type B. To reduce it to 96%, **I'd** need to be missing 8 from 200, so take away 4 more.... But **I** quickly *saw* that that wasn't the same as just taking away A's – that would be exchanging A's for B's so as to maintain a total of 200. (This didn't emerge clearly articulated: It came first as a general *sense* that unfolded into these details, but there were no *surprises* at the language that fairly quickly emerged to articulate it. It also prompted a vague and unarticulated *expectation* that the real answer should be close to 4; after all, the change was small!)	*The account begins with the witness noting an initial 'gut reaction', but keeps ringing alarm bells so that cognition is not carried away in the flow of initiated actions.* *Notice the reference to pre-articulate thoughts that 'unfolded' into the words of the account. It is often the case that when you ask someone how they got an answer they have just given, they have no idea. This is particularly the case for learners who are mathematically inclined. One reason for being reluctant to speak about the origins of insight is that "to express is to over stress" (Mason, 2002a, 2002b, p. 198). Once some aspect of a complex experience is expressed it 'becomes' the experience so that subtler aspects may be lost. As a teacher, dismissing their actions as the behaviour of good mathematical thinkers simply delays the learners' encounters with being stuck and developing personal resources for dealing with being stuck (Mason, Burton, and Stacey, 1982/2010). The great temptation as a learner is to glory in success rather than probe beneath the surface to become more sharply aware of effective and ineffective actions, dispositions, and thoughts. As a researcher, dismissing their actions as intuitions simply maintains researchers' attention on the surface phenomena; we are trying to probe beneath the surface.*
Once **I** recognized that the total would not remain the same when B's were removed, **I** noted that (at that moment), **I** couldn't intuitively and simultaneously appreciate the change in both the total and the number of A's. (Perhaps **I** didn't articulate this fully before proceeding, because it was only as **I** was writing these reflections that **I** *noticed* that the number of B's stays the same, which isn't something that **I** originally considered.) **I** decided to write a ratio based on the number of A's removed: Let x = number of A's removed $96/100 = (196 - x)/(200 - x)$	*An immediately available action (a reaction) is to cross multiply, but this was parked by a stronger desire to avoid arithmetic and to enable direct perception (seeing). What* ***was*** *available was seeing 196 as $100 + 96$ and 200 as $100 + 100$. Here we have an instance of attention in the form of gazing at the whole but not discerning fine detail. The urge to cross multiply emerges from the overall structure (something over something equal to something over something). Attending to the actual numbers and recognising a relationship amongst them opens the way to alternative actions.*

Continued

I didn't *want* to cross-multiply, because **I** was *trying* to retain an intuitive connection to the actual problem, and this got in the way. Looking at it now, it seems funny that the solution wasn't visibly obvious, but at the time, it *seemed to demand* cross-multiplication as an easier route, so **I** gave in [**to whom**?]:

$$(100)(196 - x) = (96)(200 - x)$$
$$19{,}600 - 100x = 19{,}200 - 96x$$
$$400 = 4x$$
$$x = 100 \ (HUH!?)$$

Even then, **my** first reaction was *self-doubt*: **I** *must have* messed something up in my conceptualizing of the ratio. **I** double-checked **my** thinking and **my** calculations, then deliberately released the *sense of wrong* to allow for what might be right, possibly because there was also already an emerging *sense of rightness* (quite distinct from solving the proportion) rising to consciousness. When the possibility of rightness did register loudly enough that **I** consciously recognized it, it was still very vague: just a *general sense that there was potential logic in 100* as an answer and that there might be something creating a *flawed sense of wrongness*.

Before exploring the root of that feeling, **I** wanted to test 100 in the original problem to *confirm* that it worked and in the *hope* that doing so would make the reasonableness (or not) of the answer become apparent:

$$\text{Before Removing A's}: 196 + 4 = 200 \ (98\% \text{ A's})$$
$$\text{After Removing A's}: 96 + 4 = 100 \ (96\% \text{ A's! It works!})$$

I was fully *convinced*. **I** could easily have stopped here, *confident* in my conclusion. But it still didn't *feel* intuitive, and **I've** learned that almost always, plumbing counter-intuitions yields rich rewards in terms of learning.

'Giving in' is a retrospective description of an available action as part of an adherence (desire to get the answer quickly; conceiving of oneself as quick to get answers and mathematically competent). In Kahneman's terms it is the lazy S2 not bothering to consider the offerings of S1. In Minsky's terms, the parameters for action all have default values so the frame fires.

An adherence and coordination of enaction, affect and will is seeking dominance, and achieves it. Energy now flows through different actions.

Here cognition (based on competence at algebra) is over ruled by the contrast in affect between expectation and result, leading to a sense of possible, even probable error. Over a brief period of time, the power relation shifted so that confidence in the result returns and doubt shifts to the expectation and away from the surprise. This could be seen as a shift of adherence, which influences will, namely to test the result and then to interrogate the origins of the mistaken conjecture and its emotional support.

The last ***I*** *is coming from the witness.*

Here the ***I*** *is the adherence of confident algebraic manipulation and trust in associated actions. There remains some affective-cognitive doubt that there is something missing, some way of thinking, perhaps some way of attending to the mathematical structure of the situation which will make the result obvious.*

All of this took place in the span of approximately 15 minutes, during which I was also sitting with my children as they played and therefore unable to dedicate my full attention to the problem. This is significant in that when focused and clear, solutions can present themselves that prevent quieter and more doubtful voices from being heard. At this point, **I** still hadn't created a visual conceptualization of the situation to see if **I** could find a way to *feel the solution as intuitively right*; **I** noted to myself that **I** *wanted* to do this later, but **I** was quite confident that **I** could and was more interested in seeing what **I** could bring forth from my subconscious: What was creating interference? Later in the evening, **I** attended to the constant number of B's rather than the changing number of A's. It occurred to me that **I** *could* build **my** ratios around that value, though again, the act of writing about **my** experience changed it. **My** writing is much slower and much more deliberately articulated than **my** original thinking.	*The "I could" was felt as an "I should," which adds emotional negative force to a sense of deficiency, whereas "I could" adds emotional and cognitive force to possibilities in the future, informed by current reflections. The* **I** *and* **me** *voice is partly the voice of the witness, but because it is in retrospect, largely the voice of an emotionally charged adherence of actions, emotions and cognition.*
Let $x = \text{Total}$ There are 4 B's and they must now represent 100%–96% = 4% of the total. Setting x as the total: $4/x = 4/100$ So $x = 100$ If the total is 100, then there must now be 196 − 100 = 96 A's, which is 100 less than there were originally. The fact that the two fractions emerged as identical threw me a bit here; 100 is both the base of the percentage and the actual total; 4 is both the percentage and the actual number of B's. **I** am suspicious about the choice of numbers! **I** want to know how to choose numbers so that everything works out as integers. In other words **I** want to generalize.	*After reaching* **my** *first resolution to the problem,* **I** *had noted a feeling of "potential rightness."* **I** *now decided to hold this in consciousness to see what might emerge to match it (see Metz, 2012, pp. 59–62 for a fuller description of the process of locating).*

Continued

At first, **I** just got a general *sense* of large changes in values resulting in small changes in percentages … along with *something about* one value being very small and difficult to change (though it wasn't even that clear; it's hard to express the "sense" in words, because words change it). At that point, a *feeling associated with principal and interest* came to mind (followed by the words). As **I** sat with that feeling, it evolved into *"something to do with paying down the principal on a big loan"*…. Dilution also came to mind … *something to do with needing a lot to make a difference* … but **I** didn't articulate it even that far at the time (as **I** write this, the words are calling forth *the sense of a too-cold tub that is too full to warm by adding more hot water* … and with issues *trying to compensate for having used too much red food colouring in icing or play-doh* … but **I** don't want to go there at the moment). Any of these situations may have offered insight, but it *felt as though something better was available* if only **I** could hold on to the sense or feel of it long enough and gently enough for it to rise to consciousness.

The pre-articulate is difficult to communicate, because once expressed it becomes something other than what was experienced. As Maturana (1988) said, "everything that is said is said by an observer". Complementing his observation is the fact that "to express is to over stress" (Mason, 2002a, 2002b, p. 198) mentioned earlier. As soon as something is expressed, it becomes the focus of attention, pushing what is not expressed into the background.

Notice the cognitive dimension in the use of 'feeling'

Notice how the ***I*** *changes subtly as the adherences between enaction, affect, and cognition become available.*

It did: **I** remembered being 40, pregnant with my second child, and worried about the risk of trisomy. **My** risk was 1/66, compared with 1/526 for 20-year old mothers. In particular, **I** recall **my** surprise at realizing the *seemingly obvious* fact that even at 1/66, there was still a 98.5% chance that my baby would be trisomy-free. The shape of a graph showing the exponential rise in risk as women approach 40 stood out prominently in my mind: The position of the point on that graph for 40-year old women did not seem commensurate with a 98.5% chance of a healthy baby. This situation immediately felt right: **I** *knew* I had found a match for the feeling of right **I** had experienced in relation to the percentage problem. But **I** knew from past experience that just because it matched the feeling didn't mean it would offer valid insight into the goldfish problem. **I** needed to map one onto the other.

Notice how personal metonymies trigger associations, and how cognition then has to consider whether to develop, park or dismiss them. Metonymies are predominantly to do with emotions, usually idiosyncratic associations that can be hard to trap. Also present are metaphorical resonances, and it seems that a mixture of metonymy and metaphor are operating in this account. Metaphors are to do with structural relationships and so are predominantly cognitive with some affective support.

Notice how the ***I*** *has switched to a memory and is now accessing different actions through re-emoting because of identifications with past experience.*

Surprisingly, **I** found the connection was difficult to articulate, and my hope in its ability to provide an immediate sense of justification waned. And yet … **I** wasn't quite ready to let it go.

This 'surprisingly' could be referred to as meta-affect combined with meta-cognition. It is a functioning of the witness which observes but does not initiate action.

Both situations involve large changes in value showing up as seemingly small changes in percentage.... Is that enough? What are the Type A's that are being removed? There is a large difference between 98.4848% and 99.8099%. But that's because, for example, 99.99% (1 in 10,000) is 100x times "bigger" (lower risk) than 99% (1 in 100)!	*Cognition is active in making 'sense', in integrating past experience*
This is more significant than the difference between 96% and 98%...isn't it? But wait.... If 96% and 98% were risk factors, 96% health would mean 4% risk. 98% healthy would be 2% risk. The risk doubles from 96% to 98%. The risk doubled, but the percentage still only changed by 2%. At some level, this seems so *obvious* that I'm embarrassed to write it down and admit that something's not quite working for me here. On the other hand, the fragility of this knowledge has been exposed, and I want to strengthen it.	
But again what are the Type A's in this situation? Healthy babies.... They aren't being taken away, but they aren't being born as often in comparison with unhealthy ones. But you have to "take away" a large number of healthy babies before the percentages change by large amounts.	*Perhaps witness is returning attention to the original problem* *Rapid shifts in the locus, focus and form of attention are common. Here cognition is re-initiating a cognitive function, namely to check an assumption or a calculation*
I wondered if it would be helpful to rewrite the original question in terms of trisomy risk. **My** first attempt reiterated the original misconception! 98% of babies are healthy for mothers of Age 1. 96% of babies are healthy for mothers of Age 2. In a sample of 200 babies, how many more will be born with trisomy at Age 1 than at Age 2?	*Note the meso-attention focus on childbirth, representative of an absolute of axis of concern for a young mother.* *Fischbein (1987) maintained that intuitions are not replaced, only at best overlaid, and that they are vulnerable to re-emergence in other situations. By developing an inner witness and attending closely to the multi-faceted aspects of intuitions as they re-emerge in other situations, we suspect that what is experienced as intuitive or counter-intuitive may in fact shift over time.*
Of course, here B's (trisomy babies) are being *exchanged for* A's (healthy babies) as opposed to A's disappearing. The question should have been how many fewer healthy babies would turn 98%–96%. In fact, this was *quickly obvious*. **I** include it here to point out the persistence of the thought patterns that the problem invoked.	*Notice the shift to a witnessing 'I' which is justifying itself.*

Continued

This formulation does, however, draw attention to the fact that the number of trisomy babies doubles from Age 1 to Age 2. Of course.... To double the numerator, you halve the denominator.... To double the percentage of trisomy babies, you can either double the number of trisomy cases out of 200 or halve the number of babies without changing the number with trisomy (i.e., remove healthy babies from consideration). That doesn't make sense in this context, but mathematically, it produces the same effect.

Attention was absorbed by the functioning of an adherence between action, emotion, and cognition which *blocks out or dominates over other possibilities. The visual image 'suddenly' (a shift of attention being marked and remarked upon) fits with cognition to summon up the emotional component of 'clarity'.*

Does this insight finally allow me to experience the original problem intuitively? Only if I focus on the 2% vs. the 4% of B's and to the fact that this is a doubling, not just a difference of 2; the doubling was not initially apparent, again, due to the clever choice of particular numbers. To double the percentage of B's, I must halve the total. Focusing attention on 98% vs. 96% disguises that relationship. Visualizing makes all of this even clearer, and suddenly I wonder why I semiintentionally set that aside so early on ....

A key point here is that that the existence of a clear solution does not necessarily dispel counter-intuitiveness - often, it simply overrides it, as mentioned earlier. I am suggesting that listening to that which typically gets over-ridden is usually productive, but to do this requires the witness to be awake and observing. [enaction, multiple selves, attention]

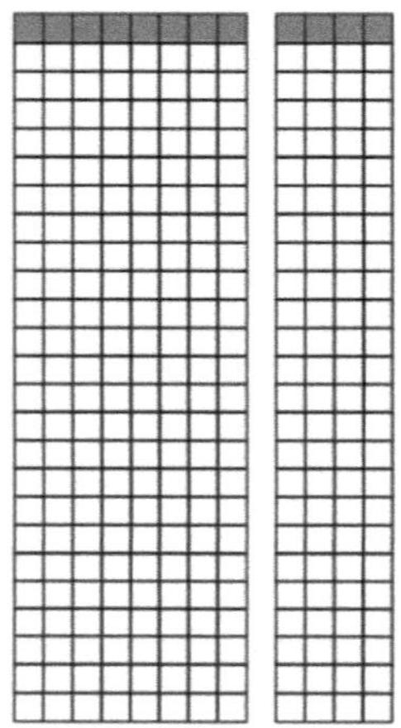

Each dark gray cell is 0.5% of the total in Picture 1 (2%) and 1% of the total in Picture 2 (4%).
Here again, the solution seems so obvious, but my attention was fixated on other actions that somehow begged for their own resolution.

	In words, my final solution may seem little different from what I wrote early on. But the sense of it is much evolved. In the end (for now), the connection that originally struck me as so meaningful (trisomy risk) did in fact turn out to be helpful. Exploring it strengthened my understanding of both situations. It is significant that the feeling of rightness that the trisomy situation prompted far preceded my ability to articulate the connection between the two problems. It is also important to note that that feeling is not always an accurate indicator of relevance.
Working on counterintuitive problems has much to do with listening to ALL the voices.... Early on, **I** had a solution to the goldfish problem that **I** found fully convincing. It would have been easy to stop there and declare a "winner" in the inner debate: Logic had declared, "**I** got this!" But what it "got" was narrowly limited. When put into conversation, both logic and intuition evolved. Counterintuitions were exposed and articulated, and logic was brought to bear on a broader field of knowing. When **I** aim for consensus and make an effort to hear quiet voices that could easily be overlooked/overheard/not noticed, **I** almost always find that it allows me to deepen my understanding much further than simply solving (or even solving and extending) would allow. It seems likely, however, that counterintuitions are deeply woven into many situations that might come to mind and become part of a self-in-the-moment.	*Notice the shifts in* **I** *between cognition and witness.*

Further Commentary and Analysis

We propose that the process of attending and responding to the counterintuitive feelings is learnable and that engaging in that process can deepen the sensitivities of both teacher and learner (Metz & Simmt, 2015). It is akin to William Blake's notion of *imagination,* which he feared was being lost in the process of industrialization (Pullman, 2015). Metz (2012) described an increasing sensitivity to what she described as a *feeling* of potential knowing that she compared to the experiences of figuring out a name "on the tip of her tongue," of feeling as though she has forgotten something, and of trying to remember where she placed a missing something. It seems appropriate to assign these inchoate "feelings" to the influence of the witness, the observer behind the scenes. Their very ineffability is cognitive, yet this is supported by an emotional desire to "keep looking," to try to "see" more clearly, and by an enactive sense of "gut-based" feeling. The combination of cognition, emotion, and action is an adherence, a self that feeds energy to the will so as to continue the search or to initiate some other action in support. The self that was trying to resolve the initial task has been relegated to the background. Notice that the emotional component of the adherences being expressed is far from "remoting." Rather it is engaging and re-engaging attention. Energy is being provided to will so as to continue rather than give up, although in many instances learners experience a tipping over into remoting as the degree of challenge experienced exceeds the threshold of sustainable actions. When possible actions dry up, frustration and negativity can take over, auctioning a flight-based adherence rather than fight-based ones.

Efforts to attend to and articulate the ineffable seem to include three stages: (a) awareness of its presence, (b) a deliberate sitting with the awareness and allowing words to come forth to describe it, and (c) recognition of the "rightness" of the words that accurately describe it. These are similar to tacit knowledge as described by Polanyi (1958) and to a description by Gendlin (1978, 2009) of a three-part role for what he terms a "felt sense" that functions in the following ways:

(a) to indicate something potentially known,
(b) to indicate incorrect and correct formulations of it,
(c) in the process of recall, because to recall something, we concentrate our attention on the "feel" of it. (p. 76)

This process was evident in the goldfish account in the search for the situation that "felt like" a right match for the goldfish problem, culminating (for the time being) in the identification of trisomy risk. It is important to re-emphasize that such identification prompted a sense of certainty in the sense that it was a certain match. But this certainty should not be confused for an *appropriate* match. Like the value of attending to the ineffable, suspicion of this certainty is *learned*. Identification of trisomy risk was the beginning of further work unpacking and exploring the suitability of previously subconscious cognition now brought into the spotlight of attention. As the process of attending and responding to such subtle indicators of possibility becomes familiar, it also becomes possible to better empathize with other learners' experiences and to act as an empathic coach (Varela & Shear, 1999) who can encourage others to attend closely to the subtle voices of "selves" that might otherwise remain unheard.

Further Exploration

One aspect of an energized will is to ask further questions, to explore the situation further. This can only happen if familiar actions are available, actions such as generalizing or asking what it is about the numbers that makes the initial reaction to the task so compelling. Questions that come to mind (that is to cognition, with varying degrees of emotional commitment) include the following:

What if the 200, the 98%, and the 96% were different?
What happens to the percentages of large and small fish as increasing numbers are removed (in some sort of systematic way, say removing half of what is left or taking away 10 at a time)?
What would occur if the question asked for how many more large fish would be needed to add in so that the small fish fell to 96%?
What numbers and percentages would give integer answers for both questions?

LESSONS LEARNED: INFORMING FUTURE ACTION

Our aim has been to show how it is possible to probe beneath the surface of reacting, re-emoting and responding to tasks by making use of a six-part structure of the human psyche, and how isolating those parts only confuses, whereas combining them together captures something of the lived experience. However, analyzing one's own experience is at best self-indulgent unless there are lessons to be learned which might inform future action.

The notion of multiple-selves or microidentities or adherences between actions, emotions, and thoughts enables observers of behavior to recognize that what appears to be happening on the surface may not fully present what could be going on underneath. By observing shifts of state, shifts of attention, and shifts of adherences of enaction, emotion, and cognition in oneself, it is possible to become more sensitive to the experience of learners. By listening to learners and interpreting what they say and do as only partial expressions of a more complex state, they may recognize a positive relationship with a caring teacher—a teacher who cares both for learners and for mathematics—and consequently develop a caring relationship with mathematics.

The English language, despite not having a word for an automatically triggered emotion (we use re-emoting), does have several words whose meaning includes some degree of each of enaction, affect, and cognition. For example, the words *sensing* and *feeling* refer to prearticulate or inchoate experiences prior to attempts to express. "Sensing" is etymologically enaction-based, yet it has affective and cognitive aspects, while "feeling" is sometimes associated with enaction (feeling one's way) but also has a strong affective and some cognitive aspects. This provides further justification for probing beneath the surface of emotions and calling upon a complex image of the human psyche. The word "knowing" in the gerundive form indicates an ongoing process. Although it is commonly linked to cognition, it has a strong link with enaction, as evidenced by the enactivist stance that "knowing is acting and acting is knowing" (Maturana & Varela, 1972). Johnson (2007, p. 53) described doubt as a "fully-embodied experience of hesitation, withholding of assent, felt bodily tension, and general bodily restriction" and noted that "such felt bodily experiences are not merely accompaniments of doubt; rather they *are* your doubt." The notion of choosing or deciding to act is fraught with difficulties because as neuroscience experiments have confirmed, many apparent "choices" are actually reactions that have become habitual and automatic (Bennett, 1964; Norretranders, 1998).

Our analysis of Martina's work on the goldfish problem suggests that close inspection of one's own experience can reveal polyphrenic shifts among sets of adherences between actions, emotions, and cognitions. Observing these in oneself can contribute to being sensitive to similar adherences in learners. This in turn can enrich the actions that are triggered and become available to be enacted, as well as awakening the inner witness so as to enable parking of actions whose parameters have default values and so are eager to be enacted. It can also make it more likely that habitual emotions can be parked and that positive rather than negative emotions can be accessed, which in turn empowers the will to initiate fresh actions. All of this can be enhanced and amplified by making use of the discipline of noticing, through imagining oneself in some future situation acting, feeling, sensing, and thinking freshly.

References

Ballard, P. (1928). *Teaching the essentials of arithmetic*. London: University of London Press.

Bennett, J. (1964). *Energies: Material, vital, cosmic*. London: Coombe Springs Press.

Bennett, J. (1993). *Elementary systematics: A tool for understanding wholes*. Santa Fe: Bennett Books.

Boaler, J. (2000). Mathematics from another world: Traditional communities and the alienation of learners. *Journal of Mathematical Behaviour, 18*, 379–397.

Boaler, J. (2010). *The elephant in the classroom: Helping children learn and love maths*. London: Souvenir Press.

Bourgeault, C. (2008). *The wisdom Jesus: Transforming heart and mind—A new perspective on Christ and his message*. Boston, MA: Shambala Press.

Calvino, I. (1980). *Italian folktales* [G. Martin Trans.]. New York, NY: Harcourt Brace Jovanovich.

De Geest, E. (2006). *Energy transactions in the learning of mathematics* [Unpublished Ph.D. thesis]. Milton Keynes: Open University.

Duffin, J., & Simpson, A. (1993). Natural, conflicting and alien. *Journal of Mathematical Behaviour, 12*(4), 313–320.

Dweck, C. (2000). *Self-theories: Their role in motivation, personality and development*. Philadelphia, PA: Psychology Press.

Eco, U. (1983). Horns, hooves, insteps: Some hypotheses on three types of abduction. In U. Eco & T. Sebeok (Eds.), *The sign of three: Dupin, Holmes, Peirce* (pp. 198–220). Bloomington, IN: Indiana University Press.

Edwards, B. (1992). *Drawing on the right side of the brain*. London: Souvenir Press.

Eliot, T. (1969). *The complete poems and plays of T. S. Eliot*. London: Faber & Faber.

Fischbein, E. (1987). *Intuition in science and mathematics: An educational approach*. Dordrecht: Reidel.

Gendlin, E. T. (1978). *Focusing: A step-by-step technique that takes you past getting in touch with your feelings—To change them and solve your personal problems*. New York, NY: Everest House.

Gendlin, E. T. (2009). We can think with the implicit as well as with fully-formed concepts. In K. Leidlmair (Ed.), *After cognitivism* (pp. 147–161). New York, NY: Springer.

Gordon, W. (1961). *Synectics*. New York, NY: Harper Row.

Guillery, R. (2014). A neuroscientist's view of action, perception and the self. *Oxford Magazine*, 10–13. Fifth week Hilary term.

Gurdjieff, G. (1950). *All and everything: Beelzebub's tales to his grandson*. London: Routledge & Kegan Paul.

Hacker, P., & Guillery, R. (2014). Letters to the editor. *Oxford Magazine*. Second week Trinity term.

Hadamard, J. (1945). *An essay on the psychology of invention in the mathematical field*. Princeton, NJ: Princeton University Press.

Hamilton, E., & Cairns, H. (Eds.), (1961). *Bollingen series LXXI. Plato: The collected dialogues including the letters*. Princeton, NJ: Princeton University Press.

Henningsen, M., & Stein, M. (1997). Mathematical tasks and student cognition: Classroom-based factors that support and inhibit high-level mathematical thinking and reasoning. *Journal for Research in Mathematics Education, 28*(5), 524–549.

Houston, J. (1997). *A passion for the possible: A guide to realizing your true potential*. San Francisco, CA: HarperSanFrancisco.

Hudson, L. (1968). *Frames of mind*. London: Methuen.

James, W. (1890). In *Principles of psychology. Vol. 1*. New York, NY: Dover. [reprinted 1950].

Jaynes, J. (1976). *The origins of consciousness in the breakdown of the bicameral mind*. Toronto, ON: University of Toronto Press.

Johnson, M. (2007). *The meaning of the body: Aesthetics of human understanding*. Chicago, IL: The University of Chicago Press.

Kahneman, D. (2012). *Thinking fast, thinking slow*. London: Penguin.

Kahneman, D., & Frederick, S. (2002). Representativeness revisited: Attribute substitution in intuitive judgement. In T. Gilovich, D. Griffin, & D. Kahneman (Eds.), *Heuristics and biases: The psychology of intuitive judgment* (pp. 49–81). Cambridge: Cambridge University Press.

Kahneman, D., & Frederick, S. (2005). A model of heuristic judgment. In K. Holyoak & R. Morrison (Eds.), *The Cambridge handbook of thinking and reasoning* (pp. 267–293). Cambridge: Cambridge University Press.

Lave, J. 1988. *Cognition in practice: Mind, mathematics and culture in everyday life.* Cambridge: Cambridge University Press.

Leron, U. (n.d.). *Application of dual-process theories in mathematics education (and vice versa)* http://www.open.ac.uk/Arts/dualprocess/abstracts/Leron_paper.pdf (Accessed April 2014).

Leron, U., & Hazzan, O. (2006). The rationality debate: Application of cognitive psychology to mathematics education. *Educational Studies in Mathematics, 62*(2), 105–126.

Locke, J. (1710/1975). Essay on human understanding. In P. Nidditch (Ed.), Oxford: Clarendon Press.

Mandler, G. (1989). Affect and learning: Causes and consequences of emotional interactions. In D. McLeod & V. Adams (Eds.), *Affect and mathematical problem solving: New perspectives* (pp. 3–19). London: Springer-Verlag.

Mason, J. (1998). Enabling teachers to be real teachers: Necessary levels of awareness and structure of attention. *Journal of Mathematics Teacher Education, 1*(3), 243–267.

Mason, J. (2002a). *Researching your own practice: The discipline of noticing.* London: RoutledgeFalmer.

Mason, J. (2002b). *Mathematics teaching practice: A guidebook for university and college lecturers.* Chichester: Horwood Publishing.

Mason, J. (2009a). Teaching as disciplined enquiry. *Teachers and Teaching: Theory and Practice, 15*(2–3), 205–223.

Mason, J. (2009b). From assenting to asserting. In O. Skvovemose, P. Valero, & O. Christensen (Eds.), *University science and mathematics education in transition* (pp. 17–40). Berlin: Springer.

Mason, J. (2011). Attending to oneself so as to sensitise oneself to learners. *The Mathematics Educator, 13*(1), 3–14.

Mason, J. (2015). On being stuck on a mathematical problem: What does it mean to have something come-to-mind? *LUMAT, 3*(1), 101–121.

Mason, J., Burton, L., & Stacey, K. (1982/2010). *Thinking mathematically.* London: Addison Wesley-Pearson.

Maturana, H. (1988). Reality: The search for objectivity or the quest for a compelling argument. *Irish Journal of Psychology, 9*(1), 25–82.

Maturana, H., & Varela, F. (1972). *Autopoesis and cognition: The realization of the living.* Dordrecht: Reidel.

McGilchrist, I. (2009). *The master and his emissary: The divided brain and the making of the western world.* New Haven, CT: Yale University Press.

McLeod, D., & Adams, V. (Eds.), (1989). *Affect and mathematical problem solving: A new perspective.* New York, NY: Springer-Verlag.

Metz, M. (2012). *Interacting with implicit knowing in the mathematics classroom* [Unpublished doctoral dissertation]. University of Alberta. http://hdl.handle.net/10402/era.28511 (Accessed 7 August 2015).

Metz, M., & Simmt, E. (2015). Researching mathematical experience from the perspective of an empathic second-person observer. In *ZDM Mathematics Education*: *47* (pp. 197–209).

Minsky, M. (1975). A framework for representing knowledge. In P. Winston (Ed.), *The psychology of computer vision* (pp. 211–280). New York, NY: McGraw Hill.

Minsky, M. (1986). *The society of mind.* New York, NY: Simon and Schuster.

Molina, M., & Mason, J. (2009). Justifications-on-demand as a device to promote shifts of attention associated with relational thinking in elementary arithmetic. *Canadian Journal for Science Mathematics and Technology Education, 9*(4), 224–242.

Norretranders, T. (1998). *The user illusion: Cutting consciousness down to size* [J. Sydenham trans.]. London: Allen Lane.

Polanyi, M. (1958). *Personal knowledge.* London: Routledge & Kegan Paul.

Pullman, P. (2015). Visions of heaven: Inspiration and the unconscious. *Prospect, 292*, 21–25.

Roberts, B., & Donahue, E. (1994). One personality, multiple selves: Integrating personality and social roles. *Journal of Personality, 62*(2), 199–218.

Scataglini-Belghitar, G., & Mason, J. (2012). Establishing appropriate conditions: Students learning to apply a theorem. *IJSME, 10*(4), 927–953.
Scharmer, O. (2007). *Theory U: Leading from the future as it emerges. The social technology of presencing*. Cambridge: The Society for Organizational Learning.
Schoenfeld, A. (1985). *Mathematical problem solving*. New York, NY: Academic Press.
Seeley-Brown, J., Collins, A., & Duguid, P. (1989). Situated cognition and the culture of learning. *Educational Researcher, 18*(1), 32–42.
Shah, I. (1970). *The dermis probe*. London: Octagon Press.
Shah, I. (1983). *The exploits of the incomparable Mulla Nasrudin*. Two volumes in one London: Octagon Press.
Shantock Systematics Group. (1975). *A systematics handbook*. Sherborne: Coombe Springs Press.
Skilling, K. (2013). *Factors that influence year 7 students' engagement and achievement in mathematics* [Unpublished Ph.D. thesis]. Sydney: The University of Sydney.
van Hiele, P. (1986). Structure and insight: A theory of mathematics education. *Developmental psychology series*. London: Academic Press.
van Hiele-Geldof, D. (1957). The didactiques of geometry in the lowest class of secondary school. In D. Fuys, D. Geddes, & R. Tichler (Eds.), *English Translation of Selected Writings of Dina van Hiele-Geldof and Pierre M. van Hiele* 1984. Brooklyn, NY: National Science foundation, Brooklyn College. Reprinted from.
Varela, F. (1999). *Ethical know-how: Action, wisdom, and cognition*. Stanford, CA: Stanford University Press.
Varela, F., & Shear, J. (1999). *The view from within: First person approaches to the study of consciousness*. Bowling Green, OH: Imprint Academic.
Watson, A., & Mason, J. (2002). Extending example spaces as a learning/teaching strategy in mathematics. In A. Cockburn & E. Nardi (Eds.), *Proceedings of PME 26, Vol. 4* (pp. 377–385). University of East Anglia.
Webref. (2013). http://quoteinvestigator.com/2013/04/11/better-light/ (Accessed August 2015).
Wenger, E. (1998). *Communities of practice: Learning, meaning and identity*. Cambridge: Cambridge University Press.
Whittier, J. (1872). *The brewing of soma*. en.wikisource.org/wiki/The_Brewing_of_Soma (Accessed June 2015).
Wikipedia 1. (n.d.). en.wikipedia.org/wiki/Blind_men_and_an_elephant (Accessed August 2015).

Further Reading

Watson, A., & Mason, J. (2005). *Mathematics as a constructive activity: Learners generating examples*. Mahwah: Erlbaum.

CHAPTER

16

On the Irreducibility of Acting, Emoting, and Thinking: A Societal-Historical Approach to Affect in Mathematical Activity

*Wolff-Michael Roth**, *Margaret Walshaw*†

*University of Victoria, Victoria, BC, Canada

†Massey University, Palmerston North, New Zealand

[The separation of] intellect and affect ... is a major weakness of traditional psychology, since it makes the thought process appear as an autonomous flow of "thoughts thinking themselves." **(Vygotskij, 1934, p. 10)**

Affect and cognition, in most theoretical approaches to mathematics learning, are conceived of as two factors that influence a student's learning. Affect, in contemporary research, is affecting cognitive activity from the outside, and that activity, in turn, influences whether or not learning will take place. In underscoring the linear sequence of events, Piaget (1981) described emotions as the gasoline that drives the structured (cognitive) engine. In this chapter, we pursue a very different approach to the relation between cognition and affect, inspired by the works of L. S. Vygotsky (Vygotskij, 2001), who, as seen in the introductory quotation, urges researchers to consider emotions holistically. In this approach, every instant of human activity, every act, has affective, intellectual, and pragmatic colorings—none of which can be considered independent. Drawing on an analogy with water and the need to conceive of it as a compound rather than individual elements, to understand its dowsing qualities, Vygotskij (1934) argues that we have to study the conduct of a whole person in the full vitality of life rather than breaking affect and cognition apart.

http://dx.doi.org/10.1016/B978-0-12-802218-4.00016-9

Our departure from standard approaches to theorizing affect and intellect in mathematical learning builds on the relational nature of Vygotsky's (e.g., Vygotskij, 2005) ontology, as embodied in the leitmotif of *experience* [perežіvanie]. The explanation of affects (emotions) in mathematics, generally, and of anxiety, specifically, does not stop at an analysis of the social or historical context or at an analysis of the cognitive or neural. Rather, in denouncing the separation of affect and intellect, the framework conceives of culture and history as embodied and embodying contexts of how students think, do, and feel about mathematics. Hence, the emotions associated with learning cannot be separated from historical, social, and cultural antecedents that are inscribed in each student, specifically, and in the current situation, generally. Thus, it follows that emotions come into being *in* and *as relation with* the material and social context of the student, both in the past and the present, with other individuals and communities. To that end, a change within a particular classroom context may also result in a change in students' long-term assessments of themselves as mathematics learners. As a consequence, a task or event within the mathematics classroom likely has a direct bearing on the kinds of mathematical identities that students might take up and the kinds of proficiencies to which they might aspire.

We begin by reviewing some of the literature on affect in mathematics, generally, and mathematics anxiety, more specifically. We then articulate an approach to affect that has arisen in the societal-historical tradition in a lineage that ranges from its founders Vygotsky through Leont'ev and the German critical psychology (e.g., Holzkamp and Osterkamp-Holzkamp). We provide a concrete example of investigating affect in elementary classrooms and end with a discussion of the affordances that the societal-historical approach provides to researchers interested in affect and mathematics.

BACKGROUND

A number of competing characterizations of affect have found their way into mathematics education research. In our selective review we note that metaanalyses (e.g., Frost, Hyde, & Fennema, 1994; Hembree, 1990; McLeod, 1994) have been particularly influential. Studies utilizing existing self-report scales or informed by studies reported within the meta-analyses are extensive and include investigations into the relationship between mathematics performance and anxiety (e.g., Maloney, Schaeffer, & Beilock, 2013; Sloan, Daane, & Giesen, 2002). The relationship of affect to behavior, cognition, and physiology (e.g., Meece, Wigfield, & Eccles, 1990; Turner et al., 2002) has been explored from the perspective of both teachers and students.

Other scholarly work builds around the notion of affect as representation (Goldin, 1987). One model of affective structures is inclusive of beliefs and belief structures, values, attitudes, emotional states, ethics, and morals; and it is constitutive of a system of communication by means of facial expression, tears, exclamations, and so forth (DeBellis & Goldin, 2006). States of feeling are deemed to interact and exchange information with the cognitive system and other internal systems through a representational configuration. In one study on affect and calculator use, for example, affect is framed as a representation of information relating to factors such as previous experiences, conditions, events, and to others within the learning setting or environment (McCulloch, 2011). From the understanding that the mind conveys information unproblematically between the interior and exterior spheres, it becomes possible to draw a causal relationship between anxiety and mathematical performance.

Socioconstructivist perspectives on affect offer a persuasive criticism of the interior formulations of affect, proposing that affective states be described as reflections upon and interactions with the environment. For example, anxious responses to the world of mathematics can be made within the outward and inward meaning systems established by the individual and through the communities and settings in which the individual engages. This kind of conceptual grounding of affect, derived from post-Piagetian work (e.g., von Glasersfeld, 1996), has had a major influence within mathematics education (e.g., Di Martino & Zan, 2010; Else-Quest, Hyde, & Hejmadi, 2008).

Discursive accounts of affect bring to the fore the idea that affective states are discourses and, within that, the idea of the politicization of affect. Theorizing anxiety in this way moves to a concern with the governance and regulation of anxiety. For example, one study investigated the way in which affective states are constituted in discourses and controlled and monitored through technologies of power (Evans, Morgan, & Tsatsaroni, 2006). It was found that a student's anxiety in a small group was associated with the hierarchical positioning made available to that student relative to the other two students' hierarchical positioning. Anxiety was normalized and regulated within the dynamic of the students' interactional practices, surreptitiously operating through specific cues operating within the social fabric of small group classroom interactions.

A recent yet not so prevalent tendency conceptualizes affect from the psychoanalytic framework. Rather than relegating the unconscious, the irrational, and nonsense to the terrain of the uncharted, such approaches are based on theories of the psyche (Lacan, 1977; Žižek, 1998) and provide instructive lessons about the way in which the anxious self is enacted into being. Drawing on psychoanalytic resources, one study found that anxiety was initiated by the introduction of a new curriculum as differences emerged the images of teaching with which the teachers chose to identify and their own capacity to meet the new curriculum demands (Hanley, 2010). Another study focused on the anxieties and tensions resulting from the discursive masculine construction of mathematics (Bibby, 2010).

Taking a different stance from these four approaches to anxiety, our theoretical pathway asks: What if affective states were always in motion? What if anxiety was a reflection of activity paralleling conscious reflection that produces visibility as a cover? The approach attends to the psychophysiological correlates of affect. It does not equate affect with talk, beliefs, or feelings. The interest for us is not in developing internal representational schemes or in finding out how the exterior world influences the individual. Neither is the focus on discursive or intrapsychic activity. Rather, the interior and exterior relevant to an individual in a specific setting are investigated in their spatial and temporal complexity by using the category of experience [pereživanie]. In our work, societal-historical theory provides us with the tools to construct something more than a "simplified picture" (Ma, 1999, p. 523), one in which anxiety is read as mutually reinforcing collaborative activities and practices.

AFFECT: A SOCIETAL-HISTORICAL, PRAGMATIC APPROACH

In its early years, psychologists in the circle of Vygotsky and A. N. Leont'ev referred to themselves using the adjective societal-historical, a term taken up in some languages such as German, to make salient the fact that society is the phenomenon that determines the human psyche. Indeed, Vygotskij (2005) and Leont'ev (1983) both hold that higher psychological functions first *were* societal relations and that personality is the ensemble of societal relations a person has been part of. Their work has led to the categorical reconstruction of fundamental aspects of the human psyche, including affect and the associated motivation (e.g., Holzkamp, 1983; Holzkamp-Osterkamp, 1975, 1976). In this approach, it is recognized that variables are external to the phenomenon and variables exhibit *extraneous* relations to each other rather than something about the essence of the phenomenon. Precisely because the subjective dimensions of affective experience are eliminated in the variable approach and because the role of individual subjectivity in monitoring the person-environment unit becomes invisible when the data are gathered into distributions representing the population, traditional psychological approaches fail to get at the function of anxiety in the unity of subjective experience (Holzkamp, 1993). In the following, we articulate an approach to affect based on a reconstruction of the human psyche.

AFFECT: A REFLECTION (MEASURE) OF THE PERSON-ENVIRONMENT UNIT

In societal-historical approaches, affect is theorized very differently from mainstream psychology (Roth, 2011). Such approaches tend to engage in a materialist dialectical reconstruction of psychological categories. In such reconstructions, affect has been shown to be an *integral* aspect of life in a nonhomogeneous medium, that is, in life after the initially existing brine where cells first emerged (Leont'ev, 1959). At that time, together with sensitivity and mobility, affect emerged as an adaptation by means of which movements and the associated food availability came to be associated with a value (increase in food = positive, decrease in food = negative). Action, re/cognition, and affect came to be aspects of the irreducible organism-environment unit. Early on, this formerly single action-intellect-affect complex was separated into two aspects. On the one hand, affect was a measure of the very existence of the organism in its relation to the environment (e.g., "threat," changing sensation with movement); on the other hand, some aspects of the organism-environment unit became activity-orienting stimuli that were oriented toward future states of the relation. That is, already very early in the evolution, two aspects of affect emerged, one as a measure for the current situation and the other as a measure of the distance to some future goal. Affect, therefore, is not something particular to the individual organism but a measure of the subject-environment relation and, therefore, an aspect of the environment as well; thus, emotional processes are, next to the cognitive processes, "functional means in and for the activities of individual persons in relation to their world" (Dreier, 2008, p. 28). Unsurprisingly, emotions are found in every form of life activity (Holzkamp-Osterkamp, 1975). Leont'ev is held to write about the affective properties of the environment. With the evolved capacity of a dual form of affect, one related to the state of the current situation and the other a projected one as the result of the animal's activity, a characteristic exists that orients the animal toward the future and become a driving force, described by the concept of *motive* in societal-historical activity theory. The animal therefore no longer is governed by the current valuation of the relation, but it acts in view of the possible results of its activity. Thus, even though there may be hardship currently experienced with negative tonality, there is a tendency to continue because of the ultimate payoffs experienced positively.

The orientation to learning activities—that is, to demands from the surrounding (school) environment—may be expansive or defensive (Holzkamp, 1993). In *expansive learning*, engagement promises an engagement of action potential, which, inherently, comes with an increase in control over the environment. To experience a particular school task as expansive, students actually have to individualize the motive of activity (i.e., "buy in"). Increases arising in expansive learning are reflected affectively in a positive manner. In such a case classical approaches to psychology might detect "motivation."

On the other hand, *defensive learning* is oriented toward the avoidance of negative consequences, associated with negative affective values for the person. This becomes clear especially when the learning of mathematics in school is theorized in situation specific terms (Holzkamp, 1993). Thus, "students learn to reckon in school not to cope in the supermarket but to cope at school, that is ... to satisfy the teacher's expectation and to avoid failure and blame" (p. 490). The individual acts in ways that achieve results—for example, by means of copying or cheating—merely to avoid consequences associated with negative affective value. Indeed, the person might become good at achieving results without ever learning what the official curriculum states as the learning outcomes. As a result, people fail to use formal mathematics not only in the supermarket but also in school. The person would be described in classical psychological approaches as "unmotivated," when it really is more a case of avoidance of failure, negative repercussions, and negative forms of affect associated with both. In that case, "rewards" are often provided externally, which essentially is a way of compensating affectively for an experience not characterized in an affectively positive way. In expansive learning, negative situational forms of affect are accepted because of the ultimate increase in agential possibilities and control available to the individual (e.g., "no pain no gain"). In defensive learning, the same negative situational forms of affect appear in the light of the fact that the outcomes themselves will not have increased the person's room to maneuver and control over conditions.

Holzkamp (1993) provides an autobiographical example where, as a student, who heretofore has received only failing grades in mathematics, encounters a tutor hired to prepare him for the comprehensive examination of academic high school graduation (Ger. A*bitur*). In the story, the author wants to avoid mathematics and desires help in all the other subject areas to be examined. The tutor, rather than trying to teach him and to give specific advice, engages the student in discussions about issues in relation to the fundamental, deep structure of mathematics (e.g., the presence of contradictions and how the removal of contradictions leads to further contradictions). The student, beginning to ask questions to know more, participates in the representation of mathematics during the talk. When they eventually focus on the examination requirements, the student easily and in a relaxed manner copes with school mathematics and ultimately achieves a B (German "2" or "good"). In this account, leaving the school-related issue about exam preparation momentarily aside, the student comes to appreciate mathematics in the course of discussions where he finds out just what it is that excites professional mathematicians in doing mathematics. Another study investigates the engagement of pioneers attending a model airplane workshop (Leont'ev, 1983). In the initial condition, participants show little interest in the physics of flying and consult the resources infrequently (mode = 5 min). When the task conditions are changed such that students are asked to build airplanes that fly a certain distance, participants start to consult the physics resources considerably more (mode = 30 min). That is, what looks from a traditional psychological perspective as "low motivation" in the first condition changes into "high motivation." It does so not by providing incentives, but by changing the task conditions.[1] Participants begin to consult physics materials because it expands their room to maneuver with respect to a result that is affectively valued positive (achieving the motive of the activity, which is crossing the target line).

EXPERIENCE [PEREŽIVANIE]: CATEGORY AND UNIT OF ANALYSIS

Societal-historical theory (Vygotskij, 2001), as pragmatic philosophy (Dewey, 1934/2008), uses experience [pereživanie] as a category of analysis. This category spans activity as a whole, irreducible unit that has near one pole the subject and on the other pole the environment; it has internal and external dimensions simultaneously. The category thereby emphasizes the unity/identity [Rus. *Edinstvo*] of person and environment. Experience [pereživanie] therefore is a unit that has practical, cognitive, and affective colorings. Indeed, it is impossible to separate out intellectual, practical, and affective dimensions. The irreducibility of the experience category is another way of saying that there is "a single *quality* that pervades the entire experience in spite of the variation of its constituent parts. This unity is neither emotional, practical, nor intellectual, for these terms name distinctions that reflection can make within" (Dewey, 1934/2008, p. 44). It is only afterwards that we may consider experience to be dominated by more of one rather than another of the three attributes (i.e., intellect, practice, and affect). We may write the analytic unit as *person-acting-in-environment*. The Russian term pereživanie may be more appropriate than the English meaning of experience because it also translates "feeling," and it is therefore inherently is associated with affect (emotion). All experience, therefore, is reflected affectively (emotionally), even if in what may appear to be dispassionate behavior. At best, if considered on their own, practical, intellectual, and affective aspects are one-sided manifestations of the experiential

[1] Critical psychologists—such as Holzkamp, 1993), Holzkamp-Osterkamp, 1975), or Nissen (2012)—see in traditional motivation psychology little more than an endeavor to find ways to make people—school students and workers especially—do what they do not want to do on their own, and, in so doing, to increase the external control over them.

whole. To understand affect, therefore, we need to study its manifestation in the activity of interest rather than asking people *about* it. The environment of the person is not the objective environment of the scientist; instead, it is the environment as it appears to the person (Vygotskij, 2001), his or her lifeworld. How a person acts, what he or she says, and what affect expresses itself are in part manifestations of the environment as it appears to the person.

AFFECT IN AN ELEMENTARY MATHEMATICS CLASSROOM

In this section, we present a method of researching affect without splitting it from intellect and then show it at work in a case study from an elementary level mathematics classroom where the students are involved in a task intended to introduce algebraic concept at an early age (Roth & Radford, 2011). The case study manifests forms of affect that might well contribute to subsequent mathematics anxiety. We draw on research on psychophysiological processes, which has established vocal correlates of emotion (Scherer, 1989a).

INTRODUCTION: ETHNOGRAPHIC AND ANALYTIC BACKGROUND

The episode derives from a fourth-grade mathematics course, where the 9–10-year-old students engaged in an experimental curriculum intended to teach algebraic notion to elementary school students. Several groups of students were recorded during each lesson. During group work, a table microphone was wired into the camera (see, e.g., Fig. 3B, center of table. Because of standard informed consent procedures, the students knew that they were recorded. Extensive fieldwork shows that the (novelty) effect of the camera wears off within a lesson (Roth, 2005).

In the episode, the students were given goblets, play chips of two different colors, and a text about a girl who had received a piggybank and, after receiving a starting amount of \$6, added \$3 each week. Students were asked to model the story using the goblets and chips for the first 6 weeks and then fill up a table of values that had already some cells filled in Fig. 1.[2] The first of the two rows required students to write 3+6, 3+3+6, etc.;

[2] The text reads, "Problem 4. For her birthday, Marianne receives a piggybank containing \$6. She decides to save \$3 per week. At the end of the first week, she says "I have \$9. Questions: (a) Model the problem until week 6 with the help of the goblets and chips; (b) fill up the following table of values."

Problème 4 :

Pour son anniversaire, Marianne reçoit une tirelire contenant 6$. Elle décide d'épargner 3 $ par semaine. À la fin de la première semaine elle se dit : « J'ai 9 $! ».

Questions :

a) Modélise le problème à l'aide des boîtes et des jetons jusqu'à la 6e semaine

b) Remplis la table de valeurs suivante :

Numéro de la semaine	1	2	3	4	5	6
Montant épargé ($)	3+6	3+3+6	3+3+3+6	3+3+3+6	3+3+3+3+6	3+3+3+3+3+6
Ou	3+6	2x3+6	3x3+6	4x3 + 6	5x3+6	6x3+6

c) Trouve le montant épargné après la 10e semaine. Explique clairement ta démarche.

FIG. 1 The task and Aurélie's solution. Highlighted are the number of weeks in the first row and the numbers or number places in the third row, which will be those of the number of weeks.

in the second row there was the opportunity to rewrite the preceding row as 1×3+6, 2×3+6, etc. Ultimately, the curriculum stated as its intended outcome for children to arrive at a statement such as "the number of the week times 3 plus 6." Whereas some students arrived at filling up the table correctly, many others did not. Although Aurélie, the student we follow below, had almost all cells correct at the end of the lesson, she had in fact copied from her classmate seated to her left.

Experience [pereživanie], as defined above, cannot be usefully decomposed into its practical, intellectual, and affective manifestations without a loss of the phenomenon. There is perhaps no better case than that of speaking, which constitutes acting to affect others in the course of getting a job done. The words spoken, in many (but not all) situations, have semantic relevance to the situation. But we all know that speaking also expresses affect—angry people tend to shout, increase pitch, and speak faster (Goodwin, Goodwin, & Yaeger-Dror, 2002; Paulmann, Titione, & Pell, 2012; Scherer, 1989b).

Research on the features used in the communication of affect has found—in addition to intensity differences, pitch, and speech rate—that the vocal expression of emotion may also involve pausing structure (Cahn, 1990), segmental features (Kienast, Paeschke, & Sendlmeier, 1999), and the duration of accented or unaccented syllables (Mozziconacci, 1998). All these features of speech recognize affect successfully, whether the affect has a positive (e.g., confident, interested, happy) or a negative (e.g., angry, stressed) valence. Sighs, cries, and inhalations also demonstrate a fidelity to the speaker's emotions (Schröder, 2000). If voice, including extralinguistic interjections, has been shown to map successfully onto the speaker's affect, the vocal expression of emotion is then a relational practice in a shared space. Rather than an intrapsychic experience, the vocal expression of emotions, as Parkinson (2011) has argued, has a communicational intent. As "active and embodied modes of engagement with the social and practical world" (p. 409), they are oriented to other people's actions and reactions, whether actual, anticipated, or imagined. In addition to voice, body movements—their rhythm and temporality—are also important ways in which affect manifests itself in intra- and intercultural communication (Roth, 2007; Roth & Tobin, 2010).[3]

Analyzing language and body movements allows us to view different dimensions of situated behavior without reducing it to any one of its manifestations. For example, in studies of psychophysiology, vocal correlates have been identified that go with specific affects (e.g., Scherer, 1989a). Among the prominent, often tested correlates include pitch (F_0), pitch range, pitch contour, and speech intensity (e.g., Table 1).[4] Although the first and second dominant frequencies (F_1, F_2) are sometimes included,

TABLE 1 Selected Affects and Their Correlates According to Published Studies

	Emotion			
	Resignation (Dejection)	Frustration		Anxiety
		Despair	Displeasure	
F0 (pitch)				
Mean	< >	>	>	>
Range	≤	>		
Contour	≤	>		>
Intensity				
Mean	≤≤	>	>	
Speech rate	≤	>		
High frequency energy	> <	>>	>	>

Note. > *is increase from normal;* < *is decrease;* ≤ *indicates the same or decrease; and* < > *indicates direction depending on preceding voice type, which exerts opposing force.*
Based on Scherer, K. R. (1989a). Vocal correlates of emotional arousal and affective disturbance. In H. Wagner & A. Manstead (Eds.), Handbook of social psychophysiology *(pp. 165–197). New York: John Wiley & Sons.*

[3] It has to be pointed out that serious scholars reject the notion of "body language," because there is no formal grammar that regulates body movements in the way that such grammars exist for language and language-like gestures (e.g., American Sign Language) and emblems (e.g., the sticky finger).

[4] In our work, we make use of the PRAAT phonetics software, which is freely available for the most common operating systems, including Macintosh, Windows, Linux, SGI, and as source code (http://www.fon.hum.uva.nl/praat/).

classroom recordings include a lot of background noise that affects especially the higher frequencies but not the information of pitch. The analysis of the extract that follows is given as an example of the plausibility and potential of analyzing voice and body movements for studying students' ongoing classroom experience. For example, in Fragment 1, Aurélie asks what they should do next (turn 05). It follows the same query on the part of her peer Mario, and a logatome[5] (i.e., "Woing"). In stating "Table," the next turn treats the statement as a query. This is accompanied by a deictic (pointing) gesture to the table of values on the worksheet, as an invitation to fill up that table as per task request.

Fragment 1

01	M:	On a vingt quat'. So maintenant quoi? [We have twenty-fo'. So what now?]
02		(0.66)
03	T:	Oing. [Woing]
04		(0.91)
05	A:	Maintenant quoi? [Now what?]
06	T:	TablEAU. [Table.] ((*Points to table of values* on worksheet))

In this situation, her voice was normal, matter of fact, asking how to proceed. The mean pitch is $F_0 = 256$ Hz, with a range from 290 to 230 Hz (Fig. 2, left). There is a falling tendency of the pitch with a modulation of the interrogative *quoi* [what] rising toward the end, which therefore is heard as a question. One might even hear a hint of resignation in her voice—especially given the fact that immediately before she has already raised a question about what they were doing and her bodily manner (with upper body oriented slightly backward and (right) hand/arm supporting head) could be perceived by participants and observers alike to indicate she is uninvolved (e.g., Roth & Jornet, 2017).[6]

On the other hand, Aurélie only a few instants later, reproduced in Fragment 2, asks Mario where he was seeing that the first week they had $3 plus $6 in their piggybank. It is a voice that manifests frustration and despair. We observe, consistent with the reported correlates for these emotions, a much higher mean pitch ($F_0 = 584$ Hz), a large variation in the pitch from a low of 239 to a high of 625 Hz, and a large contour just on the syllable *ça* [that] alone (Fig. 2, right). In addition, the mean speech intensity

[5] Literally "sense cut off," a logatome is pseudoword, a nonsense vocalization.

[6] Although bodily behavior is sometimes talked about in terms of "body language," this term is a folk psychological notion. This is so "because body movements and positions are neither structured nor used like language" (Roth, 2001, p. 368).

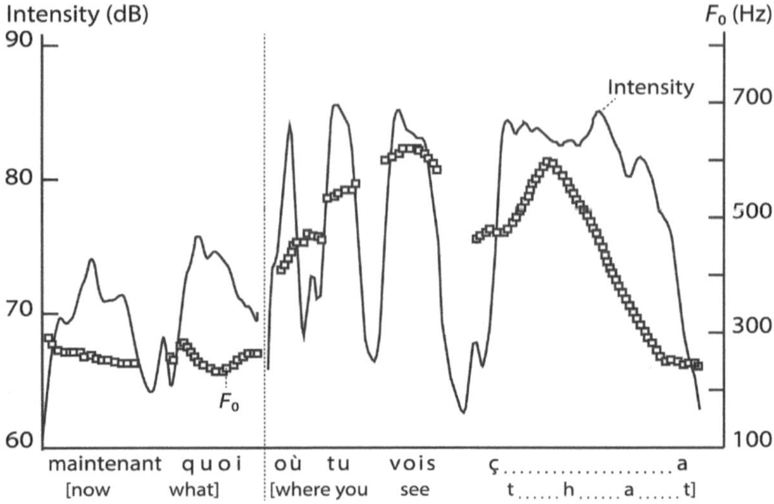

FIG. 2 Prosodic analysis including pitch (F_0, Hz) and intensity (dB) for the normal voice (left) and plaintive voice with despair and frustration (right).

has risen from $I=71.5–81.3\,dB$, which equals nearly 10 times the speech intensity with respect to the first articulation. A longish pause follows before Mario, using a confirming "Yeap," then offers up a query, "What are you doing?" Aurélie repeats, and, using her first name, directly names the addressee. The articulation of "Yep [Ouai]" exhibits a contour similar to the earlier one ($243 \leq F_0 \leq 516\,H$).

Fragment 2

01	M:	Oh. Trois plus six. [Oh. Three plus six]
02		(0.72)
03	A:	↑OÙ TU VOIS <u>ÇÀ</u>↓<u>:::À</u> [Where do you see th*at*?]
04		(1.05)
05	M:	Ouai. Qu'est tu fais? [Yep. What are you doing?]
06	A:	Ouai. Qu'est-ce que tu fais, Thérèse? [Yep. What are you doing, Thérèse?]

With respect to body movements, when a person clearly steps away from an activity, or moves in other ways that are accompanied by no longer doing what the current task demands, others in the setting perceive the behavior as disengagement. Pounding on a feature of the setting (e.g., table or desk) or kicking a door are other ways in which Aurélie's anger or disappointment manifest.

A CASE STUDY OF AFFECT IN A MATHEMATICS LESSON

In *A cultural-Historical Perspective on Mathematics Teaching and Learning,* Roth and Radford (2011) provide an extended case study of one fourth-grade student (Mario) in particular, who, although frustrated and saying that he does not understand persists and, eventually, and including an extended exchange with the teacher, comes to a point where he says, "Me, I now understand." He was frustrated; and the teacher, too, was frustrated because he did not understand and because she apparently had trouble helping him overcome his block. From the societal-historical approach taken here, Mario *had to act* because it was only in acting that the affect could change. Only through his action could he diminish the gap between his current state and the desired state of the activity, understanding the task and having learned what it was intended to teach. Even though frustration was apparent and increased at times, the actions that also expressed frustration came to overturn the frustration when Mario, after doing some of the work on his own, could see the point of his task. From a societal-historical perspective that takes inspiration from a Marxist reading of Spinoza, we might say that "affect is vanq[ui]shed by affect" (Vygotsky, 2010, p. 93, original underline). At that point, when he discovered the pattern, he also recognized the motive of the activity, that is, why he had been asked to do what he had done. In this way, because the learning task was undisclosed and required him to arrive without knowing where he was going, his frustration was the affective manifestation of the activity. On the plane of intellect, the activity was reflected in the reflexive comment, "This is dumb. I don't understand."

Roth and Radford present and analyse about 75% of the 12-min video and 171 turns at talk of the original rough transcription. However, we also have access to the other 25%, including a further 42 turns of the raw transcription. In the total of 171 turns, Aurélie contributes especially in the beginning (12 of the first 42 turns), the part that Roth and Radford do not provide; and her participation teeters out in the part that they do because there are only 3 of the last 100 turns when she talks, all of which are concerned with having finished the task. Aurélie has three one-word turns ("Yes" and twice "Madame"), none of which is associated with a reply. Ten turns are concerned with their progress in the activity ("Now what?"; "We did all this"; "Look, we already wrote this"; "We haven't even finished the first"; "What are you doing Thérèse?", "Are we supposed to do this?"; "Me, I finished"; "Finished"; and "We need to do c, d, and e").

The remaining turns pertain to the task. Especially her in/comprehension of what to do had to be worked out. Thus, for example, she tells Mario that they only have to go to 24 [chips]—which is 6 days × \$3/day + \$6—and asks him why he only adds blue chips. She twice tells Thérèse that they (the event leaves unclear whether she alone or she is with Mario) are working on a different task, or that the next item in the table is "three plus three." Then the comments about incomprehension begin and become frequent. She says, among others, "I am just going to the next one because I have no clue what [Thérèse] is doing"; "This makes no sense"; "We have no idea what [Thérèse] is doing"; "I don't understand/comprehend, and I will never understand/comprehend"; and "I don't understand." Many of those statements are marked by vocal qualities that Roth and Radford describe as "plaintive" and "lamenting." Thus, for example, Fig. 2 exhibits the sharp rises in pitch values, which more than double and then return; the figure also shows the sharp upward contour on a single syllable "ça [that]." In the course of the episode, and before she completely abandons, Aurélie repeatedly ($n = 8$) heavily hits the desk with her hand (Fig. 3A), followed by some additional movement that further manifests disengagement. For example, at one point Aurélie, after having pounded on the desk, throws herself against the backrest of the chair and, remaining there, emptily gazes into the air (Fig. 3B).

Over the first 61 turns in the raw transcriptions, the vocal expressions of displeasure and despair increase until, at about 3:40 min into the video. That crucial instant arises out of a situation when Aurélie has her head on the table, apparently not working on the task and having asked Thérèse what she is doing and twice noting that she has no idea what Thérèse is doing. Now Thérèse is writing "3×3" into Aurélie's table cell below "week 2" (Fig. 3C, left; see also Fig. 1). Six seconds later, Aurélie rises up, turns into the direction of and gazing toward Thérèse and her worksheet (Fig. 3C, right), and announces: "I don't understand/comprehend and I will never understand/comprehend." It is apparent that we may see precisely here markers of a later mathematics anxiety, a reflection of the fact that she will never understand whatever the amount of work she puts into mathematics. Moreover, because she will not understand, she will fail, which would be apparent in the marks received. A condition in which one has to engage in an activity that necessarily leads to failure, with no way out because of the compulsory nature of the school subject, will find its expression in a particular form of affect—often precisely mathematics anxiety.

After that she only makes three more brief comments about having finished and one comment querying whether they have to do the latter part of the worksheet. It is not as if she had completely abandoned working on it. When the teacher asks whether they had done parts c through e of the assignment, it turns out that the two girls have not. Following that exchange, Aurélie can be seen again writing down numbers, which turns out to be the work of figuring out how much money there would be in the piggybank after day 10 (Fig. 3D, left; see also Fig. 1, lower right). However, after working on that part for a while, Aurélie again pounds the table, pushes the worksheet away, and rests her head on the table (Fig. 3D, right).

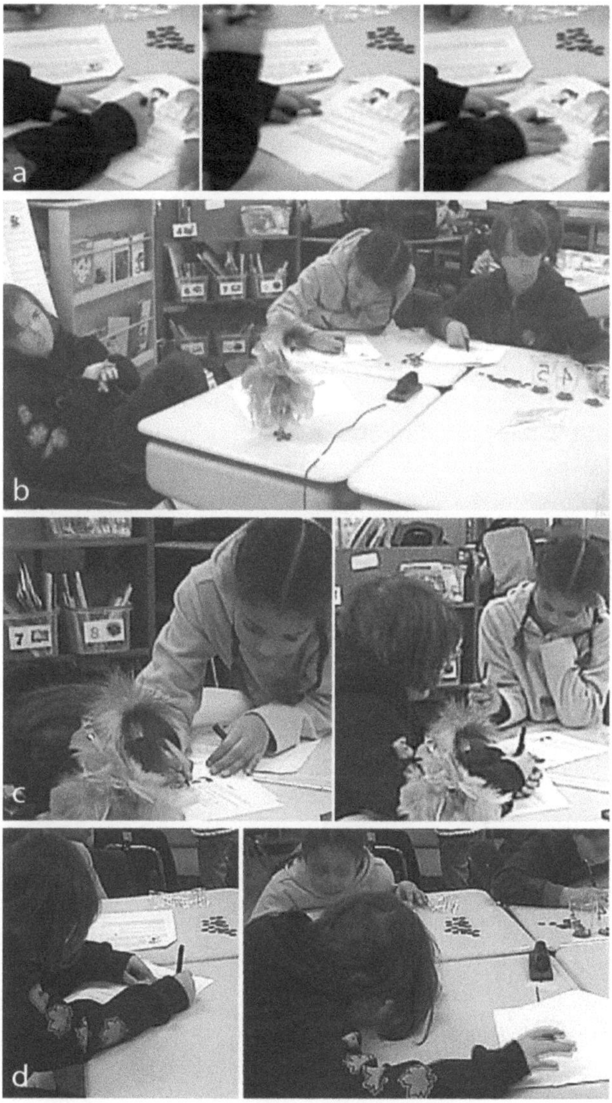

FIG. 3 Bodily manifestations of emotion in a mathematics lesson. (A) One of the repeated instances in which Aurélie pounds on the desk—manifestations of displeasure, frustration, and perhaps despair. (B) Aurélie has thrown herself against the backrest of her chair after having pounded on the desk. (C) Thérèse is "helping" Aurélie by filling in the contents of cell 2 in the first row to be filled (left); 6 seconds later, Aurélie announces, "I don't understand; and I will never understand. (D) Having given it another try at continuing with the task (left), Aurélie slaps the table once again then pushes the worksheet away and rests her head on the desktop (right) (faces have been blurred).

These are vocal correlates of displeasure and despair. But one also observes those correlates that go with resignation (dejection), which, for example, prosodically expresses itself while saying, "I have no clue what she is saying." A significant turn will have been the situation where she articulates what may be the beginning of a self-fulfilling prophecy. It is the statement pertaining to her current state of understanding of the task, "I don't understand [je ne comprends pas]," and the consequence Aurélie states for the future, "and I will never comprehend/understand [et je ne vais jamais comprendre]." The plot of the prosody parameters exhibits the doubling of the normal pitch, the wide range in pitch while talking about comprehension ($246 \leq F_0 \leq 484$ Hz), and a mean pitch of $F_0 = 368$ Hz (Fig. 4, left), which is considerably higher than normal (Fig. 2, left). The statement begins fast and then slows down (e.g., on the syllable "not [pas]," which transitions into the prediction about comprehension in the future, drawn out with small pitch range ($227 \leq F_0 \leq 330$ Hz), a mean pitch near the normal ($F_0 = 271$ Hz), and a low contour (Fig. 4, right). The spectrogram exhibits a lot of energy in the higher frequency ranges in the sound-word "understand [comprends]," and especially in the two negations "not [pas]" and "never [jamais]" (Fig. 4). The underlined syllables show that these prosodic parameters manifest themselves to the listener as emphases, which, therefore, all fall on those associated with comprehension and negation.

Following the articulation that she will never understand (comprehend), Aurélie virtually ceases all work toward the task and picks up only one more time when the teacher points out that they also had to do parts c

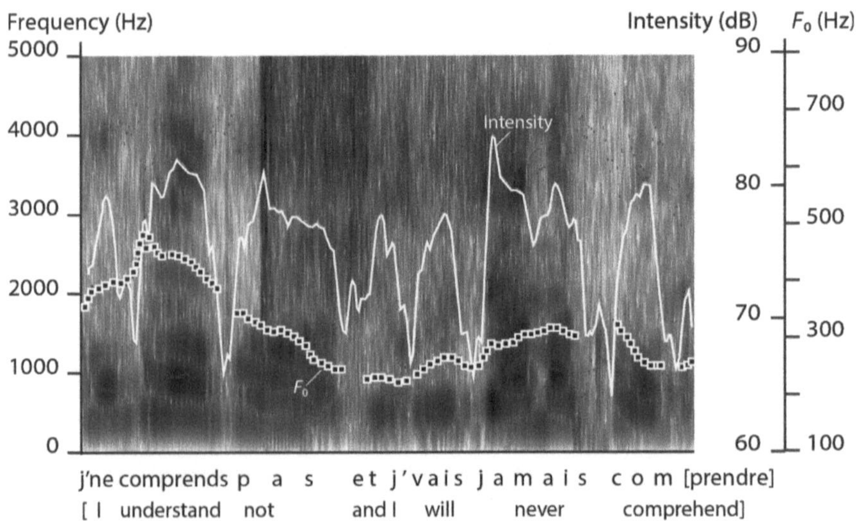

FIG. 4 Prosodic parameters—spectrogram frequencies, intensity (–), and pitch (F_0, □□)—manifest displeasure/despair, and then resignation (dejection).

through e of the worksheet. But not working on the task inherently comes with incomprehension; the continuing frustration has turned into despair and then resignation. Most immediately, never understanding pertains to this task, which she did not come to understand precisely as she said, and which is the direct result of not doing the work. In contrast to Mario, who continued despite the frustration manifested in his actions and words, displeasure (frustration) has led to despair and finally resignation for Aurélie. However, one can easily imagine that after a few experiences such as this one, for Aurélie, not only individual tasks but also mathematics more generally will be associated with negative forms of affect to the point that she might exhibit mathematics anxiety. "Never understanding" then is extended—as it does in so many instances and especially amongst girls—to mathematics as a whole (Walkerdine, 1988). As Table 1 shows, the vocal correlates of anxiety all are in the same directions as those of the affects manifested in this lesson.

CASE DISCUSSION

It is important to point out that in the foregoing, we do not separate affect from other aspects of the activity. We do not ask Aurélie what and how she feels, and we make no inferences about her internal affective states. We report those vocal and behavioral correlates that observers in everyday life treat as manifestations of certain affective states, and that are also the basis of the perceptions that establish the vocal and behavioral correlates of affect. That is, sociologists of emotion show the ways in which affect is written all over social situations (Collins, 2004); and analyses have shown that humans understand affect only because of its exteriority (Husserl, 1939), that is, when they have come to look at themselves through the lens of other people (Franck, 1981). Taking the societal-historical perspective that Vygotskij (1934) proposes, we do not separate words from their material sound carrier so that they combine into one unit the intellect and affect. If we had done so, we would not understand why the sound has those physical characteristics that go together with the manifestations of psychological states; words would have been regarded to be the result of pure mental acts. Moreover, any statement about affect, which Aurélie in fact does not provide as part of her ongoing activity, would only be an intellectualization of something that is not intellectual at all: the fact that affect is something that we are exposed to, that we are patients who are affected by something that is outside of our control. Nobody, not Aurélie or Mario and not even the teacher or witnesses of the events, can decide not to feel frustrated when they are frustrated or not to feel anxious when they feel anxious. But engagement in the task does have the potential to change our state—for better but also for worse. Without acting, however, Aurélie cannot have any hope to get out of frustration, and in resigning, she has given up hope to ever get out of a state that intellectually is described as not understanding and which is reflected affectively by the despair that comes with knowing that one cannot reach some goal.

In this episode, affect manifests itself in and through speaking and acting. It is a reflection of the specific person-environment relation; and this specificity would only become invisible if we were to measure it by means of variables put together with the results from other persons, where the specificity of affect to the situation is treated as error variance. Instead, acting, knowing, and feeling all are manifestations of the same unit (category) experience [pereživanie]; this unit includes what is commonly referred to as experience, that is, the manifestations in the different forms of consciousness reflecting the person-environment relation. This is particularly salient in those statements that do not concern comprehension or knowing what to do, such as "Yo dude, we don't have the same thing as you" or, while referring to the yellow highlighted numbers or places for numbers in the table of values, "It's written here in yellow," or while saying "the next is three plus three." There is no evidence that in addition to contributing to the group talk to move the mathematics task ahead Aurélie also reflected on her affective state and expressed it *differently* from and in addition to what she was saying or doing.

An immediate consequence of this work is that scholarly research cannot investigate and report on the words participants produce alone. Instead, if affect (emotion) manifests itself prosodically, such information is then required as evidence supporting the claims of researchers. If body movements and position are manifestations of affect that cannot be reduced to semantics, research reports then must include them as evidence (data) supporting the claims of researchers. This is so, as we see in this case study, because the semantics (intellect) is different from sound, it expresses something that is different epistemologically. Manifesting itself in prosody or body movements, there is no evidence that participants are conscious of and reflecting about their affective state in an intellectual manner. Translating what can be seen or heard into the verbal modality, however, makes it appear as if affect merely were another way of presenting intellectual content (conscious awareness). Instead, affect is a different plane of reflection than consciousness (thinking) (Leont'ev, 1983; Vygotskij, 1934); and the evidence mustered in research accounts has to take this irreducible difference into account.

GENERAL DISCUSSION

The central aim of this chapter is to contribute to the development of a theory for the work that affect generally and anxiety specifically do within the mathematics classroom. Premising the development on two key understandings—namely, (a) that affect is a vital characteristic of and intrinsic to classroom life and (b) that affect shapes, is shaped by, and may preclude engagement in mathematical activity—we are interested in understanding how affect plays into students' disposition toward and performance in mathematics. Our case study exemplifies the interrelation of verbal and bodily expression, and from that revelation, it is a small step to suggest that affect is one form of reflection of the person-environment unit that not only parallels conscious (intellectual) awareness, but also perfuses the latter. The difference between the affective and intellectual forms of reflection clearly exhibits itself in the physical manifestations (e.g., Aurélie's voice, pounding actions, and body positions), which the intellect does not require because of its metaphysical qualities. In fact, this articulation raises the very question about the connection between affect and intellect in traditional research approaches that one of the foremost thinkers formulated at the very end of his life: how can something nonphysical, such as thought or the psyche, be accompanied by simultaneous physical events in the organism unless the two are of the same kind (Freud, 1999)? If intellect and affect are but manifestations of a *unity/identity*, an irreducible whole, then the entire question of their relation becomes superfluous.

Pivotal to our conceptualization of affect are the mutual and relational effects of the society and persons within their setting, time, and space. When affect is conceived of in this way, it points to the realization that affect is a reflection of individuals interacting dynamically with layers of societally, culturally, and, in Aurélie's case, historically perceived-as-negative situations or influences pertaining to mathematics. In this framing, neither nature nor nurture are passive backdrops but are, rather, vital performative agents in cuing experience, cultivating ways of thinking, being, and doing; in short, they are individuating the individual confronted with a mathematical task toward a particular response. The anxious individual, as an example, is thus capacitated to be anticipatory so as to feel and act with emotional distress as soon as an encounter with numerical information or a mathematical task is made.

In the societal-historical approach, it is recognized that the environment *can be* articulated in an objective manner; however, *how this environment is reflected* in individual consciousness and affect cannot be known without an investigation specific to the individual (Vygotskij, 2001). Aurélie does what she does and feels the way that she feels because of her way of being in the world; the expressed affect is a manifestation of how Aurélie finds herself caught up in a situation not in or under her control. From Vygotsky's Spinozist position, the beginnings and endings of thinking, doing, and feeling always exist in the environment so that we can understand Aurélie only when we consider her irreducibly enmeshed with the environment. However, standard approaches to affect in mathematics generally and to anxiety specifically eliminate the specificity to the individual by aggregating data across individuals (Holzkamp, 1993). As a result, the unity/identity of the phenomenon—i.e., experience theorized as the person-environment unit where the practical, intellectual, and affective perfuse each other—has been lost. In the societal-historical approach, this unity/identity of person and environment is regained, for it investigates affect in the context of the individual subject's relation to the personally experienced environment, to which the subject always orients herself in ways that are grounded in her position and in the "good reasons" that underlie her actions. But research approaches that aggregate data across individuals cannot get at this phenomenon, which needs to be investigated at a different level (unit) and with the specificity arising from the person-environment transaction. As R. E. Snow, an educational psychologist interested in aptitude suggested, we cannot understand the contributions of the person to behavior unless we account for the environment, and we cannot understand the contributions of the environment to behavior, unless we take into account the characteristics of the person (Corno et al., 2002). As a result, if we give an instrument to many individuals, we find out a lot about the instrument; conversely, to find out a lot about persons, we have to follow them across many different situations.

In some research traditions, researchers might be interested in investigating affect when there are (a) fewer contextual hints or (b) when people try to consciously hide emotions. Apart from the fact that the American Psychological Association takes the position that "there is little evidence that polygraph tests can accurately detect lies" (APA, 2004), from a societal-historical approach to understand social *life* the question of concealed emotion is not of interest. What really matters is the transactional relevance of what people do, because it influences the unfolding of life in the particular event. Thus, if a certain emotion does not perceivably manifest itself, it will not affect the situation and others. It therefore does not enter the transactional dimensions that move the event ahead. But lack of emotional expression may indeed become the topic of the relation between two people, at which point the lack of emotion expressed becomes a driving moment of the unfolding event. The method does allow following affect at a collective level in classrooms, such as the levels of solidarity with a fellow student, shown in the coordination of rhythmic parameters in the ebb and flow of an argument (with the teacher) over matters of subject matter knowledge (Roth & Tobin, 2010).

The relational approach thus allows us to address the issue of decontextualization that appears to put limits on the study of affect within mathematics education. By way of explanation, words such as troubled, worried, panicky, heated and disengaged, withdrawn, frustration and despair, while associated with affect, can never fully capture the small print of the emotive register. Expressions of emotions, as in Aurélie's case, rising to a grand crescendo, will always exceed their own terms and conditions. Approaches that conflate or reduce experience to speech, as this is done in discursive approaches, draw on a model that assumes words and speech are innocent carriers of an interior semantic. The very terms in which anxiety is articulated in such approaches are haunted by terms that preclude it from confronting the problem of representation. Acknowledging the indeterminacy of speech, we have taken up the challenge to understand the fullness of emotional experience within the mathematics classroom.

References

American Psychological Association (APA). (2004). *The truth about lie detectors (aka polygraph tests)*. http://www.apa.org/research/action/polygraph.aspx. (Accessed 3 May 2016).

Bibby, T. (2010). *Classrooms and their discontents: Education and psychoanalysis*. London: Routledge.

Cahn, J. (1990). The generation of affect in synthesized speech. *Journal of the American Voice Input/Output Society, 8*, 1–19.

Collins, R. (2004). *Interaction ritual chains*. Princeton, NJ: Princeton University Press.

Corno, L., Cronbach, L. J., Kupermintz, H., Lohman, D. F., Mandinach, E. B., & Porteus, A. W. (2002). *Remaking the concept of aptitude: Extending the legacy of Richard E. Snow*. Mahwah, NJ: Lawrence Erlbaum Associates.

DeBellis, V. A., & Goldin, G. A. (2006). Affect and meta-affect in mathematical problem solving: A representational perspective. *Educational Studies in Mathematics, 63*, 131–147.

Dewey, J. (2008). Later works vol. 10: Art as experience. In J.-A. Boydston (Ed.), Carbondale, IL: Southern Illinois University Press [First published in 1934].

Di Martino, P., & Zan, R. (2010). 'Me and maths': Toward a definition of attitude grounded on students' narratives. *Journal of Mathematics Teacher Education, 13*, 27–48.

Dreier, O. (2008). *Psychotherapy in everyday life*. Cambridge: Cambridge University Press.

Else-Quest, N. M., Hyde, J. S., & Hejmadi, A. (2008). Mother and child emotions during mathematics homework. *Mathematical Thinking and Learning, 10*, 5–35.

Evans, J., Morgan, C., & Tsatsaroni, A. (2006). Discursive positioning and emotion in school mathematics practices. *Educational Studies in Mathematics, 63*, 209–226.

Franck, D. (1981). *Chair et corps. Sur la phénoménologie de Husserl* [Flesh and body: On Husserl's phenomenology]. Paris: Éditions de Minuit.

Freud, S. (1999). *Gesammelte Werke Band XVII* [Collected works vol 17]. Frankfurt: Fischer-Verlag.

Frost, L. A., Hyde, J. S., & Fennema, E. (1994). Gender, mathematics performance, and mathematics related attitudes and affect: A meta-analytic synthesis. *International Journal of Educational Research, 31*, 373–385.

Goldin, G. A. (1987). Cognitive representational systems for mathematical problem solving. In C. Janvier (Ed.), *Problems of representation in the teaching and learning of mathematics* (pp. 125–145). Hillsdale, NJ: Lawrence Erlbaum Associates.

Goodwin, C., Goodwin, M. H., & Yaeger-Dror, M. (2002). Multi-modality in girls' game disputes. *Journal of Pragmatics, 34*, 1621–1649.

Hanley, U. (2010). Teachers and curriculum change: Working to get it right. In M. Walshaw (Ed.), *Unpacking pedagogy: New perspectives for mathematics classrooms* (pp. 3–19). Charlotte, NC: Information Age.

Hembree, R. (1990). The nature, effects, and relief of mathematics anxiety. *Journal for Research in Mathematics Education, 21*, 33–46.

Holzkamp, K. (1983). *Grundlegung der Psychologie* [Founding (foundation of) psychology]. Frankfurt: Campus.

Holzkamp, K. (1993). *Lernen: Subjektwissenschaftliche Grundlegung* [Learning: Foundation in a science of the subject]. Frankfurt: Campus.

Holzkamp-Osterkamp, U. (1975). *Grundlagen der psychologischen Motivationsforschung* [*Foundations of psychological motivation research*]: *Vol. 1*. Frankfurt: Campus.

Holzkamp-Osterkamp, U. (1976). *Grundlagen der psychologischen Motivationsforschung* [*Foundations of psychological motivation research*]: *Vol. 2*. Frankfurt: Campus.

Husserl, E. (1939). *Erfahrung und urteil: Untersuchung zur Genealogie der logik* [Experience and judgment: Investigation to the genealogy of logic]. Prague: Academia.

Kienast, M., Paeschke, A., & Sendlmeier, W.-F. (1999). Articulatory reduction in emotional speech. In *Proceedings of Eurospeech 99. Budapest* (pp. 117–120).

Lacan, J. (1977). *Ecrits: A selection*. London: Tavistock.

Leont'ev, A. N. (1959). *Problemj razvitija psixiki* [Problems of the development of mind]. Moscow: Akademii Pedagogičeskix Nauk.

Leont'ev, A. N. (1983). *Izbrannye psixologičeskie proizvedenija* [*Selected psychological works*]: *Vol. 2*. Moscow: Pedagogika.

Ma, X. (1999). A meta-ana1ysis of the relationship between anxiety toward mathematics and achievement in mathematics. *Journal for Research in Mathematics Education, 30*, 520–540.

Maloney, E. A., Schaeffer, M. W., & Beilock, S. L. (2013). Mathematics anxiety and stereotype threat: Shared mechanisms, negative consequences and promising interventions. *Research in Mathematics Education, 15*, 115–128.

McCulloch, A. W. (2011). Affect and calculator use. *Journal of Mathematical Behavior, 30*, 166–179.

McLeod, D. B. (1994). Research on affect in mathematics education in the JRME: 1970 to the present. *Journal for Research in Mathematics Education, 25*, 637–647.

Meece, J. L., Wigfield, A., & Eccles, J. S. (1990). Predictors of math anxiety and its influence on young adolescents' course enrolment intentions and performance in mathematics. *Journal of Educational Psychology, 82*, 60–70.

Mozziconacci, S. J. L. (1998). *Speech variability and emotion: Production and perception*. Eindhoven: Technische Universiteit Eindhoven.

Nissen, M. (2012). *The subjectivity of participation: Articulating social work practice with youth in Copenhagen*. Basingstoke: Palgrave-Macmillan.

Parkinson, B. (2011). How social is the social psychology of emotion? *British Journal of Social Psychology, 50*, 405–413.

Paulmann, S., Titione, D., & Pell, M. (2012). How emotional prosody guides your way: Evidence from eye movements. *Speech Communication, 54*, 92–107.

Piaget, J. (1981). *Intelligence and affectivity: Their relationship during child development*. Palo Alto, CA: Annual Reviews.

Roth, W.-M. (2001). Gestures: Their role in teaching and learning. *Review of Educational Research, 71*, 365–392.

Roth, W.-M. (2005). *Doing qualitative research: Praxis of method*. Rotterdam: Sense Publishers.

Roth, W.-M. (2007). Emotion at work: A contribution to third-generation cultural historical activity theory. *Mind, Culture, and Activity, 14*, 40–63.

Roth, W.-M. (2011). Motives and emotion: A cultural-historical activity theoretic approach to motivation in learning and work. In D. M. McInerney, R. A. Walker, & G. A. A. Liem (Eds.), *Sociocultural theories of learning and motivation: Looking back, looking forward* (pp. 43–63). Greenwich, CT: Information Age Publishing.

Roth, W.-M., & Jornet, A. (2017). *Understanding educational psychology: A late Vygotskian Spinozist approach*. Dordrecht: Springer.

Roth, W.-M., & Radford, L. (2011). *A cultural-historical perspective on mathematics teaching and learning*. Rotterdam: Sense Publishers.

Roth, W.-M., & Tobin, K. (2010). Solidarity and conflict: Aligned and misaligned prosody as a transactional resource in intra- and intercultural communication involving power differences. *Cultural Studies of Science Education, 5*, 805–847.

Scherer, K. R. (1989a). Vocal correlates of emotional arousal and affective disturbance. In H. Wagner & A. Manstead (Eds.), *Handbook of social psychophysiology* (pp. 165–197). New York: John Wiley & Sons.

Scherer, K. R. (1989b). *Vocal measurement of emotion*. New York: Academic Press.

Schröder, M. (2000). Experimental study of affect bursts. In R. Cowie, E. Douglas-Cowie, & M. Schröder (Eds.), *Proceedings of the ISCA workshop on speech and emotion: A conceptual framework for research* (pp. 132–137). Belfast: Queen's University.

Sloan, T., Daane, C. J., & Giesen, J. (2002). Mathematics anxiety and learning styles: What is the relationship in elementary preservice teachers? *School Science and Mathematics, 102*, 84–87.

Turner, J. C., Midgley, C., Meyer, D. K., Ghenn, M., Anderman, E. M., Kang, Y., et al. (2002). The classroom environment and students' reports of avoidance strategies in mathematics: A multimethod study. *Journal of Educational Psychology, 94*, 88–106.

von Glasersfeld, E. (1996). Aspects of radical constructivism and its educational recommendations. In L. Steffe & P. Nesher (Eds.), *Theories of mathematical learning* (pp. 307–314). Mahwah, NJ: Lawrence Erlbaum Associates.

Vygotskij, L. S. (1934). *Myšlenie i reč': psixologičeskie issledovanija* [Thinking and speaking: Psychological investigations]. Moscow: Gosudarstvennoe social'noèskonomičeskoe isdatel'stvo.

Vygotskij, L. S. (2001). *Lekcii po pedologii* [Lectures on pedology]. Izhevsk: Udmurdskij University.

Vygotskij, L. S. (2005). *Psyxhologija razvitija čeloveka* [Psychology of human development]. Moscow: Eksmo.

Vygotsky, L. S. (2010). Two fragments of personal notes by L. S. Vygotsky from the Vygotsky family archive. *Journal of Russian and East European Psychology, 48*(1), 91–96.

Walkerdine, V. (1988). *The mastery of reason*. London: Routledge.

Žižek, S. (Ed.), (1998). *Cogito and the unconscious*. Durham, NC: Duke University Press.

Further Reading

Mozziconacci, S. J. L. (1995). Pitch variations and emotions in speech. In *Proceedings of the XIIIth international congress of phonetic sciences (Vol. 1). Stockholm* (pp. 178–181).

CHAPTER

17

Emotional Orientations and Somatic Markers: Expertise and Decision Making in the Mathematics Classroom

David Reid, Laurinda Brown[†], Tracy Helliwell[†]*

[*]University of Bremen, Bremen, Germany

[†]University of Bristol, Bristol, United Kingdom

INTRODUCTION

Learning to teach is learning to make the kinds of decisions in classroom situations that teachers make. In this chapter, we explore a theoretical basis for describing this process of coming to decide as a teacher does.

We analyze the teaching and reflections of two experienced teachers using theoretical constructs at three levels. At the most general level, we adapt Maturana's (1988b) phrase "emotional orientation" to refer to a "teacherly emotional orientation," which is the set of decision-making criteria appropriate to teaching. Having a teacherly emotion orientation simply means a person behaves as one expects a teacher to behave. Maturana and his student Varela elaborate a theory of cognition founded on the principle that cognition is an intrinsic aspect of being for living things, focused on a continual recreation of the self in interaction with a world (Maturana & Varela, 1980, 1992; see Reid & Mgombelo, 2015 for an outline of key concepts in the context of mathematics education).

At the most specific level, we name the decision-making criteria appropriate to teaching "somatic markers," adopting the concept from Damasio (1996), a neuroscientist who hypothesized somatic markers to account for the ability to make decisions quickly without conscious reflection, which is sometimes lost due to brain damage. Individuals have many emotional

http://dx.doi.org/10.1016/B978-0-12-802218-4.00017-0

orientations related to different communities, characterized by different (probably overlapping) constellations of somatic markers. For a teacher, one emotional orientation is the teacherly emotional orientation, which contains somatic markers leading the teacher to make specific decisions in classrooms consistent with that identity.

At the middle level, "purposes" group somatic markers into patterns that make sense at a conscious level, linked to actions. Collections of purposes, in turn, form emotional orientations. The notion of purposes was introduced by Brown (2004) as a tool in teacher education. They are guiding principles that structure student teachers' learning from their own experience. Purposes are significant for researchers, teachers, and teacher educators in that they provide a level of description that allows an individual to see whether they are acting effectively or not, and they can lead to changing a person's behaviors.

This chapter considers emotional orientations, somatic markers, and purposes in observations of the practices of two expert teachers alongside follow-up interviews of them. The first teacher is in flow in a familiar classroom culture that he has established over time. The second has been an expert teacher who then changes schools and was observed over 3 months as she established a new class of students who did not work in the way that she was used to. We will alternate descriptions of classroom events and selections of teachers' writing with analysis in terms of emotional orientations, somatic markers, and purposes, introducing and elaborating these theoretical ideas as needed. Our goal is developing a theoretical framework for describing the process of coming to decide as a teacher does, rather than establishing empirical results, and the structure of our chapter reflects this.

MR. HATT—WHICH ONE'S THE BEST?

Mr. Hatt is an experienced Grade 8 (ages 12–13) mathematics and science teacher in Canada.[1] He teaches two groups of students, his "homeroom" class and one other. The following episode is from his introductory lesson for a unit on using algebraic expressions to describe mathematical situations. The students are building "trains" of trapezia of ever increasing length (see Fig. 1) and calculating the perimeters.

He has asked each group to give him an arithmetic expression that describes the calculation they did to find the perimeter of a train of six trapezia, and he has written these expressions on the chalkboard (see Fig. 2).

[1] Mathematical activity in Mr. Hatt's classroom was video-recorded as part of a research project, the Psychology of Reasoning in School Mathematics, funded by the Social Sciences and Humanities Research Council of Canada, grant #410-98-0085. See http://www.acadiau.ca/~dreid/PRISM/.

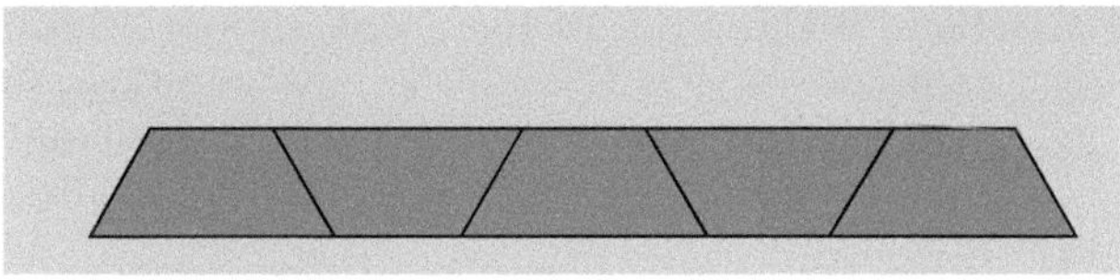

FIG. 1 Five-train of trapezia.

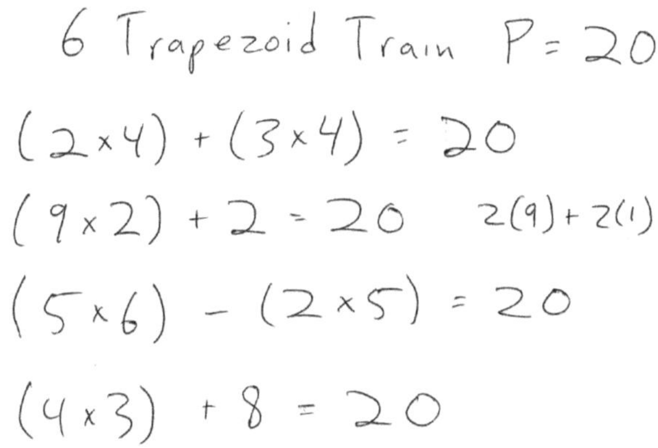

FIG. 2 Expressions for the perimeter of a 6-train.

He has also asked the groups to use their expressions to calculate the perimeter of a train of eight trapezia. He is now discussing each expression in turn and has come to the last one.

Mr. Hatt:	Which one does that kind of resemble?
Students:	The first one.
Mr. Hatt:	The first one, the first one, OK.
Student:	Which one's the best?
Students:	Ours, ours.

There are many things we can imagine Mr. Hatt doing after the student asks "Which one's the best?" He could say any number of things:

"'Best' is a moral judgement; this is a mathematics class."
"Best was the Beatles' first drummer."
"I won't tell you."
"I'm asking the questions here!"
"I'm not so sure we're at the best one yet."
"In what way does this one resemble the first one?" (ignoring the question entirely).
"The homeroom class had a better one."
"The third one is the best. Now, how does this one resemble the first one?"
"Which one's the best? Good question. Let's discuss that."

He could also have walked out of the room, saying nothing, or burst into tears. Some of these possible responses may seem less likely to you than some others. Why was this so?

Discussion 1—The Teacherly Emotional Orientation

Maturana (1988a) would call teaching a "domain of explanation," characterized by a community whose members can recognize, in others, behaviors appropriate to the community, although they probably could not give specific criteria for doing so. As teachers we recognize some possible responses to the question "Which one's the best?" as being the sort of things teachers would do or say, and other responses as being unlikely. We expect certain behaviors from teachers, and if Mr. Hatt did not act like a teacher we would not consider him to be a teacher.

> If someone claims to know algebra, that is, to be an algebraist, we demand of him or her to perform in the domain of what we consider algebra to be, and if according to us she or he performs adequately in that domain, we accept the claim. ***(Maturana, 1988a, p. 3)***

Mr. Hatt's choice must be based on something other than conscious reflection because there is no time for conscious reflection here. The decision is fundamentally an emotional one, based on preferences that have evolved over time and which are shared by other teachers.

> Whether an observer operates in one domain of explanations or in another depends on his or her preference (emotion of acceptance) for the basic premises that constitute the domain in which he or she operates. Accordingly, games, science, religions, political doctrines, philosophical systems, and ideologies in general are different domains of operational coherences in the praxis of living of the observer that he or she lives as different domains of explanations or as different domains of actions (and therefore of cognition), according to his or her operational preferences. ***(Maturana, 1988b, pp. 33–34)***

Maturana suggests the phrase "emotional orientation" to name the set of criteria appropriate to a domain of explanations. He uses "emotional" because it is a bodily predisposition rather than conscious reflection that is operating when we perceive some behaviors as appropriate and others as not.

> Members of academic disciplines and other experiential domains rely on culturally defined, recognizable patterns of relating to each other. They share beliefs and values, ways of communicating, and forms of acting. Taken together, these constitute an emotional orientation. ***(Drodge & Reid, 2000, p. 250)***

We have different emotional orientations appropriate to different domains. We observe the actions of others as being appropriate to a particular domain according to our own emotional orientation for that domain. So a mathematical emotional orientation allows us to recognize the activity of others as being mathematical, and hence to identify those people as mathematicians. A teacherly emotional orientation allows us to recognize the activity of others as being teacherly, and hence to identify those people as teachers. We have many emotional orientations because we belong to many communities, and communities can overlap, contain other communities, or subtly blend into other communities according to the behaviors different members accept as legitimate.

Mr. Hatt's Response

So, what did Mr. Hatt *do* when asked "Which one's the best?"

Mr. Hatt:	Which one does that kind of resemble?
Students:	The first one.
Mr. Hatt:	The first one, the first one, OK.
Students:	Which one's the best?
Students:	Ours, ours.
Mr. Hatt:	I'm not so sure we're at the best one yet.
Students:	Both.
Mr. Hatt:	But, I'm not sure, uh, my home room had – I think – I believe there were two different ones here – Now, the question is this – you then had to turn it up a notch and I asked you to go to-
Students:	8
Mr. Hatt:	8. Did your method work? Yeah.
Students:	Yeah.
Mr. Hatt:	Did it?
Students:	Well what was the answer? 26?
Mr. Hatt:	Yeah. The answer is 26.
Student:	Then it worked.

Mr. Hatt made a decision in the moment, to suggest "I'm not so sure we're at the best one yet" and that "there were two different ones" suggested by the homeroom class that might have been better. He then turns to checking of their expressions by calculating the perimeter of an 8-train and comparing the result to Mr. Hatt's answer.

As we noted above, Mr. Hatt could have responded in many ways to his students' question, and the decision he made was only one of many appropriate to a teacherly emotional orientation. To describe specific decision making in the moment, we turn to the concept of somatic markers.

Discussion 2—Somatic Markers

Damasio (1996) has studied the making of such decisions through the neurological characteristics of people who no longer seem able to make them. He has put forward the somatic marker hypothesis to explain what he has observed. The term "somatic marker" is used for the juxtaposition of image, emotion, and bodily feeling we have that informs our decision-making:

> Because the feeling is about the body, I gave the phenomenon the technical term somatic state ("soma" is Greek for body); and because it "marks" an image, I called it a marker. Note again that I use somatic in the most general sense (that which pertains to the body) and I include both visceral and nonvisceral sensation when I refer to somatic markers. ***(Damasio, 1996, p. 173)***

We would suggest that Mr. Hatt's response was informed by his somatic markers, and also that your somatic markers came into play when you judged some possible responses to be likely actions of a teacher and others to be unlikely. You have a constellation of somatic markers that are active in teaching situations. Another teacher's somatic markers might be different, and he or she would make different decisions than you would, but at the same time you can recognize similarities in the choices your somatic markers would guide you toward. The teacherly emotional orientation can be seen as a constellation of somatic markers shared by teachers.

In their work on teachers' complex decision-making, Brown and Coles (2000) state the following:

> Somatic markers act to simplify the decision as to which behaviour to try. Negative somatic markers mean that the behaviours do not even come to mind as possibilities for action. A positive somatic marker means that the behaviour becomes one of a number available for use. ***(Brown and Coles, 2000, p. 168)***

Seeing a teacher use a strategy effectively that you have a negative somatic marker for and have ceased to consider creates a jolt that might allow you to consider using that strategy again. So whether somatic markers are positive or negative is not a fixed state.

> Somatic markers are thus acquired though experience, under the control of an internal preference system and under the influence of an external set of circumstances which include not only entities and events with which the organism must interact, but also social conventions and ethical rules. ***(Damasio, 1996, p. 179)***

The question that informs our learning as teacher educators is, "How is it *possible* to work with people on our courses so that they become teachers?" We know that experienced teachers deal with situations where there are many different possible responses all the time, and they normally know what to do. They behave like teachers. They have a teacherly emotional orientation. Their somatic markers are well suited to acting appropriately. The students who start our courses, on the other hand, seem to lack somatic markers that would allow them to behave like teachers in many situations. How do they acquire these somatic markers, and with them a teacherly emotional orientation?

Interview With Mr. Hatt

We will return to the question of how people come to behave like teachers, but first we will explore further Mr. Hatt's response to "Which one's the best?" We cannot say much about Mr. Hatt's somatic markers based on the brief exchange we have quoted above. Somatic markers are not directly observable. We know he made a decision to respond in the way he did, and because he did so quickly he must have done so with a smooth decision-making process. Damasio's somatic marker hypothesis provides a way to account for such decision making, but not in a way that is directly applicable to our question about the process of becoming a teacher.

Looking carefully at Mr. Hatt's decisions in the rest of the lesson led us to observe some patterns that we conjectured to be linked to somatic markers. We observed that he did not want the groups to end their work too quickly. We also noted that Mr. Hatt did not always use questions to support the discussion. He also used statements (such as "The first one, the first one" and "I'm not so sure we're at the best one yet."). How aware is he of using statements to support the questioning of his students? Does he have a gut feeling that he does not want to answer the question yet? Because somatic markers are acquired through experience, we felt it would be useful to learn more about Mr. Hatt's experiences and the kinds of stories he tells to account for the way he teaches.

The interview occurred 2 years after the lesson we had analyzed, so Mr. Hatt spoke out of his recent experience rather than directly about his response to "Which one's the best?" But he had recently taught similar lessons on using algebraic expressions to describe mathematical situations. Laurinda had prepared a semistructured set of questions about supporting discussions and questioning. In his replies, Mr. Hatt made the following comments, chosen to illustrate a range of what was talked about.

- Students have to be actively engaged or they are not learning.
- They want you to tell them, but they don't learn unless they are actively engaged.
- I like to begin most things with a physical, hands on activity, then to have a lot of representations, pictorial symbolic, students can flip back and forth.
- Can they come up with expressions, then explain to the class how they obtained their expressions. And there are a number of explanations.
- They are formulating in their own minds the easy way to do it. Then they are going to ask me, but I don't answer their questions. [LB Why don't you answer?] They would like the easy way out, but they have to have time to know that it is a problem. That it isn't as simple as they

might have thought. I provide very little information until they have found a method. There are a lot of methods.

- My biggest nightmare would be if everyone did 3 *n* plus 2. I'd ask them to come up with another. And they might say "Why, I have one that works?"
- I might have my own method, but then I saw an easier way, so they write that, but others might stay with a more difficult way that they understand. In a 14-year-old mind they are not always willing to accept a different picture of what is happening.
- To them they think the formula has to work for everything, so when they find the formula is different this is a discrepant event.
- If they ask me "why?" I turn it back to the class. I think that comes up in their groups. They come and ask me why, but I want them to go work on that.
- A friend of mine who is a doctor works with Medical students and he focuses on questioning. And he says if you don't know the questions and you don't know the answers you'd better listen.
- Don't be too quick to answer questions, or to ask them. The psychiatrist doesn't answer the question "Am I mad?" He invites the patient to tell more.

These responses do little to reveal the somatic markers that might inform Mr. Hatt's decision making when teaching. This is not surprising because the whole point of somatic markers is to operate unconsciously so that the conscious mind is not overwhelmed with the complexity of simply existing. Somatic markers are not, and should not be, accessible to conscious reflection.

Mr. Hatt's responses give insight into what motivates his teaching through some strong images. For example, he expressed a belief in beginning with something physical, in looking for different solutions and in finding ways of getting students to be more engaged. An image, like that of the psychiatrist unasking the question, points to a general intention, not being too quick to answer.

Discussion 3—Purposes

We call such intentions, purposes, a word adopted by Laurinda as part of a theory-in-action (Argyris & Schön, 1978) she developed to account for the transformation that occurs in teacher education courses as people become teachers.

> Purposes emerged for me as a description of the sorts of guiding principles that student teachers found energising when learning from their own experience. Purposes seemed to be in the middle position of a hierarchy between philosophical attitudes and teaching behaviours in the classroom. ***(Brown, 2004, p. 4)***

Laurinda developed a routine when working with groups of student teachers describing the details of their practice and discussing incidents that seemed to them to be in some way similar. From these discussions, a vocabulary of what they considered to be important would develop, such as "sharing responses" or "how do I know what the pupils know?" Such phrases are examples of purposes. Discussions at the level of purposes, seemingly somewhere between philosophy and behaviors, seem to be effective in allowing a group of students to begin to collect together a range of strategies for use, to adapt their beliefs and attitudes.

Purposes are more specific than a philosophy of teaching and more general than specific behaviors. They are basic-level categories (Rosch, cited in Lakoff, 1987; Varela, Thompson, & Rosch, 1991). Rosch reported that basic-level categories are "the generally most useful distinctions to make in the world" (Lakoff, 1987, p. 49) and are linked to actions; for example, we recognize a "chair" as a "sitting-on object." Having been shown the picture of a dog we are able to recognize it as such and most people would use the word "dog" to describe the picture. This is the "basic-level category" and is linked to a range of somatic markers dependent on the individual's experience over time with dogs. One response might be running away (read as fear/flight), and for another it might be reaching out to pet the animal (read as pleasure). There would be some individuals who might know that particular dog and so use its name as a descriptor and be able to give details of that dog's individual characteristics and behaviors (interactional properties, p. 51). There are also people who would describe dog in a more general, technical way (superordinate categories, p. 51). Thinking of purposes as basic-level categories supports Laurinda's learning though experience that teachers talking in terms of purposes leads to the most easily expressed distinctions to motivate development in student teachers' teaching actions.

Purposes fit as basic level categories between teachers' philosophies of mathematics and mathematics teaching and their behaviors in classrooms, and also between the theoretical constructs of emotional orientations and somatic markers (see Table 1). Hypothesizing a teacherly emotional orientation gives a way to account for the experience of recognizing some people as teachers and others as not being teachers based on observed behaviors. Purposes provide an accessible level of description in the middle, such as Mr. Hatt describing his students as needing to be "actively engaged." Purposes group somatic markers into patterns that make sense at this conscious level, being able to be talked about. Somatic markers provide a construct that accounts for individual actions that a teacher makes in the practice of teaching, deciding between a range of possible actions. Somatic markers—the actions—can be grouped into collections that are recognizably about what teachers do.

TABLE 1 Purposes as Basic-Level Categories

	Rosch categories	Brown teaching strategies	Theoretical constructs
Abstract layer (mind)	Superordinate	Teachers' philosophies of mathematics and mathematics teaching	Emotional orientations
In-between layer (most useful distinctions in the world)	Basic-level	Purposes	Purposes
Detail layer (actions)	Interactional properties	Behaviors	Somatic markers

The intentions Mr. Hatt described when interviewed seem to be at this basic level. They might be described as a purpose of "continuing discussion of strategies." Mr. Hatt operates to support the continuing discussion of many different strategies, not rushing to fix on one alone. The actual form of the statements and range of responses is not fixed but contingent upon what happens.

This purpose suggests the presence of a pattern of somatic markers that point to the behaviors that might actually happen. As a research method we can go back to the recording of the lesson to examine what is actually done in practice toward this purpose. We can also, as teacher educators, use such a purpose as an example of the level of thinking useful for people becoming teachers, between the too general, "Behave like a teacher" and the too specific, when asked "Which one's the best" say "I'm not so sure we're at the best one yet."

MS. HUTT—CHANGING SCHOOLS

Ms. Hutt is an expert teacher who has recently moved schools. In such a move, even an expert teacher finds that teaching that was automatic at the previous school, seemingly effortless, now does not work in the new situation. Her somatic markers lead her to make decisions that do not work out well. In the previous school, the whole department was working in similar ways teaching a curriculum that supported the children "becoming mathematicians." Over time the recurrent coordination of behaviors at that school had led to everyone—children, parents, and teachers—knowing what they were doing. The children expected mathematics lessons to be like they were. They knew how to act in those lessons. Now the situation was different. Ms. Hutt needed to work with her classes to establish new ways of working. Tracy Helliwell's research examines Ms. Hutt adapting to her new school. Tracy observed Ms. Hutt's classroom teaching and interviewed her after each lesson. Ms. Hutt wrote down her reflections after the interviews. The excerpts here are drawn from that writing.

Ms. Hutt—Interrupted Automaticity

Voids

I can feel uncomfortable with voids—empty spaces, not just without voice, but without a sense of anything really happening. A period of silence is not a void unless there is also no thinking or doing involved. In my mathematics lessons before there would often be quiet times to consider, reflect, develop ideas, and formulate questions. These I would describe as quiet spaces rather than voids. For me, a void contains nothing, except perhaps confusion, bewilderment, awkwardness, no production of anything useful. When I sense a void I feel compelled to fill it; fill it with a question, further explanation, or just my voice. When I have established a collaborative way of working with a class either there are no voids or, if there are voids, I am not compelled to fill them with my voice. I am confident that the void will be filled eventually by some other means.

When I began teaching this group of 14-year-old students, I was sensing voids and of course filling them because I had not yet established a way of working with these students that I was comfortable with. A void might appear after I had asked a question or after I had offered an explanation. I could feel an empty space. It would feel awkward. I became aware of these voids, particularly in reflection after the lessons. My first awareness was of the awkwardness that the void was presenting me with followed by my reaction to this. I needed to be explicit about the ways in which I wanted the students to communicate about their mathematics. They were not talking. I do not think they were really even thinking. Asking questions was failing to create responses in the way that similar questions had done for me in the past so I could no longer work automatically. I had to force some classroom behaviors.

Questions

I like to think about the questions that I am going to ask students in the classes that I teach. When I plan my lessons I am thinking about how I can really get the students to think mathematically and part of me doing this is planning some questions that I would like to ask. In the lessons, I am aware that I am thinking carefully about what question I might ask next, if indeed a question is appropriate, and I can give this task quite a lot of attention when it is needed. In classrooms where I have established a culture, I am able to give this the attention that it needs because ways of working and managing discussion are in place. In my established classrooms, I sometimes ask questions where I do not expect a response and

I am usually explicit about the fact that I am not expecting a response, at least not an immediate one. When I really have students working in the way that I want them to be working, then it is more often than not their questions that get asked rather than my own. It is their questions that motivate them to find answers and it is their questions that open up new avenues that are worth exploring. It is their questions that offer me insight into where they are at in their thinking around a particular problem, technique, or process. When I am most comfortable is when there are many questions being asked by students for whatever purpose they might be asking them.

In the time before I had established a working culture with my year 10 group, I was aware that students were not asking many questions at all (other than maybe, "what's the title?"). I would ask questions which would often be met with some sort of void. This led to my questions becoming more focused, more closed, less probing. I was getting little in return. The students seemed nervous and not able to ask questions. Many students appeared not to have voices at all. I needed to return to my task design and my own initial questions. The scheme of work had not done me any favors with short amounts of time on topics.

What I needed was an extended block of time to work on something "rich" where students could begin to formulate questions of their own. From whom and from where the questions were coming was an important awareness that I had developed. It is important to me that I am not the "owner of knowledge" so to speak and that we, as a group of individuals, will develop our mathematics collaboratively. Offering tasks that could create a space where questions could evolve and allowing time for the students to engross themselves in the mathematics had a massive impact on who was asking the questions. The more tasks like this that I offered, the more the students would ask questions. I could refocus my energies on hearing questions and responding with questions of my own.

Discussion—Interrupted Automaticity

Some of the decisions Ms. Hutt makes when teaching require her conscious attention: "In the lessons I am aware that I am thinking carefully about what question I might ask next, if indeed a question is appropriate, and I can give this task quite a lot of attention when it is needed." As an expert teacher Ms. Hutt has somatic markers that let her quickly and unconsciously make many other decisions required in the course of her teaching, which frees her mind to focus on questioning: "In classrooms where I have established a culture, I am able to give this the attention that it needs because ways of working and managing discussion are in place."

In her new school some of her decisions do not lead to the results she is used to: "Asking questions was failing to create responses in the way that similar questions had done for me in the past so I could no longer work automatically." The fluency of her teaching was interrupted by voids, which she observed emotionally: "A void might appear after I had asked a question, or after I had offered an explanation. I could feel an empty space. It would feel awkward." This feeling is related to an immediate change of behavior: "When I sense a void I feel compelled to fill it; fill it with a question, further explanation or just my voice." But these feelings also establish or modify somatic markers, which will affect future behavior. Damasio discusses the way such feelings mark the development or change in a somatic marker: "This led to my questions becoming more focused, more closed, less probing."

Instead, Ms. Hutt reflects on her feelings: "I became aware of these voids, particularly in reflection after the lessons." She reflected not only on her feelings but also on her behavior responding to the feelings: "My first awareness was of the awkwardness that the void was presenting me with followed by my reaction to this." With her automaticity interrupted, she turned to slower, more deliberate thinking about her teaching. This led her to generate a goal: "I needed to be explicit about the ways in which I wanted the students to communicate about their mathematics." In identifying this goal, Ms. Hutt provided herself with a purpose: "creating a space where questions could evolve" to guide her decision making and to act as a counter to the possible development of negative somatic markers associated with the kind of open questioning she normally used in her teaching. This purpose could have been achieved in many ways. Ms. Hutt specifically adopted a pair-share strategy as one way to "create a space where questions could evolve" and also allowed more time for her students to "engross themselves in the mathematics."

Ms. Hutt—Diversity in Students' Work and the Mathematics

Marking students' work can be hard. It is made harder when everything you are marking is the same, like in a test. I have grown accustomed to the diversity in students' exercise books. The difference in what they record in terms of their mathematics can be considerable, making the job of marking their books far more stimulating and of course informative. When students have been working on an open task, you do not know what you are going to get. There are many surprising directions taken, different explanations, and different methods of approaching things.

The first time I marked the exercise books of the year 10 students, I became aware of the lack of differences in their work. It felt, to a certain extent, as if I was marking the same script over and over, with the same mistakes noticed, the same gaps appearing, the same lack of comments about the mathematics being made. Some students had completed more than others but this was the only real difference that was noticeable. The task of marking was tedious and to a certain extent alarming. On reflection, I can offer an explanation of sorts for this. If a class is used to being told what to do, only offered one way of doing things, not having the confidence to do something different, then of course their work will reflect this. Was this what I was doing? Was I only offering one way of working on something? Was I *telling* the students how to solve the problems they were working on? The students' work was offering me a new insight into what I needed to do differently. In an established mathematical classroom culture, it wouldn't matter if I gave the students the same set of questions to work through, they would approach the task in different ways, set their work out differently, write different notes, make different comments. Students would make their own decisions about which way was the most appropriate for tackling something. There would be flexibility in approach.

What did I do to allow this flexibility? In my written feedback, I made sure that I set students' actions and targets that were individualized, perhaps offering an explanation, or extending a problem, or making a correction using a different approach, encouraging them to have a conversation with me. This started to force some diversity and allowed me to see a student's particular response.

What about in the classroom? I began to think really carefully about the starts of my lessons. I wanted to offer the students a set of questions that they would all need to answer but that could afford different approaches. I worked on some starter questions, within the context of solving equations, where I was aware that I did not want any two questions to be the same and I wanted some questions that allowed multiple ways in. As a class we had not been through any formal approaches to solving equations:

(1) $3x-9=14$
(2) $10x+3=7x+12$
(3) $2m+3=5m-7$
(4) $14(2n-3)=42$
(5) $3m^2-1=146$
(6) $x^2+4x+3=0$

Having not taught any formal methods myself, I was interested in what the students would do and asked them to report back. I could allow all of my energy to be spent on hearing their ideas and responses, their methods and solutions and inevitably their questions. I could offer ways in, different formal methods and different images. I gave the students 10–15 minutes just to do whatever they could do for any of the questions on the board. I allowed them to get stuck, asked them to talk with each other, work together and test their answers. I encouraged them to write down as much as they could even if it did not feel like it was going to get them to a final answer. I had lots of conversations with groups of students and individuals, discussing ways in to some of the solutions. We spent a long time

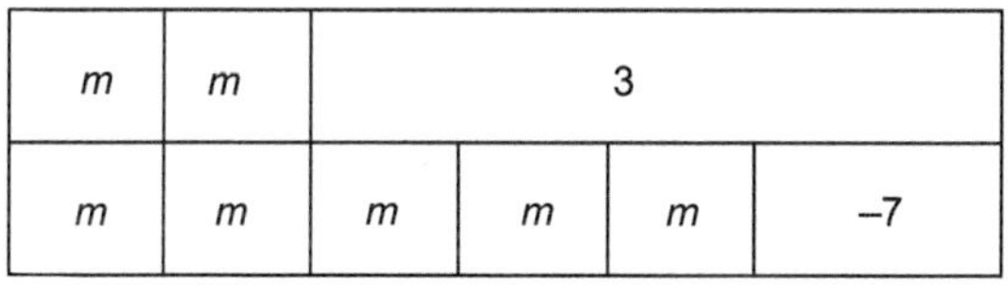

FIG. 3 Visual support for $2m+3=5m-7$.

discussing these questions, but I was careful not to spend time in the detail particularly. I was not going to get bogged down in a particular method. My aim was not to make sure each student could do each question confidently but more to make sure the students understood the importance of flexibility and diversity.

For question 3, I offered this image (see Fig. 3)—for some nothing else needed to be said.

For question 4, we talked about how to read the question, the order of operations and how this could be "undone." We looked at expanding brackets. I wrote everything down but did not insist on them writing everything down. They could choose which bits to write down for themselves. I told them they would be looking at a similar set of questions the next lesson so their notes needed to be helpful and useable. We got into factorisation for the last two questions but I did not dwell on this. As lessons progressed the students' confidence grew, as did the complexity and richness of the conversations. The responses in their books became more diverse. I could turn my attention to the mathematics, for example, when I found myself thinking that exact values are preferable to decimals, noticing that many of them are using decimals, I found myself stressing the preference for fractions. When I am able to turn my attention to this sort of awareness my classroom feels normal—the students' books become more different and I no longer feel alarmed.

Discussion—Changing Students' Somatic Markers

Here Ms. Hutt talks about another purpose, allowing flexibility/forcing diversity. This purpose informs her decision making in offering the set of equations for the students to solve and allowing them to get stuck. She is clear both about what she is not trying to do "I was not going to get bogged down in a particular method" as well as her goal, "My aim was not to make sure each student could do each question confidently but more to make sure the students understood the importance of flexibility and diversity." What does it mean to understand the importance of flexibility and diversity? Ms. Hutt has come to this purpose from observing her students all doing the same thing in their exercise books. She ascribes this behavior to their prior teaching: "If a class are used to being told what to do, only offered one way of doing things, not having the confidence to do something different, then of course their work will reflect this." In other words, the students' experiences of being told what to do and only offered one way of doing things has led them to develop somatic markers that are positive for doing what their peers do and negative for doing something different. Ms. Hutt's goal of making sure "the students understood the importance of flexibility and diversity" expresses an intention to change those somatic markers. She does this in two ways. First, she creates an environment in which there are opportunities for diverse approaches and time for diverse approaches. Second, she valued differences in their responses. There is some evidence that this effort was successful: "As lessons progressed the students' confidence grew as did the complexity and richness of the conversations. The responses in their books became more diverse."

Ms. Hutt established, in herself and in her students, somatic markers that allowed her to teach in her new school in ways that fit the teacherly emotional orientation she brought from her old school and allowed a new teacherly emotional orientation to emerge in the new context. This meant she could turn her attention to other goals. In her previous school the whole department was working in similar ways teaching a curriculum that supported the children "becoming mathematicians." They were developing some aspects of a mathematical emotional orientation. Once somatic markers consistent with her teaching were established, Ms. Hutt could turn to mathematical preferences, such as "the preference for fractions." Such preferences are related to the mathematical emotional orientation and are also based on somatic markers.

CONCLUSIONS

We have explored here somatic markers, purposes, and emotional orientations in the context of two experienced teachers' teaching. Mr. Hatt, an experienced teacher working in a familiar context, has a set of somatic markers that allow him to quickly act flexibly in sensible ways in his classroom. Purposes that group his somatic markers at another level are evident in his later reflection on his teaching. For Ms. Hutt, on the other hand, as an equally experienced teacher working in an unfamiliar context, awareness of her somatic markers emerged out of experiences of what those markers no longer allowed her to do in her new context. She was forced to slow down, to consciously reflect on her teaching goals and new context, and to decide to act ahead of time according to new purposes that led to the establishment of new somatic markers more useful in her new context. At the same time, her actions changed the somatic markers of her students, helping to bring her new context more into line with her goals and her teacherly emotional orientation. The language of purposes and somatic markers allows us to learn more about the teaching of these teachers.

References

Argyris, C., & Schön, D. (1978). *Organizational learning: A theory-in-action perspective*. Reading, MA: Wesley.

Brown, L. (2004). It's about learning: From purposes to basic-level categories to metacommenting. *BSRLM Proceedings*, *24*(3). http://bsrlm.org.uk/IPs/ip24-3/.

Brown, L., & Coles, A. (2000). Complex decision-making in the classroom: The teacher as an intuitive practitioner. In T. Atkinson & G. Claxton (Eds.), *The intuitive practitioner: On the value of not always knowing what one is doing* (pp. 165–181). Philadelphia, PA and Buckingham, UK: Open University Press.

Damasio, A. (1996). *Descartes' error: Emotion, reason and the human brain*. London: Papermac.

Drodge, E., & Reid, D. (2000). Embodied cognition and the mathematical emotional orientation. *Mathematical Thinking and Learning*, *2*(4), 249–267.

Lakoff, G. (1987). *Women, fire, and dangerous things: What categories reveal about the mind*. Chicago, IL: The University of Chicago Press.

Maturana, H. (1988a). Ontology of observing: The biological foundations of self consciousness and the physical domain of existence. In R. Donaldson (Ed.), *Texts in cybernetic theory: An in-depth exploration of the thought of Humberto Maturana, William T. Powers, and Ernst von Glasersfeld*. Felton CA: American Society for Cybernetics. (conference workbook). Retrieved from http://ada.evergreen.edu/~arunc/texts/cybernetics/oo/old/oo.pdf. Originally published in Caianiello, E. (Ed.), (1987). *Physics of cognitive processes* (pp. 324–379). Singapore: World Scientific.

Maturana, H. (1988b). Reality: The search for objectivity or the quest for a compelling argument. *The Irish Journal of Psychology*, *9*, 25–82.

Maturana, H., & Varela, F. (Eds.), (1980). *Autopoiesis and cognition: The realization of the living*. Boston, MA: Reidel Publishing.

Maturana, H., & Varela, F. (1992). *The tree of knowledge: The biological roots of human understanding*. Boston, MA: Shambhala [first edition published 1987].

Reid, D., & Mgombelo, J. (2015). Survey of key concepts in enactivist theory and methodology. *ZDM Mathematics Education*, *47*(2), 171–183. http://dx.doi.org/10.1007/s11858-014-0634-7.

Varela, F., Thompson, E., & Rosch, E. (1991). *The embodied mind: Cognitive science and human experience*. Cambridge, MA: MIT Press.

Index

Note: Page numbers followed by *f* indicate figures, *t* indicate tables, and *b* indicate box.

FSC
www.fsc.org
MIX
Papier | Fördert
gute Waldnutzung
FSC® C083411

Kolibri 360 GmbH
Ferdinand-Jühlke-Straße 7
99095 Erfurt, Deutschland
produktsicherheit@kolibri360.de